BURNED ALIVE

ALBERTO A. MARTÍNEZ

BURNED ALIVE

GIORDANO BRUNO, GALILEO and the INQUISITION

REAKTION BOOKS

Published by
REAKTION BOOKS LTD
Unit 32, Waterside
44–48 Wharf Road
London N1 7UX, UK
www.reaktionbooks.co.uk

First published 2018

Printed and bound in Great Britain
by TJ International, Padstow, Cornwall

A catalogue record for this book is available from the British Library

ISBN 978 1 78023 896 8

CONTENTS

Men of Galilee, why do you stand gazing up at the sky?

Acts 1:11

I have indeed asserted infinite particular worlds similar to the Earth, which with Pythagoras I consider a star, similar to which is the Moon, other planets and other stars, which are infinitely many.

Giordano Bruno to the Inquisition in Venice, 1592

INTRODUCTION

Most people know about Galileo. They know that he said the Earth moves around the Sun. And they know that he got in trouble with the Catholic Church for doing so. The Roman Catholic Inquisition condemned old Galileo to surrender his freedom for the rest of his life. Nowadays he is admired as a hero in the history of science.

Fewer people know about Giordano Bruno. Years before Galileo, Bruno too was put on trial by the Inquisitors. They imprisoned him for almost eight years. Then they finally condemned him to what was feared as the worst kind of punishment: the jailors gagged him, tied him to a post in a public place in Rome, and set the pyre on fire to broil and burn him alive.

Historians say that Giordano Bruno was not condemned for his beliefs about astronomy or cosmology – unlike Galileo. Still, Bruno is increasingly famous. He was featured in the remake of the popular TV series *Cosmos*. Its first episode, which dedicated much of its airtime to Giordano Bruno, aired in 2014 and was watched by roughly seven million people in the U.S. The narrator, the astrophysicist Neil DeGrasse Tyson, said that Bruno risked his life by voicing his vision of the cosmos: 'the penalty for doing so, in his world, was the most vicious form of cruel and unusual punishment.' Soon, many commentators complained that *Cosmos* had misrepresented Bruno by echoing the myth that the Inquisition killed him for his cosmological views.

This book shows, despite all expectations, that Giordano Bruno really was burned alive for his beliefs about the universe. Bruno argued that the Earth has a soul and that many worlds exist. However, historians didn't know that those beliefs were considered *heresies*, that is, crimes against God, punishable by death. By inspecting books on heresies and Catholic law, I found that these beliefs were heretical long before Bruno advocated them. Bruno didn't know this, and neither did later science writers. Furthermore, this book will show that these same censured beliefs were involved in the Catholic opposition to Galileo.

For over a hundred years, people have wondered whether the infamous trial of Galileo was connected to the Inquisition's previous trial against Bruno. This book will show that these trials were indeed linked. Some of Galileo's critics were annoyed that he seemed to support Bruno's 'horrendous' beliefs: that many worlds exist and that the Earth moves *because it has a soul.*

Christians denounced such beliefs as 'Pythagorean'. In ancient times Pythagoras was allegedly the Greek author of the theory that Earth is one of the stars and that it moves. But historians didn't know that for a thousand years prominent Christians demonized Pythagoras and his disciples as deceptive sinners and false imitators of Jesus Christ. This book will illuminate the Copernican Revolution in that neglected context. Surprisingly, Bruno's condemnation was caused mainly by his obstinate defence of such beliefs: the existence of many worlds and the soul of the world. Most importantly, this book reveals an unpublished Latin manuscript in which Galileo's most critical judge, the author of the most negative reports used by the Inquisition against him, censured Galileo and the Copernicans for those same beliefs. The Inquisition condemned the Copernicans as a heretical 'sect' of New Pythagoreans.

This story is about the roots of the notorious conflict between science and Christianity. Actually, the two have not often opposed one another.[1] In fact, Christianity has often sponsored and supported the sciences. But we still struggle to understand what exactly happened between Bruno, Galileo and the Church.

There is another reason, too, why this story needs to be told: the Roman Inquisition won. Not only did it succeed in killing Bruno and silencing Galileo, it also succeeded in preventing people from knowing Bruno's beliefs. To this day, more than four hundred years after Bruno died, most scientists and teachers are unaware that Giordano Bruno's account of the universe was far more correct than the beliefs of Copernicus, Kepler and Galileo. To this day, scientists don't know that Bruno belongs among those famous men.

This is because the Inquisition banned and burned Bruno's books. The clergymen of the Index of Forbidden Books prohibited all Catholics from reading, quoting, discussing or even mentioning any of Bruno's beliefs. They were forbidden from even writing Bruno's name. He was a heretic, and Catholics should not write about heretics. Thus they buried his role in the history of astronomy and cosmology.

But here's what really happened. Like Copernicus, Bruno too believed and argued that the Earth really does move, at a time when hardly anyone else asserted this, and long before Galileo and Kepler. Yet Copernicus, Kepler and Galileo wrongly claimed that the Sun and the stars do *not* move. But Bruno rightly said that they move. In 1613 Galileo realized that the Sun spins on its axis, yet did not realize that it also really moves through space, as Bruno expected.

Copernicus, Kepler and Galileo believed that the Sun is the centre of the entire universe. But here too they were wrong. Instead, Bruno rightly insisted that the Sun is not the centre, not at all. He alone rightly explained that our universe has no centre. Copernicus, Kepler and Galileo also believed that the stars were all arranged in a spherical heaven, but that was another big mistake. Instead, Bruno thoughtfully argued that the stars are distributed homogeneously throughout a boundless, infinite space.

And Copernicus, Kepler and Galileo thought that the Sun is not a star. To the contrary, Bruno rightly insisted that the Sun *is* a star, all the stars are suns, and many of them are larger than our Sun. He first published these claims in 1584. Only five decades later, in 1633, did Galileo briefly suggest that the stars are suns – in just one phrase in the fictitious voice of one character in his *Dialogue of the Two Chief Systems of the World*.

Copernicus, Galileo and Kepler mistakenly said that the orbits and motions of the planets were circular. But in 1584 Bruno rightly denied it, insisting that no heavenly motions are really circular. Only 35 years later did Kepler realize and prove that the orbits of planets are not really circular. Galileo apparently never agreed with Bruno's or Kepler's claims that the orbits were not circular. He mistakenly argued that all natural motions are circular.

Next, in 1584, Bruno rightly anticipated that the Moon has mountains and valleys. Copernicus, Kepler and Galileo published no such prediction or claim: yet by 1610 Galileo confirmed with his telescope that there are mountains and valleys on the Moon.

In 1584, and frequently thereafter, Bruno claimed that 'innumerably many stars' exist, stars never seen before. Copernicus, Kepler and Galileo made no such claim or prediction. By 1610 Galileo, with his telescope, detected countless many stars that nobody in human history had ever seen. He announced it on the title page of his book as the discovery of 'INNUMERABLE STARS'.

Bruno argued that the stars are surrounded by planets that are invisible to our eyes. Copernicus and Galileo made no such claim, and Kepler even denied it. Yet again Bruno was right: four centuries later, in 1988, astronomers finally detected and confirmed the existence of a planet outside of our solar system. To date they have found thousands of planets, just as Bruno expected. They are now known as 'exoplanets'. In May 2016 news articles reported that astronomers had recently confirmed the existence of many more Earth-sized planets detected with a space telescope. It is ironic that this telescope-spacecraft, designed precisely to detect exoplanets, is named Kepler, who actually denied the existence of exoplanets. It should have been called the Bruno Telescope.

Next, Bruno also specified that some planets, other than the Earth, are surrounded by moons. Copernicus, Kepler and Galileo anticipated no such thing. Yet in 1610 Galileo discovered that, indeed, there are moons orbiting Jupiter. Astronomers nowadays expect that moons orbit exoplanets too, as Bruno said.

Furthermore, Bruno argued that planets are made of the same elements as the Earth, that effectively they constitute other worlds. Astrophysicists now know that indeed the planets are made of the same elements as our world, in different proportions. And our spaceships have explored the surface of those worlds. Bruno proposed that heavenly bodies are surrounded by some kinds of atmospheres, like the Earth. Again he was right.

Bruno even anticipated that humans might walk on and live on such distant worlds, even on the closest one, our Moon. In contrast, Copernicus made no such claim. Then in 1969, human beings travelled to the Moon and walked on it. Only after Bruno's writings and Galileo's lunar observations did Galileo consider that indeed some sort of living beings might possibly live on the Moon. But he wrote very few words about this, and occasionally denied it.

When Galileo published a book in 1610, revealing his stunning, telescopic discoveries – mountains on the Moon, satellites around Jupiter and innumerable stars – such evidence did not confirm the cosmological theory of Copernicus. Instead it confirmed Bruno's claims about the universe. Kepler pointed this out.

Bruno made some predictions that haven't been confirmed. One of them, which he made repeatedly, was that many alien beings live on heavenly bodies other than the Earth. Today, many astronomers and astrobiologists believe that this too is true, but they haven't yet found any clear evidence of such alien beings, despite countless astronomic investigations. But when they finally do, they will prove once again that Giordano Bruno was right.

It is no exaggeration to say that we don't live in the cosy Sun-centred universe of Copernicus. We live in the immense universe of Giordano Bruno. Who was this man who was right about so many things? Why has he been so neglected and disdained in the history of astronomy?

Scientists and historians who know about Bruno never make the kind of accounting above. Instead they simply give Bruno a plain dismissal: they say he wasn't even a scientist. Why do scientists and teachers know so little about him? It is because of the long-lasting consequence of Catholic censorship of Bruno's works. The Inquisition won. This book will discuss why Bruno's beliefs offended the Catholic clergymen so much and how such beliefs were echoed later with hesitation by men who now are far more famous: Kepler and Galileo.

BUT FIRST, I should say a few things about my approach.

I focused on primary sources. I have also consulted scholarly works that have been very helpful, but for brevity I have abstained from long discussions of the extensive, valuable literature. Often I give quotations from primary sources rather than paraphrase. Having studied the evolution of myths in the history of sciences and mathematics, I know how much meaning can change by omission or addition of a single word. So wherever I see an interesting paraphrase in a book or article I immediately wonder: what does the original actually say? My preference for quotations reflects that sentiment. In most cases I give my own translations from the originals, mainly from Italian, Latin, Greek, French and Spanish sources. Regarding excerpts from poems and songs, I've translated each word literally rather than retaining rhyme or metre by inventing phrases.

Writers make fascinating portrayals of the past, coloured by conjectures. However, I also appreciate the modest approach in which writers take great efforts to convey the past plainly, to say clearly what happened, while trying not to get in the way, not to obstruct through mediation. How much more would we value old sources if instead of writers paraphrasing one another they had actually used quotations? How much that is ambiguous would instead be clear? In studying Bruno, for instance, I used secondary sources at first. I was surprised by how much they disagree with one another. Whenever a writer mentioned four or five Catholic accusations against Bruno, I wondered how many others were there? How do those few fit into the larger set? Only when I checked the primary sources themselves did I realize that there were dozens of accusations. Yet to convey them all is very unusual. When one finds a secondary source that does so, then finally both the trees and the forest become simultaneously visible.[2] Accordingly, I have worked to synthesize certain claims not just into paragraphs but to dissect them as informative lists in which I have tried to be faithful to the originals and to organize the evidence more clearly than in a paragraph format. Such lists convey how the most relevant particulars fit into the broad collection.

For centuries, people have interpreted the ideas of Pythagoras and his followers in many different ways. In a previous book, *The Cult of Pythagoras*, I showed how scholars have radically disagreed in their interpretations. Some said that Pythagoras was a polytheist, others that he believed in one God, or that he was an atheist. Others said that he was a pantheist, or that he imitated Jewish doctrines, or that he was a Buddhist, or that he believed in the Holy Trinity, and so forth. Some writers say that Pythagoras was an aristocrat. For others he was a poor, revolutionary exile. Others say that he was a socialist, others that he was a communist, while others have argued that he was an early pioneer of democratic thought. Owing to such an extremely wide range of interpretations, I decided to try not to invent interpretations. History is full of fascinating events that actually happened, including many that have not been discussed, so history doesn't need my speculations. This is especially true in the study of Galileo's trial, which has generated a wealth of publications that one scholar describes as 'a vast secondary literature, prone to identify speculation with fact . . . that gives us reason to recoil before the torrents of rhetoric and floods of hypotheses which continue to obscure the facts of the case'.[3]

So I have tried not to make unwarranted inferences. Finally, please note that if I don't specify something in an explicit sentence, then I'm not trying to imply it. Books are structures that generate meaning. The author doesn't control all the meanings that sentences generate in each reader's mind, which brings its own ideas and definitions. But it's a valuable interaction. I hope this book will ignite lively discussions.

ONE

THE CRIMES OF GIORDANO BRUNO

In 1616 the Inquisition in Rome ruled that Galileo Galilei should immediately abandon the belief that the Earth moves around the Sun.

An old and notorious Inquisitor, Cardinal Roberto Bellarmino, now known as Bellarmine, officially warned Galileo that he could no longer believe that, otherwise Galileo would go to prison. Right then, the Index of Forbidden Books also denounced the theory that the Earth moves, calling it 'the false Pythagorean doctrine'.[1] Galileo too had credited it to the ancient philosopher Pythagoras.

Years earlier, in the 1590s, some of the same Inquisitors, including Cardinal Bellarmine, had interrogated Giordano Bruno, who was a prisoner of the Inquisition in Rome. And they censured his belief in the Earth's motion. Yet Bruno had boldly defended his theory that the Earth moves because it is animated by a soul. To his Inquisitors, he too called this belief 'the Pythagorean doctrine'. This is the missing link between the infamous trials of Bruno and Galileo.

What a word: 'Pythagorean'. Most people who write about Galileo don't even quote that word. But then we miss something. Namely, that the ancient Greek man whom schoolteachers portray as a famous mathematician, Pythagoras, was something else. He lived in southern Italy, where he became a philosopher or a religious leader. He died around five hundred years before Jesus of Nazareth was born. And, actually, there is no evidence that Pythagoras did what centuries later made him famous: there's no evidence that he proved anything in geometry. Instead, in ancient times most people knew him for other reasons. In particular, Christians viewed Pythagoras as a pagan religious leader, a rival of Jesus Christ. Reportedly, Pythagoras taught that when you die your soul travels out of your body and is reborn in another. People also attributed to Pythagoras strange beliefs about the Earth: that it is a heavenly body and it moves.

The followers of Pythagoras were a scattered and mystical group. They practised secrecy and silence, like a cult, and were bound by oaths and ancient religious beliefs. For centuries curious philosophers sought the secrets of Pythagoras. In particular, Giordano Bruno praised 'the School of Pythagoras'. Yet Catholics and Protestants both excommunicated Bruno. Like a good Christian, Bruno wanted to eat the host – the body of Christ. But they forbade him from doing so. In 1592 the Inquisition arrested and imprisoned him in Venice. He denied having believed any heresies and he defended himself as a philosopher. But there were some beliefs that he refused to renounce, beliefs about souls and many worlds: that there exist 'infinite particular worlds similar to the Earth, which with Pythagoras I consider a star, similar to which is the Moon, other planets and other stars, which are infinitely many'.

Bruno was also accused of having said blasphemies about Christ, the Virgin Mary and more. The Pope and the Roman Inquisition proclaimed 'all these propositions are heretical, and not now declared so for the first time, but by the most ancient Fathers of the Church and the Apostolic Chair.'[2] On a Thursday in February 1600, Giordano Bruno was burned alive. He was barely 52 years old.

Less than ten years later, Galileo used an innovative telescope to look at the Moon. He made a sensational discovery: he saw mountains and valleys on the Moon. He pondered whether this was visual evidence 'to revive the ancient Pythagorean opinion that the Moon is like another Earth'.[3] Perhaps now the Moon and the planets seemed to be not just luminous bodies, but *other worlds*?

But Galileo did not mention that the most vocal, recent advocate of the theory that the heavenly bodies are worlds was a dead man one should never name: the heretic Giordano Bruno.

There are mysteries here. What were the ancient 'heretical propositions' that the Inquisitors said were not original in Bruno but had been denounced by 'the most ancient Fathers of the Church' and the Apostolic See?

Likewise, in Galileo's trial by the Inquisition in 1633 there is the question: why were some Church officials so extremely upset by the 'false Pythagorean doctrine' of the Earth's motion? Even Galileo's former supporter Pope Urban VIII became furious. The Pope complained that Galileo 'had dared entering where he should not have, and into the most severe and dangerous matters that could be stirred up at this time . . . matters, which involve great harm to religion and more awful than were ever devised'. The Pope denounced it as 'the most perverse subject matter that one could ever handle.'[4]

But why?

Some Jesuits said that Galileo's book of 1632 was 'more harmful to the Holy Church than the writings of Luther and Calvin'.[5]

But *why*?

The Catholic Church had lost half of Europe because of the influence of Martin Luther and John Calvin. How could a scientific question seem more dangerous than the deeds of Protestant heretics? How could the 'Pythagorean doctrine' of the Earth's motion involve the 'most perverse subject' imaginable?

I think the problem was that the theory of the moving Earth was associated with radical pagan heresies. I will link the trials of Bruno and Galileo in a neglected historical context: the pagan beliefs of the cult of Pythagoras. Historians know that the Pythagoreans were a secretive religious group, yet there was no historical account of how the Christians criticized their evolving beliefs. How did this ancient religion relate to astronomy? How did it clash with Christianity?

The Pythagorean thread is remarkable because it explicitly shows up in the works of many important figures in the Copernican Revolution, including Copernicus, Bruno, Kepler and Galileo. This book will explain some well-known events by setting them in the context of others that are virtually unknown.

The cult of Pythagoras became notorious for anti-Christian beliefs: souls being reborn in many bodies, divination, many worlds, many gods, the infinity and eternity of the universe, and denials of the uniqueness of Jesus. The 'false Pythagorean doctrine' of the Earth's motion, censured by the Inquisition since the 1590s, was connected to heretical beliefs that had appeared in works on philosophy, magic and demonology for centuries.

Occasionally a few historians have briefly sensed the implications of denouncing the Copernican theory as 'Pythagorean'. For example, at the end of an article on Bruno, Jole Shackelford said:

> The association of Copernicus's ideas with the ancient central fire cosmology of Pythagoras was more than a dismissal of the antiquity of heliocentrism; it was especially damning, inasmuch as it implied other shared heresies, such as the Pythagorean belief in the transmigration of souls. Such teachings were not to be tolerated in post-Tridentine Rome.[6]

Still, investigators have not traced such implications until recently.[7] We will see that such heresies were of utmost importance in Bruno's deadly trial in the 1590s. Later, the Inquisition's proceedings against Galileo involve striking continuities.

Some historians have done valuable research on the history of the belief in many worlds, throughout the centuries.[8] But now, the belief in many worlds will acquire a surprising, new importance in understanding the Inqisition's proceedings against Bruno and Galileo. Because this belief was *heretical*. It was a crime against God.

Pythagoras and Copernicus

To understand the conflicts in the Renaissance, it is necessary to first trace some earlier conflicts. Early Christian theologians, such as Hippolytus, Chrysostom, Jerome and Augustine, had formulated notions that were echoed much later by Catholic theologians. In the times of Bruno and Galileo, such ancient notions were not forgotten. On the contrary, those were the authoritative judgements that were followed by the Inquisitors and the censors of the Index. Hence the ancient theology is directly relevant to the later conflicts.

According to late accounts, Pythagoras was born on the Greek island of Samos in the Aegean Sea, around 570 BCE, and he became a philosopher in southern Italy. Nowadays astronomers and historians refer to Pythagoras as a forerunner of Copernicus, but this is actually a mythical identity. There is no reliable evidence that Pythagoras was an astronomer or even a mathematician.[9] Nowadays, many school textbooks claim that Pythagoras was the first person to prove the Pythagorean theorem: the square on the diagonal of a right triangle equals the sum of the squares on the other two sides. However, ancient sources don't say this. I have found no evidence that anyone in antiquity claimed that Pythagoras *proved* the theorem.

The earliest such instance I have found is not from a century or five after Pythagoras died, but from roughly two thousand years after his death. In a work on Euclid's *Elements* published in 1557, the author Jacques Peletier commented about the hypotenuse theorem, that 'Pythagoras proved a general demonstration of it.'[10] Thus similarly many other myths about Pythagoras' alleged achievements in mathematics have arisen over time.

In addition to there being no writings by Pythagoras himself, the extant texts of early Greek writers on mathematics or astronomy, by Plato, Aristotle, Archimedes, Apollonius of Perga and many others, make no claim whatsoever that Pythagoras worked on mathematics or astronomy. Various ancient writers mentioned him, including Heraclitus of Ephesus, Plato, Herodotus, Heraclides and Isocrates. The sum of their various brief comments gives the impression that Pythagoras was a wise man, a popular religious leader who argued that the human soul is born repeatedly, even in animal bodies. It seems he taught his followers to live in a disciplined way, including some dietary restrictions, such as not eating beans or animals, at least of certain kinds. His followers modified his teachings, sometimes creating opposite beliefs. Aristotle criticized 'the so-called Pythagoreans' for saying 'that the Earth is one of the stars' and that it moves around a central fire.[11] Still, we have no claims by Aristotle about Pythagoras himself.

Pythagoras was most famous for allegedly teaching that human souls can be repeatedly reborn, even in the bodies of animals. According to some mythical accounts, Pythagoras was the son of a god, either Apollo or Hermes.[12] Allegedly, the soul of Pythagoras was originally born as Aethalides, and Hermes granted him the gift of remembering his past experiences, even in future lives. His soul was reborn in plants and animals, and even travelled to Hell. Later he was reborn as the warrior Euphorbus, next as Hermotimus, followed by Pyrrhus the fisherman, and fifth as Pythagoras, who still remembered his past lives.[13] According to the poet Ovid, Pythagoras occasionally departed from the Earth and travelled to the stars.[14]

For more than a thousand years famous Christian theologians vilified the Pythagoreans for sins and blasphemies. They complained that Pythagoras practised 'ten thousand kinds of sorcery'. Some of his followers said that Pythagoras was the son of a god and a human mother. Whereas Jesus had died and resurrected only once, Pythagoras had survived death multiple times. Allegedly his soul had been reborn in other bodies. St Hippolytus of Rome (*d. c.* 236 CE) criticized the 'alliance between heresy and the Pythagorean philosophy' and he denounced the 'enormous and infinite heresies' of the 'disciples not of Christ but of Pythagoras'.

The cosmology of the Pythagoreans annoyed the Christians. The Pythagoreans said that the infernal regions begin with the Milky Way, from where souls fall to the Earth. Others said that human souls come from the soul of the Earth. According to one book, allegedly by Plutarch, Pythagoras taught that, 'when bodies die, souls join the soul of the world.'[15] Others said that the Earth is alive, that therefore it moves. Stars and planets were other worlds. They said that souls live in those other worlds, and that there are demons living on the Moon. They said that Pythagoras came from Jupiter and had lived on the Moon as a demon.

Whereas Jesus had been dead for less than three days before his resurrection, Pythagoras allegedly spent two centuries in Hell, and yet returned to life. Diogenes cited claims that Pythagoras was the god Apollo who spent 207 years in the underworld, and saw all the men who ever died: 'he saw the soul of Hesiod tied to a column and gnashing its teeth; and that of Homer hanging from a tree with serpents around it, as a punishment.' Diogenes added that Pluto, god of the dead, ate only with the Pythagoreans.[16] Some said that Pythagoras taught that there exists not just one God, but many.

Similar to the apostles of Jesus, some of the disciples of Pythagoras allegedly resurrected the dead and exorcised demons. But unlike the apostles, allegedly some Pythagoreans even returned to life after death.

The Church Fathers complained that all of this was indecent madness, perverse falsehoods most deserving of ridicule, evil imitations of the miracles of Jesus. St John Chrysostom declared, 'this is the snare of the Devil.' Using magic and divination, the Pythagoreans seemed to control nature and gain advice from demons. The Christians complained that the Pythagoreans presumed to be equal to God. One of the admirers of Pythagoras,

Porphyry, authored a scathing attack on Christianity, *Fifteen Books against the Christians*. These books were banned by the first Christian Roman emperor. There was extensive censorship, and the Christians seized and burned the most offensive Pythagorean works until only scattered quotations remained.

Because of his extensive, historical-critical attacks, Porphyry became known as one of the greatest enemies of Christianity. His name became synonymous with blasphemy. St Augustine later said that Porphyry hated Christ and was 'the most bitter enemy' of the Christians. Augustine criticized Pythagoras as a necromancer who tried to divine the future by contacting the dead, or demons pretending to be gods.

Throughout the centuries the Christians rejected the 'poisonous doctrines', 'devilish' deceptions and perverse blasphemies of the Pythagoreans. And they declared, 'Philosophers are the patriarchs of the heretics.'[17]

In 1244 Thomas Aquinas joined the Dominican order, which had been recently founded to preach the Gospel and to combat heresies. Thomas eventually became famous for authoring extensive commentaries on scriptures and theology as well as on Aristotle's works. Thomas's opinions became so authoritative that they were held in the highest esteem not only by the Dominicans but by other Catholic orders such as the Jesuits, founded in 1540. Hence his judgements on Pythagorean topics are especially relevant.

In 1594 the Fifth General Congregation of the Jesuits issued several rules requiring that their theologians and teachers should follow St Thomas. The Jesuits explained:

> by unanimous consent of all the deputies, it was decreed that our professors must follow the doctrine of St Thomas . . . Our teachers in scholastic theology are to follow the doctrine of St Thomas. Only those who are well disposed toward St Thomas are to be promoted to chairs of theology. Those who poorly understand him, or who are not in sympathy with him, are to be relieved of the office of teaching.

Their rules also warned that professors of theology 'should not commit themselves or their students to any philosophical sect'.[18] So what did Thomas say about Pythagorean beliefs?

From 1265 to 1274 Thomas Aquinas worked on his *Summa Theologica*. Among many topics he mentioned Pythagoras. Thomas complained that Pythagoras construed the number 'one' as a kind of undivided being, and since numbers are composed of units, Pythagoras believed that numbers were the substance underlying all things.[19] It seems that Thomas was echoing Aristotle, who had argued that the Pythagoreans construed numbers as the origin of all things. Yet Thomas attributed this notion to Pythagoras himself.

Thomas also denied the notion that there exist multiple worlds. Like Hippolytus, he attributed this false claim to those who did not acknowledge the ordering wisdom of God. Thomas declared: 'Those who posit many worlds do not believe in any ordaining wisdom, but in chance, as Democritus, who said that this world and infinitely many others came from a concourse of atoms.'[20]

Citing Aristotle's *On the Soul*, Thomas denied the claim that the human soul can pass from body to body. Aristotle had said that the Pythagoreans were wrong about the transmigration of souls, yet Thomas attributed this doctrine to Pythagoras himself. In an article on 'Whether the Fire of Hell is Underground', Thomas noted:

> Pythagoras truly posited the place of punishment in the sphere of fire, which he said is in the middle of the whole world: and he called it the prison of Jupiter, as Aristotle relates (*On the Heavens and World* II). However, it is more in keeping with Scripture to say that it is beneath the earth.[21]

Again, Aquinas repeated the habit of attributing to Pythagoras beliefs of some so-called Pythagoreans.

Despite the great fame of the wise men of ancient Greece, for many Christians they could never compare to the Church Fathers, the saints, or the most pious clergymen. This was because the Greek philosophers seemed to be essentially ignorant of the most important theological truths and had indulged their misguided imaginations. For example, in Dante's *Inferno*, completed after about 1315, he named Orpheus, Democritus, Empedocles, Anaxagoras, Plato, Ovid and Virgil among the inhabitants of 'the first circle of Hell'. Dante did not mention the mystic Pythagoras, although he fits well in such company. In a separate work, which was published posthumously, Dante wrote: 'Pythagoras and his followers said that this world was one of the stars.' But Dante insisted that the 'glorious' Aristotle had refuted this opinion as false.[22]

Over the centuries Christian critiques against Pythagoras were supplemented by critiques against his advocates. Apollonius of Tyana, the reputed prognosticator and exorcist, was repeatedly denounced for allegedly performing exorcisms and resurrecting a man and a dead bride. Niceforo Callistos and Pico della Mirandola argued against divination and denounced Apollonius for practising demonic magic to imitate Christ's miracles. Porphyry, who advocated for belief in many gods, had celebrated Pythagoras as a moral and divine superhuman, more eminent than anyone. In one of his earliest works Porphyry had claimed that the Sun god Apollo 'exposed the incurable corruption of the Christians, saying that the Jews, rather than the Christians, recognized God'.[23]

But the Catholic Church did not always oppose pagan culture. In the 1330s the Italian poet Petrarch initiated a revival of interest in the ancient classical cultures of Greece and Rome. He realized that instead of studying only heavenly or worldly things, one should thoughtfully contemplate humans themselves. He argued, compellingly, that the secular study of human nature was compatible with an authentic Christian relationship with God. Petrarch's works inspired many Christian intellectuals to participate in what became known as Renaissance humanism. Hence the Church and its adherents sought ways to incorporate certain aspects of ancient Greek philosophy into Christian culture, and even into theology. It was not easy, because often the ancient pagan ideas were clearly in conflict with Christian dogma as traditionally construed.

Eventually humanist culture came to dominate the still fundamentally Christian countries and kingdoms of Europe. In this new intellectual climate, trials of heresy became progressively milder, and discussion and debate on religious and philosophical topics became more tolerated. By the 1480s the works by the Catholic priest Marsilio Ficino, especially his translations and commentaries of works by Plato and the legendary 'Hermes Trismegistus', greatly influenced humanists and the Catholic Church. Ficino even used the philosophy of the Pythagoreans and Plato to try to elucidate and refine biblical doctrines and Christian theology.

However, the situation changed dramatically by the 1520s. Following the efforts of Luther and Calvin to reform Christianity in northern Europe, the Roman Catholic Church became increasingly intolerant of unorthodox beliefs. In 1520 Pope Leo x issued his papal bull against Martin Luther. The Pope condemned the heretical writings of Luther as 'blinded in mind by the father of lies' – the Devil. The Pope denounced Luther by saying 'a new Porphyry has arisen.' This shows Porphyry's extremely inimical role for Christians at the start of the Protestant Reformation.

Meanwhile, in 1510 a 37-year-old Nicolaus Copernik became a canon of the cathedral chapter of Frombork (Frauenburg) in Poland. Privately, he drafted a manuscript, *Commentariolus*, on the theory that the Sun, planets and stars do not circle the Earth. Instead, Copernicus argued that the Sun is at the centre of planetary motions, that the Earth is a planet. Some of his friends, including the Cardinal of Capua and the Bishop of Culm, encouraged him to publish, but he resisted. Why? He did not publish his theory until more than three decades later, when he prepared to print *De Revolutionibus*. He dedicated the book to Pope Paul iii, telling the Pope that it was good to 'follow the example of the Pythagoreans and certain others, who used to transmit philosophy's secrets only to kinsmen and friends, not in writing but by word of mouth, as is shown by Lysis' letter to Hipparchus'. In this preface, Copernicus also took the opportunity to criticize those who stupidly argued about mathematical topics without understanding them. In this connection, he briefly criticized Lactantius, an ancient Christian authority, noting that although he was a celebrated writer, he was not a mathematician. Copernicus complained that Lactantius spoke in a childish way about the shape of the Earth, in saying that it is 'ridiculous' that its shape is spherical.[24]

In the ancient works of Cicero and 'Plutarch' (the *Placita philosophorum*), Copernicus found reference to various Pythagoreans, including Philolaus, who claimed that the Earth spins or moves.[25] The *Placita* stated: 'Some insist that the Earth is immovable; but Philolaus the Pythagorean says that it moves circularly around the central fire, in an oblique circle like the Sun.'

Copernicus also discussed the allegedly divine significance of the Sun. He noted that Plato had argued that to become godlike one must know the Sun, the Moon and the heavenly bodies. Copernicus added that some people called the Sun the ruler of the universe, that 'Trismegistus called it a visible god, and Sophocles' Electra, the all-seeing. Thus indeed, as though seated on a royal throne, the Sun governs the family of planets revolving around it.'[26]

Copernicus was so impressed by the Pythagoreans that he had planned to include the ancient Pythagorean letter from Lysis to Hipparchus in *De Revolutionibus*. Copernicus translated it from the Greek and included it in a draft, but for some reason he deleted it before the book was printed, though it was later found in his manuscripts. There, Lysis allegedly wrote:

> I would never have believed that after Pythagoras' death his followers' brotherhood would be dissolved. But now that we have unexpectedly been scattered hither and yon, as if our ship had been wrecked, it is still an act of piety to recall his godlike teachings and refrain from communicating the treasures of Philosophy to those who have not even dreamed about the purification of the soul. For it is indecent to divulge to everybody what we achieved with such great effort, just as the Eleusinian goddesses' secrets[27] may not be revealed to the uninitiated. The perpetrators of either of these misdeeds would be condemned as equally wicked and impious . . . That godlike man prepared the lovers of philosophy . . . divine and human doctrines were promulgated by him.

This letter would have ended Book 1 of his *De Revolutionibus*. The life of Copernicus concluded with the end of his secrecy: he died in 1543 just hours after he received a copy of his printed book.

To understand originality it is common to compare innovators to their predecessors. Long before Copernicus some astronomers such as Aristarchus had argued that the Earth circles the Sun. But Copernicus hardly knew about the theory of Aristarchus, since its main account came to light in 1544, months after Copernicus died, when *The Sand-reckoner* by Archimedes was first printed in Europe. Copernicus had vaguely heard about it, since he mentioned it very briefly in a manuscript: 'Philolaus believed in the mobility of the Earth and some say that Aristarchus of Samos was of that opinion.'[28] But Copernicus did not specify this in his book; he omitted the sentence along with his translation of the 'Letter from Lysis'.

Another account had appeared in 1440 when the German theologian Nicholas of Cusa had argued that the Earth moves and 'is not the centre of the world'. But he did not say that the Earth orbits the Sun, just that the Earth is a star and moves the least of all the heavenly bodies, as it orbits a centre that is essentially God. He also spoke of the inhabitants of Earth and the stars:

> It cannot be said that this place of the world is inhabited by men & animals and vegetables that in degree are more ignoble than the inhabitants of the region of the Sun & other stars. For although God is the centre & the circumference of all the stellar regions, & although from Him there arise in each region inhabitants of diverse natural nobility, so that not all the celestial & stellar places are empty, & that perhaps lesser beings do not inhabit this Earth alone; still, given the intellectual nature that inhabits this Earth & its region, it does not seem that, according to the order of nature, there could be a more noble or more perfect nature, even if there are other kinds of inhabitants in the other stars: man does not desire another nature, except only perfection in his own [nature].[29]

In contradistinction, Copernicus voiced no such discussions about the inhabitants of other stars or planets. In 1484 Nicholas of Cusa became a cardinal, yet his arguments about the Earth's subtle motion did not convince or grab the attention of astronomers. The substantive, mathematical work of Copernicus, however, generated plenty of attention. More than Nicholas of Cusa's theory, that of Copernicus seemed to resemble the ancient Pythagorean claim that the Earth orbits a central fire, if at least one misconstrued that central fire as the Sun. Indeed, what educated readers widely knew, thanks to the famous Aristotle, was that some Pythagoreans wrongly said that the Earth moves. Hence Copernicus seemed to revive or refine the theory of the Pythagoreans.

Still, it is inappropriate to call Copernicus a Pythagorean because the beliefs that he explicitly shared with them, at least overtly, were few.[30] Yet it seems that he practised their secrecy as he painstakingly developed the theory that he attributed to them. And in writing, by explicitly referring to the Pythagoreans, he effectively connected the new and radical astronomy to the legendary pagan cult. In turn, the critics and admirers of Copernicus both linked his theory to the Pythagoreans.

In 1542 Pope Paul III established the Supreme Sacred Congregation of the Roman and Universal Inquisition. Its official function was 'to maintain and defend the integrity of the faith and to examine and proscribe errors and false doctrines'. It served as the final court of appeal in trials of heresy.

Soon after Copernicus died – that is, promptly after his book was printed – the main censor in Rome decided to condemn the book. His name was Bartolomeo Spina and as Master of the Sacred Palace, chief theologian for the Pope, it was among his duties to sanction the printing of all religious books, and to censure heretical publications.

We know about his objections only because, at the time, the Dominican friar Giovanni Maria Tolosani specified in a manuscript that Master Spina had intended to condemn the work of Copernicus but was impeded by illness. Tolosani criticized Copernicus's book at length, for idiotically disregarding Aristotle's critiques of the Pythagoreans and for contradicting the Bible. Tolosani rejected 'the false imagination' of the Pythagoreans for having said that 'the Earth, existing as one of the stars', circles around a central fire. Friar Tolosani warned:

> the contrived Pythagorean opinion has long deserved to be extinct yet once again it rises though directly contrary to human reason and against the sacred writings, from which easily there can arise quarrels between the Catholic expositors of divine Scripture and those whose obstinate mind adheres to false opinions. This small work of ours has been written to prevent such a scandal from happening.[31]

But Tolosani's manuscript, dated 1546, remained unpublished, and his friend Spina died that year. Other important deliberations occupied the clergy, especially the Council of Trent, which transpired from 1545 to 1563. With it, the Catholic Church reintroduced the medieval persecution of heresies in a severe but highly bureaucratic way.

In the late 1550s the Roman Catholic Church also established the Index of Forbidden Books, to either prohibit or censure offensive works in order to prevent the spread of heresies, including Protestant beliefs. The precise definition of heresy became a subject of debate after the Council of Trent had led to the systematic establishment of the Inquisitional courts and the Index of Forbidden Books. Officially proclaimed heresies justified the death penalty; erroneous or distasteful opinions instead usually concluded by requiring that the culprit voice an abjuration. Nevertheless, there was an initial period in which the Inquisitors themselves tried to carry out a policy that favoured toleration and peaceful persuasion. However, as ever more Protestants abandoned Catholicism, the Popes at the Vatican enacted increasingly severe measures of punishment, including the death penalty.

By the 1590s and early 1600s various officials of the Catholic Church became increasingly hostile to pagan beliefs, just as they rejected Protestant beliefs. In that period cardinals and consultors of the Roman Inquisition as well as members of the Congregation of the Index of Forbidden Books examined and denounced various Pythagorean beliefs. Such developments echoed the previous conflict in which early Church Fathers, from roughly 150 to 550 CE, criticized and denounced Pythagorean beliefs as false and heretical. Once again, prominent Christians criticized the infectious speculations of philosophers. Pythagoras appeared in books on astrology, divination and demonology. But soon Galileo's telescopic discoveries seemed to suggest that maybe, just maybe, Pythagoras was right: there were landscapes on the Moon; it was another world.

The trials of Giordano Bruno and Galileo Galilei were caused partly by their bold, personal efforts to uphold the humanist culture of dialogue and tolerance in a period when both parts of Europe, the Catholic and the Protestant, were becoming rigidly dogmatic and intolerant.

If I may compare the Copernican Revolution to a large marble sculpture, the issues to be discussed can be compared to thin, deep stains that cut through the marble, at some points emerging in its surface as a fracture. This analogy should serve as a reminder of the approach taken: tracing inconspicuous connections between ancient notions and Copernican developments. I stress the importance of this analogy because this is not a reinterpretation of the Copernican Revolution. Instead I expose a neglected chain of events within it. It is not an account of the rise of the theory that the Earth moves; it is a supplement to such discussions. I will set the works and concerns of Copernican astronomers in the context of forgotten pagan heresies.

The Moving Earth and the Fugitive Friar

The one admirer of Copernicus who most earnestly held Pythagorean beliefs was the Italian philosopher and Catholic friar Giordano Bruno. Famously, Giordano Bruno was a martyr for the freedom of expression. Despite being imprisoned and interrogated by Catholic Inquisitors for almost eight years, Bruno refused to recant his eccentric opinions. In February 1600 the Roman Inquisition declared him guilty of heresies. The Roman city officers gagged him, took him to a public square where a crowd watched, tied him to a post and burned him alive.

Countless books and articles have been written about Bruno. Let us discuss a neglected but important dimension: how did his beliefs about the cosmos relate to his trials and execution? To trace the Pythagorean thread in his life, consider first his education.

Bruno was born in 1548 in the village of Nola, Italy, at the foot of the ancient volcano Vesuvius. At an early age, in 1565, he entered the Dominican monastery in Naples, like Thomas Aquinas, to lead a life of devotion and prayer. As a novice, however, Bruno once discouraged another novice from reading a book on Marian devotion. Someone also saw Bruno discard some images of saints and reported him. For such reasons young Bruno was denounced to the Neapolitan Inquisition, but he was not processed. Having completed his courses in theology, focused on the doctrines of Thomas Aquinas, Bruno was ordained priest in 1573.[32]

However, Bruno cultivated objections against Aristotle's philosophy, the doctrine of the Trinity and other points of theology. He thought that some heretics had proposed interesting ideas that were not necessarily wrong. Bruno also secretly managed to read some forbidden books. Hence in 1576 he was again denounced to the Inquisition for doubting the Trinity, rejecting the cult of saints and sympathizing with Arian heretics.

To avoid the confrontation, at the age of 28 Bruno fled from the monastery and travelled to Rome. Meanwhile, his superiors found that he had hidden prohibited books in the latrine of his cell.[33] The books included texts by saints Jerome and John Chrysostom that had been edited by Erasmus of Rotterdam, the Catholic priest who had become a social critic and humanist, and whose works had been banned by the Index of Forbidden Books.

Erasmus sought wisdom from many sources and he discussed many ancient ideas and sayings. Discussing philosophers, he sometimes referred to Pythagoras. Like others, Erasmus praised Pythagoras as a sage who voiced insightful pronouncements about life. Apparently Pythagoras had argued that friends should share all their possessions as common property. He described life as a solemn gathering in which some people fought, others did business, while others were spectators (philosophers who peacefully contemplate people and nature).[34] Erasmus portrayed Pythagoras as an admirable role model. He claimed that the apparent 'superstitions' that were taught by Pythagoras were just meant to train youths in ways that would facilitate learning. As examples, Erasmus mentioned the prohibition against eating meat, the praise of vegetables, the five-year restriction of silence and the oracular precepts. He argued: 'Our modern monks, who are monks in name only, seem to have imitated some of this man's teachings, and those who control their bellies and their tongues are the ones who sin least.' So according to this one priest, Pythagoras was above most, 'for no one among the pagan philosophers is more saintly than this man.'[35] Erasmus asked Christians to carry out Pythagoras' advice to review one's actions and duties before going to sleep every night.

Of course, Erasmus also discussed the Pythagorean idea of the transmigration of souls: in *Praise of Folly*, for example, Erasmus referred to Lucian's *The Dream, or the Rooster*, in which a rooster spoke and identified himself as having formerly been Pythagoras. Erasmus celebrated

> that never-sufficiently-to-be-praised Pythagorean rooster, who in his own person had occupied many shapes, as a philosopher, a man, a woman, a king, a lowly subject, a fish, a horse, a frog, even I think a sponge – after which he concluded that no animal was more wretched than man because all others were content with the limits imposed by nature.[36]

Furthermore, in other writings Erasmus said that Pythagoras construed the mind or the soul as being imprisoned in the body.[37] He also discussed the precept against eating flesh. Erasmus discussed what the Pythagoreans viewed as the 'monstrous' crime that had become very common:

> taking the carcasses of slain animals for food, to tear dead flesh with their teeth, to drink the blood and suck the gore, to 'stuff their entrails with other entrails' as Ovid says. That crime, however monstrous it might seem to gentler minds, was nevertheless sanctioned by usage and convenience. It has even become a pleasure to see, among the luxuries of the table, the semblance of a corpse. Meats are covered under crusts like buried corpses, they are embalmed with scents.[38]

Erasmus did not focus his discussions on Pythagoras or his followers, but these passages just illustrate his willingness to engage unusual intellectual perspectives, slight departures from Catholic doctrines.

Since his fellow monks knew that Bruno had read some of Erasmus's forbidden works, plus he had committed other questionable acts, it seemed that Bruno should be required to meet with an Inquisitor. But Bruno had fled in 1576, so the proceedings could not be carried out.

Impressed by the theory that the Earth moves around the Sun, Bruno studied ancient texts on astronomy, including the *Placita* allegedly by Plutarch, in which the author or authors discussed 'Pythagorean' ideas that greatly impressed Bruno: that stars are worlds, the universe is infinite, and there exist other worlds similar to Earth. Bruno also continued to study texts that had been prohibited by the Congregation of the Index. Following the Pythagoreans, he pondered their major religious belief: that human souls are reborn, even in animals. He also pondered unorthodox ideas about Jesus Christ, such as whether others could perform similar feats. Moreover, in his writings Bruno argued as if knowledge by reason is superior to knowledge by faith.

At the time numerous writers discussed magic and divination. Some of them cited references to Pythagoras and his legendary powers. For example, in 1558 Christofo de Cattan had published a popular book, *Geomancy, with the Wheel of Pythagoras*, in Italian.[39] The 'Pythagorean Wheel' was commonly called the Wheel of Fortune, which included, along its circumference, rows of numbers and the alphabet, and sometimes signs for the Sun, the Moon and the planets, all to divine a person's future arithmetically. Cattan said the Wheel could be used to detect lies, to predict who will die first, to predict the sex of a child, whether a prisoner will be freed, whether a disease will kill and so on.

Reputedly, Pythagoras was a master of divination: numerology, hydromancy, geomancy and onomancy (deciphering the future on the basis of numbers, water, bits of dirt, or words, respectively). In another work, *On the Trickery of Demons*, another author, Johan Weyer, described how magicians colluded with demons to effect seemingly impossible tricks, for example, that Pythagoras appeared in two distant cities at once, and that Apollonius travelled instantly from one city to another. Pythagoras appeared as someone who had travelled to many distant lands in search of arcane knowledge. Echoing Philostratus, Apollonius was described as a magician who allegedly once resurrected a young bride in Rome, and who summoned the soul of Achilles the warrior, from Hell, back into his body.[40] By the 1590s the Vatican launched an attack on magic and the occult arts. Apollonius was viewed as a demonic magician and great enemy of the Church, who conspired with the Devil to overthrow Christianity.

Meanwhile, a few Christian writers defended the theory of Copernicus. In 1576 Thomas Digges argued that Copernicus correctly said that the Earth circles the Sun. He echoed Copernicus in writing that '[Hermes] Trismegistus called him [the Sun] the visible god. Thus doth the Sun like a king sitting in his throne govern his courts of inferior powers.' But while Copernicus imagined a bounded sphere of stars surrounding the planets, Digges argued that the Sun-centred universe is infinite, with infinitely many stars throughout it. He repeatedly described the sphere of the stars as 'that Orbe immoveable garnished with lights innumerable', and again, 'that fixed Orbe garnished with lights innumerable and reaching up in Sphærical altitude without ende'. But at one point he backed down a bit, echoing Copernicus: 'But whether the worlde have its boundes or bee in deed infinite and without boundes, let us leave that to the Philosophers.'[41]

On the first page of his account Digges included an image of this Sun-centred universe, and he titled it 'A perfect description of the Celestial Orbs, according to the most ancient doctrine of the Pythagoreans.' He asked that this 'so ancient doctrine' that Copernicus had 'revived', should be investigated further, 'and not rashly condemned for phantasicall'.[42] Digges was one of the leading

Statue of Giordano Bruno at the Campo de' Fiori in Rome. Sculpted by Ettore Ferrari, 1889.

mathematicians and scientists in England at the time, so his book was reprinted multiple times, helping to blend the Copernican and Pythagorean notions.

In 1579 Giordano Bruno travelled to Geneva, seeking to live in freedom and safety from Catholic persecution. He later said that he had not gone to Geneva to adopt the religion of Calvin, yet he often listened to preachers and read works by Calvin and other Protestants. Soon Bruno became embroiled in an argument against a local evangelical pastor, Antoine de la Faye, and was detained. The Calvinists felt antagonized by Bruno. The Consistory admonished him for calumnies, demanded that he repent and forbade him from taking the sacrament of communion. Apparently Bruno then appealed that they undo their judgement, but he soon departed from Geneva. In 1583 he travelled to England, where, incidentally, the 'Pythagorean' account by Digges had been published. That year Bruno published a book titled *The Seal of Seals*, in which he listed the greatest men ever, the best 'inventors, teachers, leaders, and pastors of their people', starting with Pythagoras, then Moses, followed by Jesus.[43]

In 1584 Bruno published *The Ash Wednesday Supper*, a post-Copernican dialogue about the structure of the universe. In it Bruno described himself in the third person as 'the Nolan', and he used a character named Theophil, a philosopher, to convey his views. Through his mouthpiece Bruno argued that there exist innumerable worlds, similar to Earth, and that the Earth is actually a kind of animal. Like Thomas Digges, Bruno argued that the universe is infinite. He even bragged about this finding, as it gave evidence of the majesty of God's infinite power: 'Thus we shall advance to the discovery of the infinite effect of the infinite cause, the true and living evidence of the infinite vigour.' And he added that 'it consists of an infinite ethereal region', the same claim that Aristotle had attributed to the Pythagoreans.

There is no clear evidence that Bruno read Digges. However, both of them read Marcello Palingenio Stellato's *Zodiacus vitae* (1543), which argued the universe is infinite but has a hierarchical structure: life and misery exist only on the Earth, while the unchangeable heavens are illuminated by spiritual and immaterial lights, unlike the Sun.[44]

In the *Timaeus*, Plato had asserted that the sensible world must be finite and spherical. Similarly, Aristotle insisted that the universe was finite. Hence Copernicus too portrayed the universe as finite and spherical. According to Digges, however, the sphere of stars extended infinitely outward and was the Celestial Court inhabited by God and the chosen ones. For Digges, the Sun was not a star, and only the Sun was surrounded by planets. In contradistinction, according to Bruno the universe was infinite and *homogeneous*, such that the Sun was not the centre, and countless other stars were also suns, surrounded by planets. For Bruno the universe was not spherical or hierarchical. He argued that our Sun, Moon, planets and Earth were actually different kinds of 'stars', what astronomers now call heavenly bodies. Unlike Digges, Bruno construed God to be equally present in all regions of the universe.

Moreover, Bruno rejected the common notion that life and death exist only on Earth. Bruno made explicit references to having read the *True Story*, in which Lucian of Samosata had described a voyage to the Moon as an inhabited world. We might well point out that in Lucian's account Pythagoras appeared as having lived in that region of space, in some of the 'islands in the air', as his soul transmigrated through different bodies. But Bruno griped that Lucian sought merely to mock the philosophers, and hence suffered from blind ignorance of the truth of such matters. Bruno insisted that the stars and the Earth are living things, which give life to other living beings. He added that each of the heavenly bodies has a soul, which is the cause of its motions, and that such souls are both sensitive and very intelligent. Bruno further argued that although the Earth has flesh, bones, blood, veins, nerves, organs and even senses, all are somewhat different to those of humans.

Bruno repeatedly opposed Aristotle. He criticized Aristotle for using overly refined abstractions and abstruse logic. Bruno also scorned Ptolemy's astronomy. He attacked Aristotle's physics, he criticized the notion of a finite universe and he claimed that circular motions do not really exist.[45]

Bruno further developed his theory of many worlds in another book of 1584, *On the Infinite Universe and Worlds*, apparently printed in Venice. Bruno gave reasons why the universe is infinite, and he quoted passages from Lucretius' *On the Nature of Things* (*c.* 50 BCE), in which the author elaborated the views of Democritus and Epicurus. Like them, Lucretius had remarked that 'it is in the highest degree unlikely that this Earth and sky is the only one to have been created.'[46] Bruno's book was a dialogue, and the character that spoke for him voiced this argument about the stars:

> the universe being infinite, there must ultimately be more suns: because it is impossible that the heat and light from a particular one can spread throughout immensity, as Epicurus could imagine, if what others report of him is true. Therefore it necessarily follows that there exist innumerable suns, of which many are visible as sorts of small bodies: but such an apparently minor star can be much larger than the one that seems the greatest.[47]

Bruno argued that despite their heat, such suns might be inhabited by some beings or animals. Such beings would 'vegetate' by virtue of the surrounding cooling bodies, analogously to how those in our world depend on the Sun's heat. The Earth seemed to be a relatively opaque star, whereas some other stars are bright. He also argued that some stars are moons. He argued that all stars are worlds and, furthermore, that stars are rational, moving animals, which, like the Earth, have waters that move within their caverns as blood through veins. He spoke about a soul that embraces and infuses all things. He referred to the Earth as a 'divinity'. Furthermore, Bruno praised the views voiced in Plato's *Timaeus* and by Nicholas of Cusa. Against Aristotle, Bruno argued that the accepted concept of the order of the elements and the heavenly bodies was a vain fantasy.

Bruno portrayed Pythagoras as a role model, denying such praise to Plato. Bruno echoed Cicero's claim that Pythagoras coined the word 'philosopher', and then Bruno called himself a philosopher. He expressed his allegiance by referring to 'the school of Pythagoras and our own'. He praised the 'saintly words of the philosopher of Samos'. Bruno defended the theory of the transmigration of souls (metempsychosis), and argued that soul is in things, that it survives death, by existing in atoms. He voiced his desire to create a new science based on Pythagorean principles. Furthermore, in his writings Bruno referred to some famous Pythagoreans, even Apollonius, whom he called a magician. Bruno alluded to Apollonius' ability to know things at a distance.[48]

Meanwhile, in 1584 the Spanish theologian Diego de Zúñiga published a commentary on the Book of Job, defending the Copernican theory. Zúñiga noted that in the Bible the characterization of God as 'He who moves the Earth from its place, and its pillars tremble' seemed to say that the Earth moves. He claimed that this passage could be explained by 'the statement of the Pythagoreans' and of Copernicus.[49] Seemingly contrary passages, such as 'Generations will come, and generations will pass away, but the Earth remains forever' (Ecclesiastes 1:4), argued Zúñiga, referred not to the Earth unmoving but to its durability. He credited the Pythagorean opinion of the Earth's motion to Philolaus and Heraclides (citing the *Placita*), followed by Numa, and Plato when old.

In a work of 1585 Bruno described a dialogue between a donkey and a man called 'the Fool'. This book presented abundant sarcasm against the Christian religion. At one point the wise donkey brings up Pythagoras and the notion of transmigration, by saying:

> Don't be so proud, O Fool, and remember that your Pythagoras teaches not to disdain any thing found in the bosom of nature. Although I am an ass in form at present, I may have been and may become again the form of a great man; and although you are a man, you may have been and may yet become a great ass, according to what will seem expedient to the distributor of clothes and abodes, and the dispatcher of transmigrant souls.

Another character in the dialogue, Onorio, makes several statements about transmigration. One of his companions referred to Onorio as 'living now in the terrestrial, now in the infernal, now in the celestial abodes'. Such statements and others seem to suggest that Bruno favoured the notion of the transmigration of souls.[50]

Like Copernicus, Bruno also praised the legendary discipline of silence of Pythagoras:

> the most profound and divine theologians say that God is honoured and loved more by silence than by words; as can be seen more by closing one's eyes to the representing species than by opening them: thus the negative theology of Pythagoras and Dionysius [the Areopagite] is much more renowned, above the demonstrative one of Aristotle and the scholastic doctors.[51]

In 1588 Bruno moved to Prague. There the Holy Roman Emperor Rudolf II was nominally Catholic, though he tolerated the Protestants and was interested in mysticism and alchemy. Bruno promptly dedicated a work to Rudolf, titled 'One Hundred and Sixty Articles Directed against the Mathematicians and Philosophers of the Day'. Bruno voiced hopes for religious unity and complained about intolerance. He wanted to transform Catholic Christianity. He complained that 'authority usually binds and deceives in countless ways', such that thinkers should free themselves from 'subjection'.[52] Still, Bruno left Prague to go to Helmstedt where a new university had recently opened. He matriculated in January 1589 and dedicated much time to reading and writing about magic, that is, about how to effect occult but natural feats. However, Bruno encountered religious controversies again. The chief pastor of the Lutheran church in Helmstedt excommunicated him. The charge was a public dishonour, as if he were a heretic. Bruno's ideas seemed so unorthodox that some Protestants did not tolerate them. So he left the city and headed to Frankfurt in mid-1590.

In 1591 Bruno published a work building upon the ancient Pythagorean notion of 'monads', as the constituent elements of things and numbers.[53] Echoing Philostratus, Bruno noted that: 'Apollonius, thanks to the virtues of numbers, resurrected a girl, after she heard her name.'[54] Was it the case that only Jesus Christ and his apostles could perform resurrection? Bruno also again seemed to advocate transmigration. In another book, Bruno argued:

> According to how a soul behaved in one body, so too it disposes to depart from it, as said by Pythagoras, the Sadducees, Origen, & others and many Platonists . . . Therefore some [souls] travel among human bodies, others enter the bodies of heroes, while others are driven into the degraded ones.[55]

Prisoner of the Inquisition

In 1591 Giovanni Mocenigo, a Venetian aristocrat, sent Bruno a personal letter inviting Bruno to teach him. Bruno went to Venice and began teaching philosophy to Mocenigo, but in May 1592 Mocenigo sent a written complaint to the Venetian Inquisition, with serious accusations against Bruno. I will focus on the accusations pertaining to Pythagorean beliefs, but in order to see how the subsequent accusations against Bruno related to this earliest set, we should list all of Mocenigo's claims. I quote them in the sequence they appear in Mocenigo's original complaint, but add numbers to itemize them. I have separated each distinct claim rather than group them according to the semicolons; and I have omitted a few superfluous phrases, adding ellipses:

> *First Accusations against Bruno, by Giovanni Mocenigo*
>
> (1) that he said that it is a great blasphemy of Catholics to say that bread transubstantiates into flesh [of Christ];
> (2) that he is an enemy of the [Catholic] mass;
> (3) that no religion pleases him;
> (4) that Christ was a wretch, who did wicked deeds to

> seduce people, that one could well predict that he should be hanged;
> (5) that he saw no distinctions of persons in God [the Trinity], and that this would be an imperfection in God;
> (6) that the world is eternal,
> (7) and that there are infinite worlds,
> (8) and that God creates infinity continually . . .
> (9) that Christ did [merely] apparent miracles and that he was a magician . . .
> (10) that Christ showed himself unwilling to die, and that he tried to escape inasmuch as he could;
> (11) that there is no punishment for sins,
> (12) and that the souls created by the work of nature pass from one animal into another;
> (13) and that just as brute animals are born from corruption, so too will humans, when after the deluges they will be born again.
> (14) He showed himself wanting to be the creator of a new sect under the name of the New Philosophy;
> (15) he said that the Virgin could not have given birth,
> (16) and that our Catholic faith is full of blasphemies against the majesty of God . . .
> (17) that all are asses, and that our opinions are doctrines of asses;
> (18) that we don't have proof that our faith has merit with God . . .
> (19) and that he is amazed by how God tolerates so many heresies by Catholics.
> (20) He says that he wants to study the arts of divination, and that he wants to run all over the world;
> (21) that St Thomas and all the doctors knew nothing compared to him . . .[56]

Thus Mocenigo accused Bruno of many heresies and blasphemies, saying that he suspected that Bruno was 'possessed by a demon'. At the time many denunciations to the Inquisition did not generate any proceedings against a suspect. Yet Mocenigo's accusations were so significant that the Venetian Inquisition promptly arrested Bruno, imprisoned him and began a process of interrogation.

The main function of the Inquisition was to condemn heretics. When someone was accused of holding a belief that clashed with Catholic dogma, the Inquisitors tried to compel that person to recognize the mistake. If that person refused to abandon such beliefs then he or she could be condemned to execution if such beliefs had been classified as 'formally heretical'. Other lesser beliefs could be deemed 'erroneous': they might be classified either as 'offensive', 'temerarious' (rashly opposed to theological consensus) or 'nearly heretical'. If a questionable proposition seemed theologically debatable, its proponent could be asked to merely 'retract' it rather than 'abjure' it.[57]

The Venetian Inquisition subjected Bruno to six interrogations over ten days. Bruno argued that he had not criticized the Catholic faith, but had mainly spoken philosophically. He denied various accusations. I now highlight only the relevant points concerning Pythagorean beliefs. In the numbered list of initial accusations, above, there are four arguably Pythagorean claims: (6) the world is eternal, (7) there are infinitely many worlds, (12) animal souls transmigrate into other bodies, (13) human souls are also reborn. By 'Pythagorean' I do not mean that I regard them as such, since we don't even really know the beliefs of Pythagoras; instead, I mean that philosophers and theologians in the Renaissance had attributed such notions to the Pythagoreans. These four claims do not seem to stand out among the other presumably more offensive heresies against Catholicism, but as we will see, they became increasingly important in Bruno's trial.

In his defence, Bruno clearly denied practically all of the offensive, embarrassing and dangerous statements that he had allegedly voiced against God, Christ, theologians and Catholicism. The historian Thomas F. Mayer grossly misrepresented Bruno's compliance: 'Unlike Galileo, Bruno did not need accusers. He was perfectly happy to incriminate himself.'[58] But no, Bruno's depositions actually show that he was very concerned with the accusations and denied nearly all of them; acknowledging only the few that he construed as not opposing Catholicism.

Still, Bruno voiced other opinions that had questionable theological relevance. In particular, in his third deposition Bruno explained his philosophical views. Again he said, 'I have not taught anything against the Catholic Christian religion.' But he acknowledged that philosophical arguments could be misinterpreted as being indirectly opposed to the Catholic faith. He said that he had articulated his philosophical views clearly in his books. He then summarized his views:

> I affirm an infinite universe, which is the consequence of the infinite divine power, because I regard it as unworthy of the divine goodness and power that, being able to produce in addition to our world another and infinitely others, it would produce one finite world. I have indeed asserted infinite particular worlds similar to the Earth, which with Pythagoras I consider a star, similar to which is the Moon, other planets and other stars, which are infinite; and that all those bodies are worlds and numberless, which thus constitute the infinite universality in an infinite space; and this is called the infinite universe, in which are innumerable worlds.[59]

Despite his intentions, Bruno's 'philosophical' opinions had theological significance because they referred to 'the divine power' and the nature of God's Creation.

When he was interrogated by the Inquisitors, he admitted that 'regarding the divine Spirit as a third person I have not been able to grasp the sense in which it ought to be believed, but instead according to the Pythagorean way, in accord to which way I understand it as the soul of the universe, as shown by Solomon'.[60] Bruno quoted a biblical line attributed to Solomon:

> 'For the spirit of God fills the Earthly orb: and therefore he who contains everything', which entirely conforms to the Pythagorean doctrine explained by Virgil in the sixth [book of] the *Aeneid*:
> 'In the beginning the spirit nourishes within
> the sky, and lands, and regions of water,
> and the shining globe of the Moon, and the Titanic stars,
> totally infused through the limbs
> a mind agitates the mass . . .'[61]
> and that which follows.
>
> Thus from this spirit, that I have called the life of the universe, I say in my philosophy, there originates the life

> and the soul of each thing that has soul and life, which I understand to be immortal; as all other bodies.[62]

Bruno immediately added: 'As for their substance, all [bodies] are immortal, death being nothing but division and congregation.' He quoted the statement in Ecclesiastes 1:9 that 'there is nothing new under the Sun.'

To explain himself, Bruno thus referred to book 6 of Virgil's *Aeneid* (*c.* 19 BCE). In it, Virgil had written about how the hero Aeneas descended to the underworld, where he then spoke with the phantom of his dead father, who revealed a secret vision to his son: countless souls (or ghosts) crowd around a dark river, waiting for new mortal bodies. The ghost explained that a great fiery spirit fills and animates all the world, in earth, water, air, heaven, the Sun, Moon and in every star. This spirit gave rise to animals and humans, as their souls originate in Olympus but are imprisoned in earthly bodies, until those bodies die and the freed souls return to the heavens. But each body somehow stains each soul; therefore each soul must undergo a painful penance, to purge sins and crimes. Once purified, the chosen few souls gain blissful Elysium, while the rest flow back to the world, wanting to gain new bodies.[63]

By itself this episode of Virgil's *Aeneid* might not necessarily seem to refer to the Pythagorean religion in particular, but to popular pagan beliefs, if only because Virgil did not name the Pythagoreans, at least in the *Aeneid*. However, Bruno did.

Bruno had published his claim about the great Spirit in Virgil's *Aeneid* in his book *On the Cause, Principle and One* (1584), where he wrote:

> The universal Intellect is the intimate, most real, peculiar and powerful part of the soul of the world. This is a single whole that fills everything, illuminates the universe and directs nature to the production of suitable species: it is concerned with the production of natural things as our intellect with the congruous production of rational kinds. It is called by the Pythagoreans the motive force and mover of the universe, as said the poet [Virgil]: 'Mind agitates the mass and intermixes itself with the great body.'[64]

Accordingly, historian Dilwyn Knox remarks: 'Bruno, of course, recognized that the World Soul was an essential feature of Plato's and Platonic cosmology . . . For Bruno, however, the doctrine was not Plato's invention, nor was it quintessentially Platonic. Rather, it was Pythagorean.' Knox recognized that ancient Stoics developed such notions, although Bruno was unaware of it.[65]

In his manuscript *On Mathematical Magic* Bruno had presented Virgil's words as being by Pythagoras himself. Bruno there argued:

> The heavens and the world are animated, and likewise heavenly bodies that are visible, as the most noble poets and the wisest philosophers have granted, and that one universal spirit is ingrained in the universal machine, one mind, infused through the framework, moves the universal mass, as said by Pythagoras.[66]

Much of Bruno's book *On Mathematical Magic* was copied from various sources, but this passage was not copied verbatim from any source. Furthermore, in another work Bruno overtly identified the universal spirit with God. Again he attributed this notion to the Pythagoreans, and he related it to polytheistic notions about the heavens, the stars and the Sun:

> The ancients and those more recent attribute Gods to the Heavens, where by Gods they really meant stars, from which none can be the first God; although among those that are in our sphere, the Chaldeans had the Sun rising in the centre of the stars as God's great temple, and for the Pythagoreans it is an infinite spirit that penetrates everything, comprehending and vivifying.[67]

Finally, to complete this compelling chain of evidence, in his work *The Heroic Furor* Bruno again said that the heavenly bodies are composed of superior powers (that abide with the divinity) along with inferior powers that abide with the 'mass' and 'vivify' the body, and sustain 'the living things in that world'. In that book Bruno referred to Virgil as 'the Pythagorean poet'.[68]

Regarding the theory of transmigration of souls from one body to another, Bruno testified to the Venetian Inquisitors, apparently with less certainty, that 'if it is not true, then at least it has verisimilitude following the opinion of Pythagoras.'[69] In sum, Bruno recanted or denied practically all charges of heresy, *except* the doctrine of the universe and the doctrine of souls. He said that he was 'reasoning as a philosopher'. These three points are striking because they confirm his Pythagorean beliefs, that is, he willingly denied every heresy he had allegedly said, except his Pythagorean convictions. To be sure, Bruno's beliefs were not merely or exclusively Pythagorean; he had developed his own idiosyncratic, original and subtle theories, yet his views involved well-known Pythagorean notions.

We should consider Bruno's phrase 'the Pythagorean doctrine'. It is an important expression especially because the cardinals later used it, in 1616, to censure Copernican works. Expressions such as 'the doctrine of Pythagoras' appeared in some ancient works, such as Philostratus' *Life of Apollonius*, where it referred to transmigration of souls, the memory of past lives, abstinence from eating animals and the importance of silence. In 1576 Thomas Digges praised the 'Doctrine of the Pythagoreans', a supposedly perfect description of the celestial orbs that included an immovable and perpetual orb of stars extending infinitely in all directions. In 1592 an Italian book on emblems referred to the most secret 'Pythagorean doctrine' according to which numerical proportions correspond to the elements and the theory of the world, leading to knowledge of 'the Real Kabbalah'.[70] Also in the 1590s a few works on government, in Italian, used the expression 'dottrina Pittagorica' as the name for the astrological theory that distinct numbers can be associated to the planets and the signs of the Zodiac to make predictions.[71] A 1594 work by a Spaniard, advocating the establishment of a new Catholic Council, included discussions about Pope Clement I, in which he disdained this late first-century Pope's father as 'a pagan man, and educated by the Pythagorean doctrine, not Christian or Mosaic', for having opposed Christian and Apostolic teachings.[72] The author alluded briefly to 'Pythagorean errors', in the context of religious heresies. This contrast seems to echo the ancient dichotomy posited by Hippolytus, who had criticized some heretics as 'disciples not of Christ but of Pythagoras'.

Thus the phrase 'Pythagorean doctrine' had been used to refer to mystic or pagan beliefs, numerical astrology and, in Bruno's case, the soul of the Earth and the soul of the universe. The phrase had also been used in literary elaborations of classical works. Ovid's ancient *Metamorphoses* featured Pythagoras but did not refer to him by name, only as the wise 'man from Samos'. It did not even say 'the Pythagoreans'. Yet later editors and translators added such names. In 1584 Giovanni dell'Anguillara published a translation of the *Metamorphoses* into Italian that greatly expanded it. For example, in the original Latin version, in Book 15, Ovid wrote 74 lines of verse before saying that 'the man from Samos' rejected the killing and eating of animals; yet in dell'Anguillara's version there are instead 272 lines of verse until that point. Dell'Anguillara referred to Pythagoras repeatedly by name, and he also referred to the teachings of Pythagoras as 'the new Pythagorean doctrine'.[73] Again, in the *Metamorphoses* Ovid had portrayed the Samian sage as revealing sacred knowledge: that everything changes, but that souls are immortal and, not keeping the same form, they can transmigrate into other human bodies or even into animal bodies; that the soul of Pythagoras had lived before; and that Earth is a living animal. Bruno actively defended all of these beliefs.

The Venetian Inquisition summoned more witnesses. Two were booksellers: Giovanni Battista Ciotti, also known as 'Ciotto', and Giacomo Bertano, also known as Jakob van Brecht from Antwerp. Both had first met Bruno in Frankfurt. Ciotti, from Siena, had introduced Bruno to Mocenigo. Ciotti and Bertano had hardly any accusations against Bruno. Andrea Morosoni testified that Bruno seemed to be a good Catholic. Among other witnesses was Domenico da Nocera. (It is unclear whether Celestino Arrigoni participated in the Venice proceedings or only in Rome.) Overall, they testified that Bruno regularly read forbidden books. Indeed, he was a fugitive friar who had consorted with Protestants in Geneva, London and Wittenberg, while publishing multiple books on controversial topics. Some of his books had false imprints, for example purporting to have been published in Venice when in fact they had been published in London. Such activities reeked of heresy.

Bruno then begged for forgiveness, saying 'I now detest and abhor' any and all errors committed in regard to the Catholic faith and life.[74]

The Venetian proceedings were interrupted by the Inquisition of Rome. The central office of the Inquisition determined that the case was so important that it should be handled directly. They argued that Bruno was not a citizen of Venice, but of Naples. The 'Supreme Inquisitor' in Rome, Cardinal Giulio Santori of Santa Severina, ordered that Bruno be transferred because he was not merely a heretic, he was something worse: a *heresiarch*, someone who leads and converts Catholics into heretics, and can become their intellectual patriarch.[75]

They complained that this 'popular heresiarch' had fraternized with heretics, he had written heretical books, he published them 'in heretical places', and he had praised royal heretics, such as Queen Elizabeth of England. His heresies were 'extraordinarily grave', such as doubting the incarnation and the Trinity, and leading a 'diabolical life'.[76] Pope Clement VIII himself wanted Bruno to be extradited and put on trial in Rome. For months the Venetian authorities resisted the extradition order. Historian Thomas Mayer explains that almost none of the detainees requested by Rome were allowed to leave Venice, 'unless the Venetians had decided on their own that they had outlived their usefulness'. Finally in February 1593 they transferred Bruno to the Inquisition's prisons in Rome. This was very unusual, because unlike other jurisdictions the Venetian authorities 'almost never allowed the extradition of those accused of heresy'.[77]

The Roman Inquisition continued the proceedings instead of beginning anew. In Rome a fellow Venetian prisoner, a friar called Celestino Arrigoni from Verona, confirmed some of Mocenigo's accusations and added new ones. This is very important because otherwise the Inquisition essentially had just one free witness against Bruno, namely Mocenigo, since his principal accusations had not been confirmed by the other witnesses. According to the practice of the Inquisition, one witness was equivalent to no witness.[78] Friar Celestino accused Bruno of blasphemies about Christ (that Jesus had sinned), that Bruno had erroneous opinions about Hell, that he criticized Moses and the prophets, that he denied Church dogmas and the Catholic book of rites, that he disapproved of praying to the saints, that he uttered blasphemies, that he planned to burn down a monastery and said that Cain was a good man for killing his brother Abel, in addition to the previous claims that stars are worlds, and about the transmigration of souls from body to body, and from one world to others.[79]

In turn, Bruno again defended himself and gave various clarifications, for example, that 'I was primarily joking when I told him

The building of the Holy Roman Inquisition, 'S. Inquisitio', near the southern wall of Vatican City, south of St Peter's Basilica. Map by Antonio Tempesta, 1593.

that Cain was a Pythagorean, that is, by having a Pythagorean zeal, because Pythagoras abhorred the slaughter of animals.'[80] In Genesis 4:4, Abel offered an animal sacrifice to God, while Cain did not.

Still more witnesses accused Bruno too. They reported that he had argued that Jesus sinned; that Bruno uttered blasphemies; that he doubted dogmas; that he said the world is eternal; that many worlds exist, and so on. There transpired ten more interrogations and depositions (for a total of sixteen), but Bruno continued to deny charges and renounce alleged heresies, except the plurality of worlds and the Pythagorean doctrine of souls.

Six witnesses against Bruno were fellow prisoners, which greatly undermined their credibility. Also, roughly half of Mocenigo's accusations were unconfirmed.

Censured Propositions in Bruno's Books

In early 1595 the Inquisition decided that its theologian consultors should systematically inspect Bruno's books in order to pinpoint heretical statements that should be censured. Over the following months they managed to gather only some of his books, not all. The identity of the theologians is unknown, but it seems plausible that Robert Bellarmine joined them after February 1596, when he became a consultor for the Inquisition.

By late 1596 the list of censured propositions from Bruno's books was nearly ready.[81] On 16 December the Inquisitors visited Bruno in his cell and ordered 'that he be interrogated as soon as possible about the propositions excerpted from his writings and about the censures'.[82] Apparently the original list has not survived, but any need to speculate about its contents diminished when a summary was discovered.

In 1925 Angelo Mercati became Prefect of the Secret Vatican Archives, and soon learned that in 1887 a document had been discovered regarding Bruno's proceedings. Back then, however, Pope Leo XIII had ordered that it be sent immediately to him and that he 'absolutely did not want these proceedings to be given to anyone'. Mercati began an extensive archival search that continued for fifteen years, until finally in November 1940 he found the document hidden in the personal archives of Pope Pius IX. The document, titled 'The Summary of the Trial of Giordano Bruno', consists of 261 paragraphs, including 34 articles of accusations, embedded in summaries of Bruno's interrogations in Venice and Rome. Importantly, the long document ends with ten sections, under the final heading, undated, titled 'Summary of the Replies of Brother Giordano to the Censures made about Propositions that were extracted from his Books'.

The Summary includes ten censured topics or propositions. The last two seem to have been added later than the rest, for at least three reasons: that by comparison to the first eight, the last two are very brief; they do not include Bruno's replies; and the last proposition includes a cited page number, 'fol. 139', which is out of sequence with all the other cited page numbers. It seems that after the first eight propositions were composed and presented to Bruno, and after his replies were recorded, the Inquisitors added two more.

To discuss the relation of these ten propositions to Pythagorean notions, we must first list them. The shortest section (the tenth) occupies only three lines of print in Mercati's published edition, while the longest (the fourth) occupies 53 lines of print plus 55 lines of footnotes. Hence, I will abridge the ten sections by quoting only the initial lines of each, which include the brief statement of each censured proposition, our main interest. I will also summarize some of Bruno's replies. There exist various scholarly accounts, with some differences. The 'Summary' that Mercati discovered consists of ten propositions. Luigi Firpo reproduces all ten, but then adds two others from a later account by Gaspar Schoppe.[83] Maurice Finocchiaro closely follows Firpo, but interpolates yet another: 'the individual immortality of the human soul is a questionable proposition.'[84] I find no evidence for this interpolated proposition. Instead, Bruno repeatedly claimed: 'I hold that souls are immortal.'[85] It is therefore important to analyse the 'Summary' carefully, because it has generated puzzlement and speculation. In a classic monograph, Frances Yates frankly commented: 'I find this document very confused and confusing.'[86] The following analysis will explain it.

> ***Summary of the replies of Brother Giordano to the censures made about Propositions that were extracted from his books.***[87]
>
> [*1st censured proposition*] 'Concerns the generation of things since he admitted the eternal existence of two real principles from which all things are done, and which are the soul of the world, and primal matter; Interrogated whether they are eternal.'[88]
>
> [*2nd censured proposition*] 'Concerns the conditional, namely, that the nature of God is finite, if it does not in fact produce infinity, or the infinite . . . [and Bruno insisted] that as a

consequence of my philosophy, since God's power is infinite it must necessarily produce effects that are equally infinite.'[89]

[*3rd censured proposition*] 'Concerns the mode of creation of human souls stated by these words: Deriving from that universal principle, from general to particular.'[90]

'Concerning that [*4th censured*] proposition, indeed. In this world nothing is generated, or corrupted in substance, unless we want to refer to alteration in this way; a product, whatever may be its alteration, it always maintains the same substance . . . [Bruno replied:] As Solomon does not contradict this, now saying "a generation passes away, and a generation arrives, [but the Earth remains forever]" [Ecclesiastes 1:4] and now also with "Nothing new under the Sun" [Ecclesiastes 1:9], that is, what is now is what was.'[91]

[*5th censured proposition*] 'Concerning Earth's motion states: First, I generally say that the manner and cause of the motion of the Earth, and of the immobility of the firmament and heaven are to me produced with the proper reasons and authority, which are certain and not harmful to the authority of divine Scriptures . . .'[92]

'Likewise folio 292 face 2 states [*6th censured proposition*] that Stars are Angels, in these words: the Stars are actually Angels, animated rational bodies, which while they praise God, and reveal the power and greatness of that by which light, its writings etched in the firmament, "the Heavens declare the glory of God"; Angels do not mean anything other than the messengers and interpreters of the divine voice, and of nature, and these are sensible Angels, visible, while others are invisible, and insensible.'[93]

'Likewise, folio 293 posits [*7th censured proposition*] that the Earth is alive, not only with a sensitive soul, but also rational, and God expressly attributed a soul to it, by saying: "Earth produce a living soul", which is how animals are constituted with the body from a part of its body, and its

> [Earth's] universal spirit comes to animate each particular one of them, by transmitting to them from its spirit . . . That it [the Earth] is a rational animal is manifest by its rational intellectual acts, which are seen in the regularity of its motion around its own centre, another around the Sun ...'[94]
>
> 'Likewise on folio 294 [*8th censured proposition*] states that the intelligent soul is not form, in these words: "I do not intend to support by my philosophy the ordinance that the soul is form, as in no place of divine Scriptures is it so called, but the spirit which is now in the body as the inhabitant in his house, a traveller in his pilgrimage, as the inner man in the outer man, as a captive in a prison . . ."'[95]
>
> [*9th censured proposition is that*] 'Likewise, he denies that individuals' true being, what they are, is but vanity, according to the dictate of Solomon, "I saw everything that is done under the Sun, and all is vanity" [Ecclesiastes 1:14], but that true substances are species of primal nature, that truly are that which they are.'[96]
>
> [*10th censured proposition*] 'Again, he posits many worlds, many suns, necessarily containing similar things in kind and in species as in this world, and even men, as in folio 139 and the subsequent long digression.'[97]

Again, these last two propositions are brief: they don't have Bruno's replies. The tenth cites a folio number out of sequence with the rest. As noted, apparently Bruno replied to eight propositions, then the censors added two more.

Do the ten propositions form a coherent group? Some match Pythagorean beliefs, such as (10) that many worlds exist. At first it seems that several are not Pythagorean – (2), (3), (4), (6), (8), (9) – but we will see that Bruno's transgressions were essentially 'Pythagorean'. The term does not necessarily mean that Pythagoras or his early followers held such views. Instead, it means that, throughout the centuries, such views had been variously attributed to the Pythagoreans.[98]

Bruno's first censured proposition was that things arise from two principles: the world soul and primal matter – both eternal. 'The world' means either Earth or universe. The *Placita philosophorum* said that Pythagoras believed that the world has a soul. It said that he thought the universe could never be destroyed.[99] Porphyry – enemy of Christians but advocate of Pythagoras – also asserted the universe's eternity. He denied Creation and spoke of the 'soul of the world'.[100]

St Hippolytus complained that Pythagoras said the world is eternal.[101] In the 1270s Étienne Tempier, Bishop of Paris, condemned the proposition that 'the world is eternal' as heretical, with authority granted by Pope John XXI.[102] Bishop Tempier argued that members of the faculty of the University of Paris transgressed the limits of philosophy to speak erroneously about theology. Anyone stating the world's eternity would be excommunicated and subject to the Inquisition. Tempier also deemed heretical 'That the substance of the soul is eternal'.[103] Yet Bruno said that souls are made from the eternal substance of the world soul.

In 1553 the Divinity Faculty of the University of Paris censured the claim 'That the World was never made'.[104] By the 1580s a tome on heresies specified: 'One heresy is to say that elementary Matter, from which the world is made, was not made by God, but is coeternal with God.'[105]

Justin Martyr quoted Pythagoras as saying God is the 'animating soul of the universe'.[106] The Pythagorean philosopher Apollonius also said the world has a soul.[107] This notion derived popularity from Plato's *Timaeus*. Likewise, Marcus Varro said the true gods are the soul of the world and its parts. St Augustine attacked this in *City of God against the Pagans*. He denounced the polytheistic notion of a world soul as contradictory. He complained that pagans who believed such notions were 'possessed by many demons'.[108] St Jerome too rejected 'the error of heretics, who suppose that everything is animated'.[109]

To further understand the first censure against Bruno, we must discuss the French theologian Peter Abelard. He was also a famous philosopher and logician, who became infamous for having an affair with a young woman, which, since it did not end in marriage, angered the woman's uncle so much that he castrated Abelard. Subsequently, Abelard became a monk in Paris.

Even though Abelard was convicted of misinterpreting the Holy Trinity in 1121, he later argued that ancient philosophers called the Holy Spirit the 'soul of the world', a Spirit that infuses all and 'vivifies creatures'.[110] He cited Salvian who 'quoted' Pythagoras: 'A soul is intermixed or diffused in all parts of the world, from which all animals when born receive their life.'[111] Abelard discussed the World Soul not as strictly real, but as a metaphor for the Holy Spirit. But just like Giordano Bruno centuries later, Abelard repeatedly quoted Virgil's passage about how the 'Spirit nourishes within'.[112]

In 1139 Abelard's claims about the Holy Spirit angered William of St Thierry, who denounced Abelard to the Bishop of Chartres and Bernard, Abbot of Clairvaux, complaining that Abelard endangered faith in the Trinity, constituting a growing evil that threatened to become a serpent. Among the thirteen 'monstrous doctrines' listed by William was the proposition 'That the Holy Spirit is the soul of the world'.[113] Soon after, in 1141, Bernard convened the Council of Sens in which the attending bishops condemned Abelard's doctrines as heretical.[114] Pope Innocent II promptly confirmed their ruling and gave an order on Abelard's 'erroneous books': 'wherever you find them, burn them in fire.'[115] Abelard died the next year, in 1142.

Henceforth some treatises on heresies included the heresy that the Holy Spirit is the world soul. For example, the Franciscan theologian Alfonso de Castro enumerated heresies about God, including 'The eleventh heresy: that the Holy Spirit is the soul of the world.'[116] Castro's treatise was published in fourteen editions until 1578. Another book on heresies also rejected how Pythagoras defined God: a soul that permeates the world and vivifies animals.[117]

Giordano Bruno wrote about God, who 'for the Pythagoreans is an infinite spirit that penetrates everything, comprehending and vivifying'.[118] Interrogated by Venetian Inquisitors, he said that he didn't understand the Trinity or 'the Holy Spirit as a third person', except 'by following the Pythagorean way' as soul of the universe. He quoted the Book of Wisdom, 'Solomon: "For the spirit of God fills the Earthly orb: and therefore he who contains everything", which conforms entirely to the Pythagorean doctrine.'[119]

Bruno's first censure was Pythagorean: the world soul as a generating principle. It also referred to primal matter, discussed below.

The second censure was Bruno's claim that, since God is infinite, the universe He created is infinite too. Bruno had argued: 'So great is God's excellence, that it is manifested in the greatness of his empire: it is not glorified in one, but in innumerable suns: not in one Earth, one world: but in ten-hundred thousand, I say in infinite.'[120] In another book he ridiculed a finite universe as a stupid dream of a confused imagination.[121] His infinite universe was Pythagorean in that they believed that stars are worlds in an infinite ether.[122] The notion of an infinite (but still hierarchical) universe was articulated by the Christian poet Marcello Stellato in his *Zodiacus* of 1543. That book was placed on the Index of Forbidden Books, though it is unclear whether the author's views on the universe contributed to the censorship.[123] Subsequently, the Protestant or Puritan writer Thomas Digges too said the Copernican universe is infinite, crediting the Pythagoreans.

The third censure was that each human soul is *derived* from a universal principle, the world soul. Aristotle denied the Pythagoreans' claim that any soul can be clothed in any body.[124] Justin Martyr rejected the claim that souls are not created, when explaining how he left Pythagoras and Plato for Christ.[125] Hippolytus and Origen criticized the Pythagoreans for saying that souls come from the stars or fall from heaven. According to the *Placita*, Pythagoras taught that the eternal rational part of human souls is derived from an eternal Deity.[126] Porphyry and Macrobius said that Pythagoras taught that souls, immortal, come from the Milky Way.[127] Tertullian, Lactantius and Epiphanius said that Pythagoras misunderstood the soul's immortality.[128] St Jerome said that 'among the Greeks Pythagoras was the first to find that the soul is immortal, and transits from some bodies into others. That which Virgil followed in the sixth volume of the *Aeneid*.'[129] Thus the Pythagoreans thought that human souls are *derived* from a previous state, whether fragments of the world soul, or immortal souls in the heavens. Instead Christians said that God creates a soul for each person.

Bruno's fourth proposition was that nothing is created: things are only *transformed*. Was it Pythagorean? Other ancient writers, such as Heraclitus, emphasized change. Hippolytus discussed Xenophanes' belief that 'nothing is generated or perishes.' He also quoted Epicurus: 'nothing was generated, except from atoms.'[130] Yet in Bruno's self-education, Pythagoras became the most prominent proponent that nothing is new. Interrogated by Venetian Inquisitors, Bruno quoted Ecclesiastes 1:9.[131] In the Bible it is about the past and the future:

> What has been will be again,
> what has been done will be done again;
> there is nothing new under the Sun.

But Bruno spoke about the past and the present, as a personal motto:

> Solomon and Pythagoras.
> What is that which is? That which was.
> What is that which was? That which is.
> Nothing new under the Sun.[132]

He often said 'nothing new under the Sun.' His words echo the Pythagorean concern for memories of past lives, instead of the Christian afterlife. Ancient tales said that the divine Pythagoras had lived many lives.

The *Placita* stated: 'Matter is that first being which is substrate for generation, corruption and all other alterations. The disciples of Thales and Pythagoras, with the Stoics, are of opinion that matter is changeable, mutable, convertible and sliding through all things.'[133] Ovid portrayed Pythagoras as the master of transmutations, teaching that everything in the world transforms, and the soul 'passes into ever-changing bodies'.[134]

Porphyry wrote that Pythagoras 'taught that the soul was immortal and that after death it transmigrated into other animated bodies. After certain specified periods, the same events occur again; that nothing was entirely new.'[135] St Jerome too said that Pythagoras taught immortality and that 'those things which had existed, after certain revolutions of time, came into being again; so that nothing in the world should be thought of as new.'[136]

Bruno's fifth censured proposition was that the Earth moves. It was a Pythagorean idea – Aristotle complained that the Pythagoreans believed it.[137] In the *Placita*, 'Philolaus the Pythagorean' and 'Ecphantus the Pythagorean' said that the Earth orbits a central fire. The *Placita* also referred to 'Hicetas of Syracuse'. Hippolytus noted that Ecphantus said that Earth moves.[138] Porphyry argued that the world soul is selfmovable and can 'move the body of the world'. Porphyry claimed that the world soul 'is adapted to be moved in a beautiful and orderly manner, and also to move the body of the world according to the most excellent reasons'.[139] Diogenes and Joannes Stobaeus also credited a moving Earth to Philolaus.

Copernicus, Digges and Zúñiga attributed the theory of the Earth's mobility to the Pythagoreans. Similarly, in his *Ash Wednesday Supper*, Bruno listed some of the proponents of the Earth's motion: 'Copernicus, Niceta the Pythagorean from Syracuse, Philolaus, Heraclides of Pontus, Ecphantus the Pythagorean, Plato in the *Timaeus* . . . and the divine Cusa.'[140] Interestingly, the ancient Church Fathers did not condemn claims about the Earth's motion, so why was it flagged by the censors of the Inquisition when they read Bruno's books? It clashed with a literal interpretation of scriptures. But more importantly, the pagan thesis of the Earth's motion was linked to the other accusations against Bruno. For example, for Porphyry the Earth's motion was a direct consequence of the notion that it has a soul.

The sixth censured proposition was that stars are angels: animated rational bodies that convey God's voice. The Pythagoreans did not write about angels, so this was not their belief. However, Digges framed the 'Doctrine of the Pythagoreans' as involving stars as angels: 'glorious lights innumerable. Far excelling our Sun both in quantity and quality the very court of celestial angels devoid of grief.' Bruno briefly referred to 'angels, which are stars, announcing the divine infinite majesty'. He also wrote that angels flow into heavenly bodies, the elements flow to the heavens and then 'into daemons or angels'. Bruno echoed Agrippa, but Bruno's identification of 'daemons or angels' and the infusion of angels into stars were his own interpolations. He used the terms 'angels' and 'daemons' synonymously, as Christian and pagan terms: 'How to attract both good and bad angels: Good daemons are enticed in diverse ways.'[141]

To the Pythagoreans, daemons mediated between humans and gods. Some portrayed Pythagoras as a daemon. Philostratus said that a daemon guided Apollonius. St Hippolytus complained that Pythagoras said the souls of animals come from the stars. Pythagoras theorized that 'the Sun, and the Moon and the stars were all gods', according to Laertius. Likewise, Porphyry said that stars 'are visible Gods'.[142] But many astrological books construed stars as divine beings. So Bruno's sixth proposition is compatible with Pythagorean beliefs, but it cannot be described as distinctly Pythagorean: and Bruno did not attribute it to them, at least in writing.

Was it a heresy? Yes. In 553 CE the Fifth Ecumenical Council at Constantinople issued 'Anathemas against Origen'. The sixth states: 'If anyone says that the heaven & Sun & Moon & stars and the waters that are above the heavens, are animated [have souls] & material powers, he is anathema', that is, a heretic. Origen's alleged heresies were well known in the 1590s. These were 'confused dreams & wives' tales to repudiate', 'pagan errors' for 'imbecile minds', as noted by Cardinal Cesare Baronio.[143]

The seventh censured proposition was that Earth is a living animal with a rational, sensitive soul that embodies a universal spirit that animates animals. The belief that Earth is alive appears in *Timaeus*: 'the world became a living creature truly endowed with soul and intelligence by the providence of God . . . one visible animal comprehending within itself all other animals of a kindred nature.'[144] Some commentators thought that Timaeus was a Pythagorean and that Plato based his book on impressions about Pythagoras.[145] The *Placita* spoke of the 'respiration' of the Earth.[146] Diogenes said that Pythagoras taught that the world 'is endowed with life, and intellect'. Plotinus (Porphyry's teacher) argued that Earth's soul transmits growth to its parts.[147] Apollonius reportedly thought that Earth is a soulful animal.[148] Iamblichus too argued that it is a great animal.[149]

Some of the Stoics advocated similar beliefs.[150] But St Augustine denied the Earth is a living being.[151] He said that belief in the divine soul of the living Earth led to false belief in many gods. Bishop Tempier condemned the notion that the heavenly bodies are moved by a soul 'like an animal'. So this too was a heresy, forbidden under penalty of excommunication.[152]

In Ovid's *Metamorphoses*, Pythagoras declared 'the Earth is of the nature of an animal, living.'[153] In 1584 an Italian translation discussed in great length this 'Pythagorean doctrine'.[154] Bellarmine, who served on the Index of Forbidden Books, wrote in 1592 that the original *Metamorphoses* could be 'tolerated' as a model of proper Latin, but its Italian editions should be censored.[155]

The eighth censure opposed Bruno's statement that soul is not 'form'. He thus rejected Aristotle's theory of souls, dismissing it as incomprehensible. In 1312, however, the Council of Vienna had decreed that anyone who denies or doubts that 'soul is the form of the human body' is 'a heretic'.

To determine whether Bruno's heresy was Pythagorean, we must specify his definition of soul. He claimed that souls are immortal intellective substances that inhabit bodies that are human or other. Under interrogation he explained:

> Speaking as a Catholic, they [souls] do not pass from body to body, but go to Paradise, Purgatory or Hell. But I have reasoned deeply, and, speaking as a philosopher, since soul is not found without body and yet is not body, it may be in one body or another, and pass from body to body. This, if it be not true, seems at least verisimilar, according to the opinion of Pythagoras.[156]

In several books Bruno wrote about transmigration.[157] He beseeched God: 'We beg that in our transfusion, or transit, or metempsychosis we shall receive contented genii [daemons]: since however inexorable He be, we must attend him with pleas, either to be kept in our present state, or to enter into a better one, or a similar one, or a little worse.'[158] Bruno argued that death does not affect souls, so that following the saintly Pythagoras we should not fear death, which was really a transition.[159]

Aristotle denied the 'Pythagorean myths; that any soul could be clothed upon with any body – an absurd view, for each body seems to have a form and shape of its own. It is absurd as to say that the art of carpentry could embody itself in flutes; each art must use its tools, each soul its body.' Justin Martyr too denied transmigration. Tertullian said that Pythagoras used shameful lies to advocate transmigration, which corrupted Christianity. Epiphanius criticized Pythagoras for proclaiming 'the wicked, extremely impious doctrine of the immortalizations and transmigrations of souls and the dissolution of bodies'. Augustine rebuked Porphyry on transmigration. Thomas Aquinas rejected it too.[160]

In Bruno's eleventh deposition, the Inquisitors asked him whether 'the soul of one man can transmigrate from one body into another'. Bruno replied that he posited '*la transmigratione*', not as a fact, but only as possible.[161] Under interrogation, Bruno said that the false claim that the soul is the form of the body is not found in scriptures. Instead he said that the soul, 'in its pilgrimage', inhabits the body 'as a captive in a prison' and controls it like a ship.[162] He said the soul subsists outside the body, against Aristotle (who said that souls are inseparable from bodies, but that 'it is uncertain whether the soul is the actuality of its body as a sailor of a ship').[163] Bruno's analogy to a 'prison' was not rare but the Pythagoreans used it.[164] Lactantius and Jerome attributed it to Pythagoras.[165] St Irenaeus (*c.* 180 CE) denounced that doctrine as false, saying that its insane believers relied partly on the Devil to supposedly help transfer their souls into new bodies, new 'prisons'.[166] Apollonius reportedly said: 'all the time we men are in a prison we call life; our soul, attached to this perishable body, suffers much.'[167] St Basil quoted Pythagoras on people who eat a lot: 'Can you not stop planning for a heavier prison?'[168] Erasmus too said that Pythagoras construed the soul as imprisoned in the body.[169]

So Bruno's eighth proposition was Pythagorean. The Fifth Ecumenical Council denounced Origen's heresy that souls are 'condemned to punishment in bodies' like prisons.[170] They also condemned: 'That the intellect is not a bodily form, except as a captain is the form of a ship.' Tempier also condemned the thesis 'That the intellect can transit from body to body'.[171] In 1553 theologians of the University of Paris censured claims 'That the transmigration of souls is an argument for the Resurrection'.[172]

The ninth censure was Bruno's denial of Solomon's statement that individuals' true being is vanity. Consider Bruno's *De la causa*:

> Every production, of any kind, is an alteration, while substance always remains the same, since there is only one substance, as there is but one divine, immortal being. Pythagoras, who did not fear death but saw it as a transformation, reached this conclusion . . . Solomon inferred this as well, saying, 'there is nothing new under the Sun, but what is, has already been'. You see, then, how the universe is . . . eternal, while every aspect, every face, every other thing is vanity and nothingness.[173]

Bruno's proposition was Pythagorean because he credited it to Pythagoras. In another book he wrote that mutable things do not exist, but are vanity.[174] For Bruno, primal substance was not vanity, it was eternal like God.

The Inquisitors disagreed. Ecclesiastes 1:14 states: 'I have seen all that is done under the Sun, and behold: all is vanity and an affliction of the spirit.' To the Inquisitors it meant that *every thing* is vanity. They censured Bruno's belief 'that true substances are species of primal nature, that truly are what they are'.

Finally, the tenth proposition: many inhabited worlds and suns exist. Historians have argued that the accusation of many worlds was 'too vague to be defined as formally heretical', so instead, maybe it was merely 'erroneous', 'scandalous', 'injurious' and so on.[175] But no, I'll show that it was heretical. I found that many theologians, jurists, bishops, archbishops, one emperor, three popes, five Church Fathers and nine saints had denounced this notion.[176]

Aristotle had complained that the Pythagoreans said that 'Earth is one of the stars.'[177] Lucian wrote that Pythagoras lived in such starry worlds. The *Placita* said that the Pythagoreans thought the Moon and the stars are worlds, with soil, air and animals. And Bruno told the Inquisitors: 'I have indeed asserted infinite particular worlds similar to the Earth, which with Pythagoras I consider a star, similar to which is the Moon, other planets and other stars, which are infinitely many.'[178]

Was it a heresy? Hippolytus ridiculed the doctrine of infinitely many suns, moons and worlds, some inhabited.[179] Around 260 CE Pope Dionysius of Alexandria wrote a tract against the Epicureans, mainly to criticize their theory that all things are composed of atoms without divine Providence. In it, Pope Dionysius dismissed one of the corollaries of that theory, that atoms clash and combine by chance 'and thus gradually form this world and all objects in it; and more, that they construct infinite worlds'.[180]

Most significantly, however, in 384 CE the belief in many worlds was categorized as *heretical* by Philaster, Bishop of Brescia. In his *Book on Heresies*, Philaster wrote: 'Another heresy is to say that worlds are infinite and innumerable, following the asinine opinion of the philosophers, whereas Scriptures say that the world is one and it teaches us that it is one.'[181]

In 402 St Jerome complained that one of 'the most heretical' among Origen's transgressions was his claim that 'worlds are innumerable, succeeding one another eternally.' Subsequently St Augustine composed an authoritative list of 88 heresies, including 'that worlds are innumerable' as the 'Seventy-seventh heresy'.

In the early seventh century Bishop Isidore of Seville also listed it as heretical.[182] Isidore's encyclopaedic work was so admired that it was printed in at least twelve editions between 1470 and 1595; he was belatedly canonized by Pope Clement VIII in 1598.

There is evidence that the denial of the plurality of worlds was taken seriously. In 748 the abbot of St Peter's monastery in Salzburg so annoyed Pope Zacharias that he denounced the notion that more than one world exists. The abbot was an Irishman named Feargal, known in Latin as Virgilius. Apparently he said that the Earth is spherical and is inhabited everywhere, even in points diametrically opposite, known as the 'antipodes'. (Plato mentioned the antipodes in *Timaeus* and these were later discussed by Aristotle and others. Hence Diogenes said that Alexander Polyhistor found that the 'Commentaries of Pythagoras' asserted the existence of the antipodes.)

But Virgilius's superior, Archbishop Boniface of Mainz, primate of Germany, misunderstood whatever Virgilius said about the Earth and the antipodes. He complained to Pope Zacharias that Virgilius argued that another world exists and that it is inhabited. Pope Zacharias replied that Virgilius

> has lied to himself: out of the perverse doctrine, which is spoken against the Lord and his own soul too, namely that there is another world and other men beneath the earth, another Sun and Moon; if he is convicted by the summoned council of the Church for confessing this, he will be deprived of the honour of the priesthood.'[183]

Apparently there are no records of how Virgilius replied to such accusations, but he seems to have defended himself convincingly, because he was not defrocked. Presumably he had to explain that Boniface had misunderstood his remarks. Virgilius later became Bishop of Salzburg and was eventually canonized in 1233.

Nevertheless, the complaint by Pope Zacharias is very significant. It shows that the notion that there is more than one world, Sun and Moon seemed intolerable and was addressed severely by the highest authority of the Church. Moreover the Pope's complaint was set in writing and disseminated, so henceforth clergymen had another precedent, in addition to objections by other authorities, now that another pope had personally denounced that idea.

The heresy of many worlds appeared in books for centuries, for example, in a *Thesaurus of the Christian Religion* (1559) and in an edition of Philaster's *Book of Heresies* (1578).[184] St Jerome too was an authority in the Renaissance, because at the Council of Trent the Catholic Church had selected his translation of the Bible, known as the Latin Vulgate Bible, as the official version to use and trust. Writers also cited Augustine. In 1591 the Jesuit theologian Gregorius de Valentia cautioned: 'There is no need of heretics among Christians, who follow Democritus and other ancient philosophers in saying that there exists not one but innumerable worlds, as the divine Augustine reports in his book on heresies, namely the seventy-seventh heresy.'[185]

Other theologians too cited this heresy for centuries.[186] They explained the problem: 'we cannot assert that there exist two or many worlds, since neither do we assert two or many Christs.'[187] Aristotle had insisted that many worlds cannot exist since it would require more than one First Cause. The belief in many worlds seemed to entail belief in many gods.

Above all, however, the highest authority proves that Bruno's belief in many worlds was heretical. In 1582 Pope Gregory XIII issued a compilation of laws of the Catholic Church and ordered that it be used in all church courts and schools of canon law. Expanded in 1591, the *Corpus of Canon Law* includes long discussions of what are heresies and who shall be considered a heretic. Echoing Isidore, its list of heresies includes 'having the opinion of innumerable worlds'.[188] The *Canon Law* embodied the fundamental laws of the Catholic Church. All inquisitorial laws and courts had to comply with the Pope's canon laws. Even the Roman Inquisition's condemnation of Bruno refers to the 'sacred Canons' three times.[189]

In his depositions Bruno did not mention Democritus or other philosophers associated with belief in many worlds. He only credited Pythagoras. Finally there is a link between many worlds and transmigration in Bruno's motto about 'Solomon and Pythagoras'. Centuries before Bruno highlighted 'nothing new under the Sun', others had linked that phrase to claims about transmigration and many worlds. This had been done by an early theologian accused of heresies, Origen, who argued, 'The fact is that prior to this world there have existed others, as Ecclesiastes says: "What is it that has been done? the same as is the future. And what is it that has been created? the same that will be created: and there is nothing new under the Sun".'[190] Origen said that omnipotence led God to create worlds prior to ours, and after it. He said that our souls might live in future worlds. Jerome condemned Origen's notion of innumerable worlds as heretical. Others stressed the link between many worlds and transmigration. Rupert of Deutz spoke of Ecclesiastes 1:9 by denying that many worlds exist. He said that 'heretics' who posit worlds are deathly wrong to think that Pythagoras' soul transmigrated to many bodies.[191]

Remarkably, Bruno's claims echoed Origen's heresies: eternally many worlds and transmigration. Had he read Origen's censored book *On Principles*? Yes – and Bruno praised him as 'the only theologian, who like the great philosophers' dared to voice 'after the reproved sects' the truth of revolutions and eternal change.[192]

HAVING ANALYSED Bruno's ten propositions, we see that they are a coherent set of interconnected beliefs. Five propositions were clearly heretical: the first, sixth, seventh, eighth and tenth. These involve at least six formal heresies: the eternity of the world; the world soul; that the heavenly bodies are animated; transmigration; that the human soul is not form; and many worlds. The other propositions could be considered 'errors' or 'nearly heretical'. Any such proposition, once identified, could constitute a heresy if the accused persisted in not abandoning it. And Bruno did defend such propositions. Such heresies do not challenge the divinity of Christ. They are not about magic or esoteric Hermetic beliefs.

Yet without exaggeration they can be described as *Pythagorean heresies*. Only the sixth censured proposition, about angels, is not quite Pythagorean, though it has some relevant aspects. The rest – nine out of ten – of Bruno's crimes as a writer were Pythagorean propositions inasmuch as either ancient or recent writers attributed such ideas to Pythagoras or his followers, Church Fathers denounced those beliefs as Pythagorean, or that Bruno himself presented such notions as Pythagorean.

Bruno learned many of these notions from the *Placita*, Lucian, Virgil, Erasmus and Ovid. Bruno appreciated the Pythagorean tradition as one of the alternative, ancient traditions that challenged the philosophy of Aristotle. Bruno did not discuss Pythagoras just as a philosopher, but also as a contributor to theology: 'the negative theology of Pythagoras and Dionysius [the Areopagite] is much more renowned, above the demonstrative one of Aristotle and the scholastic doctors.'[193]

ON 24 MARCH 1597 the Inquisitors and theologians visited Bruno in his cell. This meeting is significant for two reasons. It is the first extant record that shows the participation of 'Robertus Bellarminius', the famous Jesuit theologian who would exert important influence in the Inquisition's proceedings against both Bruno and Galileo. Robert Bellarmine was a highly accomplished professor of theology for twenty years. He had become *the* authority in the analysis and persecution of heresies, especially against the Protestants. He claimed that 'heresy is a more serious perversity than all other crimes and atrocities, just as the plague is more formidable and fearsome than common diseases'.[194] Bellarmine became the personal theologian to Pope Clement VIII. On 5 February 1597 Bellarmine made his vow of secrecy as Consultor to the Roman Inquisition.[195] Weeks later Bellarmine served as a theologian consultor in the trial against Bruno.

The second important point about this meeting on 24 March is that the record states a single demand against Bruno: 'He was admonished to thus abandon his delusions [*vanitates*] of diverse worlds, and ordered that he should be interrogated sternly. Afterward he was given the censure.'[196]

In this connection, Bellarmine's presence is significant because he knew that certain heresies were rooted in *pagan* notions. For example, in a book of 1594, titled *On the Need for Caution When Reading Pagan Philosophers*, its author noted that 'Father Robert Bellarmine' had advised that 'certain Catholics who are too pleased with a burgeoning Pagan sense should be subjected to censorship, by admonishing the Author.'[197]

At that point, the official written 'proceedings' of Bruno's trials, in Venice and Rome, spanned 256 sheets of text, that is, 512 pages. Therefore, an officer of the Inquisition composed the summary of the long proceedings, finished before 1 March 1598, and discovered centuries later by Mercati. And after it was written, the Inquisitors confronted Bruno in a 'Seventeenth Deposition'.

We know about this important deposition because brief notes about it were added in the margins of the 'Summary' of Bruno's trial. First, in the account of Bruno's third deposition, in which he had discussed the Trinity by describing 'the Pythagorean doctrine' that the Holy Spirit is the soul of the world and quoting Virgil's *Aeneid*, an employee of the Inquisition now added, in the margin of that page: 'About these replies by him, he was interrogated [again] in the 17th Deposition, folio 257, where he affirmed the same replies in which he relapses [*reincidit*].' Second, in the immediately following paragraph, in which he said that he had some doubts about some expressions about the persons of God, again the margin bears an annotation: 'About these words he was interrogated in the 17th Deposition, fol. 257*v* up to fol. 261, and he gave replies which relapse [*reincidunt*] into the same thing'. Third, also in the summary of that third deposition, in which Bruno had said that he agreed 'with Pythagoras' that the Earth is a star similar to the Moon, again someone added in the margin: 'About this reply he was interrogated in the 17th Deposition, fol. 261, but does not seem to satisfy, because he relapsed [*reincidit*] into the same reply.' The word *reincidit* was a technical term for Inquisitors. It was part of the definition of someone who has 'relapsed' into heresy.[198] This shows that Bruno's beliefs about the soul of the world, the persons of God and the plurality of worlds were *heretical*, and that he obstinately repeated them.

By March 1598 the Congregation of the Holy Roman Office (the Inquisition) consisted of several cardinals, several coadjutors or consultors, an Assessor of the Holy Office, a General Commissary and the official notary. Father Alberto Tragagliolo was the General Commissary of the Holy Office. Tragagliolo and Bellarmine agreed to produce another list of propositions that would synthesize Bruno's principal heresies, and on 14 January 1599 Tragagliolo and Bellarmine presented 'eight heretical propositions collected from the books and trial' of Bruno.[199] The Inquisitors agreed that Bruno would be required solemnly to abjure such propositions. Unfortunately, the official document is not extant, or researchers have not found it despite many efforts.

Some scholars have tried to guess the eight propositions on the basis of the alleged transgressions in the Venetian and Roman proceedings against Bruno. In 1886 Felice Tocco speculated that the propositions were about: (1) the distinction of persons in God; (2) the Incarnation of the Word; (3) the nature of the Holy Spirit; (4) the divinity of Christ; (5, 6, 7) the necessity, eternity and infinity of nature; and (8) the transmigration of souls. These questions are theological, but in view of the present analysis of the ten censured propositions, several of Tocco's conjectures now seem Pythagorean too: the Holy Spirit as soul of the universe, the eternity and infinity of nature, and the transmigration of souls. Likewise, in 2002 Finocchiaro speculated, 'it is likely that this list [of eight heresies] dealt primarily with the questions of the temporal, spatial and plural infinity of the universe, and with issues of the nature, immortality and individuality of the soul and its relationship to the body, the world and material substance.' Such conjectures are plausible, but I won't speculate about Bellarmine's list of eight heresies.[200]

One proposition, however, *seems* to have been specified in a later document: 'That you had said that it is a great blasphemy to say that the bread transubstantiates into flesh, etc.' Was this the first of Bellarmine's 'eight propositions'? It was the first accusation that Mocenigo had raised against Bruno. From this Ingrid Rowland speculated that all of the eight propositions were extracted from Mocenigo's initial accusation.[201] I disagree. The document in question is a 'partial copy' of Bruno's final condemnation, which omits material with an 'etc.'

I suspect that the original, long list began with Mocenigo's accusations, plus others, and concluded with the last eight propositions selected from Bruno's books and trial. Indeed, the preceding sentence refers to the accusations from 'eight years ago' (by Mocenigo), and it is followed by a paragraph that mentions 'these eight propositions', clearly meaning the ones by Bellarmine.

It doesn't make sense to have a denial of transubstantiation as one of those eight propositions because, under interrogation, Bruno repudiated Mocenigo's accusation of denying transubstantiation, saying, 'I have not doubted this sacrament.'[202] Neither did he deny transubstantiation in his books or depositions. Therefore this could hardly be one of the 'propositions collected from the books and trial'. Rowland's guess stemmed from her arbitrary interpretation of the 'partial copy' of Bruno's final sentence. I cannot imagine, however, that an expert on heresies, Bellarmine, would simply echo the accusations of a Venetian layman, Mocenigo.

Moreover, Rowland missed the important fact that two of the eight propositions were mentioned in a later document, and that these are *not* among Mocenigo's accusations. Therefore, Bellarmine's list was not a copy of Mocenigo's.

Apparently, though, Bruno didn't know the eight propositions were heretical, telling the Inquisitors 'that if the Apostolic See and the Holiness of Our Lord held these eight propositions as definitely heretical, and that His Holiness recognizes them as such, or by the Holy Spirit he defines them as such, then he [Bruno] would be willing to revoke them'.[203]

The list of eight propositions was not final. The minutes of the Decree specifying that Bruno be presented with them concludes with a statement to 'Seek other heretical propositions from the proceedings and books.'[204] A notary document about the list makes a key point, the answer to the question Bruno had raised:

> Our most Holy Lord decrees and ordains that it be intimated to the apostate brother Giordano Bruno of Nola by the theologian Fathers, namely Father Bellarmine and the Father Commissary, that these propositions are heretical, and not only heretical as now declared, but by the most ancient Fathers of the Church and the Apostolic See.

Two more documents echo this claim, one of them adding that these propositions were 'heretical and contrary to the Catholic faith' and that the ancient Church Fathers had 'rejected and condemned' them.[205]

So Bruno's worst heresies were old. They had been denounced 'by the most ancient' Church Fathers. From the previous historical context, it seems fitting that some of Bruno's censured propositions appeared in ancient works by several advocates of the Pythagoreans, including Apollonius, Porphyry, Iamblichus and Philostratus. And, ancient Church Fathers such as Irenaeus, Tertullian, Athanasius, John Chrysostom, Jerome, Augustine and Isidore had denounced claims about transmigration, the eternity of the world, many worlds and Pythagorean denials of the uniqueness of Jesus.

Bruno had spent six years in the Inquisitions' prisons. Yet they then gave him forty more days to recant the heresies in Bellarmine's list. Time passed, but Bruno did not recant.

In March 1599 Bellarmine was promoted to Cardinal and Inquisitor. On 30 April one of the early consultors of the Inquisition in the proceedings against Bruno, Ferdinando Taverna, became governor of Rome. Taverna soon became infamous for being ruthlessly cruel towards criminals. He personally extracted the confession of a young woman who killed her incestuous father: Taverna's officers beheaded her and her mother, then they bludgeoned her son and tore his body to pieces. People rioted because of such brutally cruel executions, claiming that the governor and the Pope did not care for justice, but Taverna crushed them.[206]

MEANWHILE, FAR from Italy, Catholics in India reasserted their denial of transmigration. In June 1599 the Christians of the Malabar Coast in India formally united with the Catholic Church. At the Synod of Diamper, they denounced three 'common Errors' of infidels, foremost among which was the false 'transmigration of souls, which after death go either into the bodies of beasts, or of some other men'. The third error they denounced also echoes one of the accusations against Bruno, that everyone may be saved, 'a manifest heresy' since Salvation was conditional; it could only transpire through Christ. Members of all other sects would burn in Hell.[207] I do not suggest that the decrees at Diamper influenced the Inquisitors in Rome, but the converse: such 'errors' were condemned in India because they were repugnant in Rome.

In August 1599 a Decree of the Inquisitors noted that Bruno had 'clearly revoked in writing' most of the the eight propositions presented to him in April – with the exception of two: 'namely the first, about the Novatian heresy, and the 7th, which treats the soul being in the body as a sailor in a ship'.[208] Neither heresy was explained in that Decree but we discussed the latter, which was the eighth proposition censured by consultors. Bruno had asserted it in at least three of his books.[209]

The other heresy is mysterious. Novatian was a Roman bishop around 240 CE who denied that the Church could forgive the sin of idolatry. He argued that a newly elected pope, Cornelius, was too forgiving of Christians who denied their Christian faith during persecution. So Novatian proclaimed himself Pope. For this and other reasons he was deemed a heretic. As summarized by Bishop Isidore, around 630, 'The Novatians (*Novatianus*) originated from Novatus, priest of the city of Rome, who, in opposition to Cornelius, dared to usurp the priestly chair. He established his heresy, being unwilling to receive apostates and rebaptizing the baptized.'[210] What was meant by accusing Bruno of 'the Novatian heresy'?

One meaning for such an expression was the accusation against Novatian's claim that some sins are so grave that the Church cannot forgive them, only God himself can.

Trying to explain the mysterious 'Novatian heresy', historian Luigi Firpo speculated that 'maybe Bruno insisted beyond measure that there are difficulties in being admitted to penance, leading him to talk about an impossibility of his absolution and thus engendering suspicions in Bellarmine of Novatianism.' I disagree with Firpo's conjecture because it sounds distant from Bruno's words. Actually, Bruno said he wanted to do confession, that he frequently asked God for forgiveness, and that he appreciated the Sacrament of Penance.[211]

Thus Bruno did not voice heterodox views on penance and forgiveness. He acquiesced to the Inquisitors' requests on this. I see no clear reason to think that his 'Novatian heresy' was about this.

In 2006 historian Lucia Boschetti analysed this puzzling accusation, comparing editions of Novatian's works with Bruno's expressions, and found that Novatian's notion of the Trinity, especially as interpreted in the 1590s, resembles Bruno's. As Boschetti explained: 'The concept of God expressed by Novatian had been explicitly developed following Virgil, although in this Christian theologian [Novatian] we do not find a literal citation.'[212]

Boschetti compared lines from Novatian's *On the Trinity* with lines from Virgil's *Aeneid*. Virgil wrote: 'Mind agitates the mass and intermixes itself with the great body.' Novatian described God as the 'Mind that generates and completes everything'. Similarly, Bruno wrote that 'one mind, infused through the framework, moves the universal mass, as said by Pythagoras.'[213] As I've explained, Bruno quoted Virgil's words as 'the Pythagorean way' of construing the Holy Spirit. This was one of the questionable beliefs Bruno boldly defended. Boschetti detected the pervasive Pythagorean thread in Bruno's arguments about this matter.

To check Boschetti's claims, I read Novatian's work *On the Trinity* in the original Latin and can confirm that there are striking similarities. For example, consider this passage Boschetti did not quote:

> Above everything he [God] himself contains all . . . always intent on his work, pervading through everything, & moving all, & universally vivifying, and beholding the totality, & connecting discordant materials in a harmony of all the elements . . .[214]

Here, as in Virgil's *Aeneid*, Novatian portrayed God as somehow infusing, vivifying and moving everything. Novatian also said that 'the warm nature of an interior spirit was added to everything.'[215] Recall Virgil's 'Spirit nourishes within . . .', which Bruno quoted to the Inquisitors.

As for the Holy Spirit in particular, it seems peculiar that Novatian spent thirty chapters discussing God and Jesus Christ but he discussed the Holy Spirit in merely one chapter. In another chapter Novatian seems to say, briefly, that God is the same as the Spirit: 'But when the Lord says that God is a Spirit, I think that Christ spoke thus of the Father.' And Novatian there explained that just as God is not really 'love', 'fire' or 'light', God likewise is not really 'a spirit' in substance, because all spirits are creations, and because such biblical appellations are just efforts to convey what humans cannot express: what God really is.

Boschetti fairly noted that since Novatian did *not* criticize the misidentification of God the Father as the Holy Spirit, later Catholic commentators surmised that Novatian *heretically* implied no distinction between the Father and the Holy Spirit. Such commentators complained about the 'nefarious heresy' of construing the Holy Spirit as a 'breath', 'air' or 'wind', so they rejected any interpretation of it as 'vital spirit'.[216]

In 1579, for example, Jacob Pamelius published an annotated edition of Novatian's *On the Trinity*, in which he criticized Novatian's brief claim that 'Every spirit is a creation.' Here, Pamelius interpreted 'spirit' as meaning even the Holy Spirit, so he warned 'it is not surprising that Novatian erred in this,' because Novatian had not received the Holy Spirit through the sacrament of Confirmation 'and therefore he easily fell into heresy; for nothing like this had been defined by the Church about the Trinity'.[217] In this context, it seems that Bruno's 'Novatian heresy' was indeed his denial of the Catholic notion of the third person, the Holy Spirit.

I might mention another point: the question of penance (the usual meaning of 'Novatian heresy') is linked to the interpretation of the Holy Spirit as the universal soul. Virgil's *Aeneid* involved a pagan concept of penance, a punishment for sins not forgiven on Earth. Here Virgil argued that during a soul's bodily imprisonment the body somehow stains the soul, so after death the soul must undergo a painful penance to purge sins and crimes before becoming purified.

Along with the belief that soul travels in bodily vessels, Bruno refused to recant his Novatian heresy – whatever it was. In September the prosecutor, the commissary (Tragagliolo), the assessor and three consultors of the Inquisition all voted that Bruno should be tortured. Two of them (including Tragagliolo) said he should be tortured severely, and another said that he should be tortured more than twice. Only one of them specified a reason why: 'concerning foremost the holiest Trinity torture him.'[218] This confirms that one of Bruno's heresies was his interpretation of the Trinity. The Pope was present at that meeting, along with Bellarmine and the other cardinal Inquisitors.

The extant documents do not specify whether in fact Bruno was tortured. Pope Clement VIII was patient, so more months passed. In December the cardinals of the Inquisition, including Bellarmine, confronted Bruno again. They asked if he would recant and recorded his answers: 'He said that he neither wants to nor should repent, and has nothing to repent, nor has matters of repentance, and does not know about what he should repent.'[219] So they gave him even more time to recant. They submitted his books to the Index of Forbidden Books, for heresies and errors.[220]

Fire and Smoke

On 20 January 1600 a Decree of the Congregation of the Holy Office reported that Bruno had been asked to 'recognize, detest and abjure' the 'heretical propositions that were contained in his books and that he had voiced in his depositions'. However, the Decree reported: 'Bruno did not want to agree, asserting that he never said or wrote heretical propositions, but that the excerpts were done wrongly by the ministers of the Holy Office, conveying the opposite.

Because, he was prepared to give an account of all his writings, to defend them against any theologians.'[221] Since he did not recognize his heresies, the Decree stated that he would be sentenced and delivered to the secular Roman Court for punishment. On 8 February the Inquisitors issued their sentence against Bruno, including the statement that 'we condemn, reprove and prohibit all your books and writings mentioned above and others as heretical and erroneous and containing many heresies and errors.'

They ordered that his works be publicly burned in St Peter's Square and listed in the Index of Forbidden Books. In that document the cardinals condemned Bruno as an 'obstinate and impenitent heretic'. Nine cardinals, including Bellarmine, approved the final judgement, and it was sent to the governor of Rome, Ferdinando Taverna. On the same day the Congregation of the Index duly noted that Bruno's books and writings would be prohibited.

Gaspar Schoppe, a virulent critic of the Protestants, described Bruno's final days. Schoppe was a young German humanist, almost 24 years old, who had quickly gained prominence in Rome thanks to his very boastful conversion to Catholicism. In a letter to his former teacher, the jurist Konrad Rittershausen, Schoppe said that he was present at the court of the Inquisition on the day when Bruno faced his Inquisitors on his knees, to receive his sentence. Schoppe noted that first they gave an account of Bruno's life, studies and his doctrines. Schoppe said that many years ago Bruno had begun 'to doubt, or even deny' the doctrine of transubstantiation. Next he had doubted the virginity of Mary. Then he left his monastery and travelled to Geneva and other places. The Inquisitors described their efforts 'to reclaim him' to the Catholic faith, against his pertinacious impiety. Then they degraded him, excommunicated him and turned him over to the secular Roman magistrates to be punished. Schoppe said that the Inquisitors 'begged that he be punished very gently and without profusion of blood'.

Right then, Bruno said just one thing: 'Maybe you have more fear in executing your sentence on me, than I in accepting it.'[222] Here again Bruno spoke like a Pythagorean, one who did not fear death. Was he convinced of the immortality or rebirth of the soul? In his dialogue *On the Cause, the Principle and the One*, Bruno had referred to the 'saintly' Pythagoras as one 'who did not fear death, but expected the mutation'.[223] In the end, Bruno refused to recant

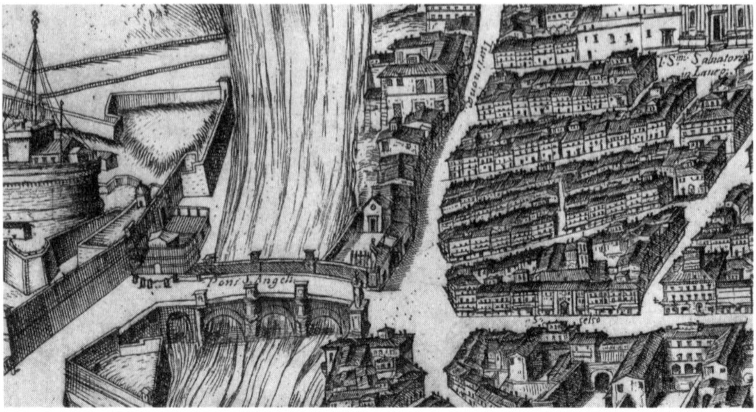

The prison of Tor di Nona (centre top), on the edge of the Tiber River, across from the Castel Sant'Angelo. Map by Antonio Tempesta, 1593.

his views. They declared him guilty of heresy and sentenced him to death.

Bruno was transferred to the Roman authorities and locked up in the Tor di Nona prison.[224] The Roman notary Giuseppe de Angelis wrote that Bruno was detained by Governor Taverna. The Roman judge Giovanni Battista Gottarello was assigned the task of carrying out the sentence imposed by the court of the Inquisition.

Death sentences for heresy usually consisted of hanging, drowning, mutilation or burning. But before being mutilated or burned, some convicts were strangled or decapitated to reduce the suffering and horror. Being burned alive was regarded as the worst punishment, for the gravest offences.[225] That was the punishment imposed on Bruno. It was a rare event in Rome. Schoppe commented that he had not heard about any heretics being burned at the stake there.[226] Although we do not know the total number of trials by the Roman Inquisition, there are some figures about executions. For example, the Archconfraternity that accompanied victims to their deaths kept records, including the names of 97 heretics executed in Rome between 1542 and 1761.[227]

On 12 February an official Notice of Rome announced that the 'obstinate heretic' from Nola would be brought to justice. It summarily stated that Bruno had been sentenced for authoring 'various enormous opinions' that he had refused to recant. It also noted some of the cities where he had lived and added that 'they say in Germany that he had disputed several times against Cardinal Bellarmine.'[228]

Did that really happen? What did Bellarmine think about Bruno's eccentric claims? Since Bellarmine wrote many works, I suspected that somewhere he commented on such beliefs. Indeed he did, years later, as we will see.

A document recently discovered in a Roman archive and published in 2010 seems to note the importance of the heresy of the plurality of worlds. It seems to have been written on 12 February 1600, the same day that the official Notice of Rome announced Bruno's execution. It states:

> The Proceedings were burned, those in which the curate [parish priest] already noted had specified so many ☉ [suns] and for which he was burned, and soon will be burned an obstinate *relasso* [someone who relapsed into heresy] called Tadeo Bruno of Nola, a famous writer who for 3 years has been at the Holy Office; who recently said to the Cardinal of Santa Severina that he knew more about doctrine and especially about philosophy than St Thomas, and that with greater happiness he had accepted his sentence to be burned, by comparison to the bitterness and displeasure with which they read it, but that he was quite disturbed when the Cardinal told him that he would be burned in a place where he would be publicly seen. And this hardened beast of a philosopher bears a resemblance to those of antiquity. They say that he is ill, but slightly, with phlegm.[229]

This document is very important because it seems to say that someone else had been burned before Bruno for the belief that there are many suns. It does not specify the name of that curate. Federica Favino conjectures that it was actually Celestino Arrigoni, one of the Venetian inmates who testified against Bruno and was executed on 16 September 1599, shortly before Bruno, both at the Campo de' Fiori.[230]

The document is also significant because it discusses Bruno immediately after specifying the objectionable claim of many suns, as if that too were his main crime, and because it compares the obstinate Bruno to the philosophers of antiquity. We should note that the tenth proposition that had been censured in Bruno's books, during his Roman trial, was precisely: 'Again he posits many worlds, *many suns* necessarily containing similar things in kind and in species, as in this world, and even men.'

A few days later, the 'Journal' of the Archbrotherhood of St John the Beheaded (S Giovanni Decollato), in Rome, notes that the 'impenitent heretic' brother Giordano Bruno was punished for having 'finally always stood in his cursed obstinacy, wandering the brain and intellect with a thousand errors and vanities'. Bruno was sentenced to public execution. When executions were public instead of private, the intention was often to publicize the particular heresies, as a kind of public warning.[231]

Sixteen years earlier Bruno had anticipated his fiery death at the end of his *Ash Wednesday Supper*, his book about Copernicus:

> if you do not want to accompany him [Bruno] with fifty, or a hundred torches (which, even if he should march at the middle of the day, will not be lacking if he is to die in the Roman Catholic land), then at least accompany him with one of those, or even if this seems too much, accompany him with a lantern with a tallow candle inside.[232]

Likewise, in his *Expulsion of the Triumphant Beast*, he had written about himself in the Dedication:

> Come now, come now, as a citizen and servant of the world [Bruno], son of the father Sun, and of the mother Earth; since he loves the whole world so much: let us see what it is to be hated, blamed, persecuted and expelled from it. But in this moment, he is not idle, nor wrongly busied during the expectation of his own death, of his transmigration, of his transformation.[233]

The entry in the Journal of the Brotherhood describes what finally happened to Bruno: 'And he persisted so much in his obstinacy, that from the ministry of justice he was taken to the Campo di Fiori, and there stripped naked and tied to a pole.'[234]

A crowd thus witnessed the execution on the funeral pyre, in front of the ruins of the ancient Theatre of Pompey. In the letter he wrote that day Schoppe said he saw Bruno's execution: 'Therefore, today, on being led to the funeral pyre, as he was dying they showed

The Campo de' Fiori, where Bruno was taken to be publicly burned alive. Map by Antonio Tempesta, 1593.

him an image of the crucified Saviour, he rejected it by throwing back his fierce face, so he perished miserably.'

The records of the Brotherhood of St John the Beheaded state that the clergymen sang as Bruno died: 'he was burned alive, accompanied throughout by our Company singing the litany, and the comforters up to the very end comforting him to let go of his obstinacy, with which finally ended his miserable and unhappy life.'[235]

Among those present was Cardinal Santori as well as the Roman notary Giuseppe de Angelis. A few years ago De Angelis's document was discovered in a Roman city archive. The most stunning thing about it is a simple but extraordinary drawing. Before that discovery there were no known depictions of Bruno during his life. De Angelis summarily described the event and drew a small sketch of Bruno, with a thin beard and wearing a tunic, as he burned on the funeral pyre, his hands apparently behind his back, while flames and smoke rose around him.[236] It is astonishing that this solitary image should precisely record the moment of Bruno's death. The ink, which has decomposed across the centuries, has created cracks and holes in the yellowed paper as if the paper itself had been burned by the fire that killed Bruno.

There exists another account. Johann Wackher von Wackenfels was an imperial counsellor who had been a patron of Bruno in Prague in 1588. He was also a patron of Schoppe. On 19 February 1600 Schoppe sent a letter to Wackher, telling him:

Giordano Bruno as he died in the fire. Drawing by the notary Giuseppe de Angelis, on 17 February 1600. Original document at the Archivio di Stato di Roma.

> Lately, on Thursday Febr. 16 [*sic*], Giordano Bruno was adopted into the family of the Baron of Atoms. When the pyre was about to start, an image of the Crucifix was offered to him to kiss it, [but] he rejected it with a scowling face. Now, I think, he shall announce to the worlds of the innumerable and Simonians how things are done in this one [world] of ours.[237]

The Simonians were a heretical Christian sect that followed Simon Magus, a reputed magician who was rebuked by the apostles Philip, Peter and John (Acts 8:9–24). The Simonians were some of the earliest Gnostics and predecessors of the Valentinians, who were denounced as Pythagoreans. The Simonians were denounced as heretics who believed in transmigration.[238]

The same day Schoppe wrote that letter, 19 February, a 'Notice of Rome' was issued:

> Thursday morning at the Campo di Fiore a Dominican friar from Nola was burned alive, of whom it was already written: obstinate heretic, and having by his caprice formed various dogmas against our faith, and in particular against the most

Holy Virgin and Saints, he chose to die obstinately in those crimes; and he said that he died as a martyr and voluntarily, and that his soul would rise with the smoke into paradise.[239]

Why the Romans Killed Bruno

We should summarize all the accusations against Bruno. Luigi Firpo counted 34 charges. Maurice Finocchiaro composed a longer list: forty charges, including the plaintiffs for each. However, Finocchiaro's list includes a few repetitions and omissions, as well as one interpolation, so I have prepared a more comprehensive list. It consists of 54 accusations, including self-incriminations, and it specifies more clearly each alleged or actual transgression. Whereas Finocchiaro described each charge briefly, for example that Bruno 'had spoken ill of Moses', I specify the actual accusation: what did Bruno say about Moses? In addition, how did Bruno defend himself? I have also grouped the accusations in five major categories. In the endnotes, I specify which accusers voiced each accusation.

1. Bruno's Alleged Departures from Catholic Practices and Rituals

1. Allegedly Bruno spoke ill of Catholicism, saying 'that the Catholic faith is full of blasphemies against the Majesty of God', that it is 'full of doctrines of asses', and that he would create a new sect of 'Giordanists' in Germany called the New Philosophy. But in his fourth deposition, Bruno denied having ever spoken against the Catholic faith, and he denied having created a religious sect.[240]

2. Reportedly, Bruno spoke ill of the breviary, the Catholic book of rites, by saying that it is not worthy of being read by good men, that it contains many confused and profane things, and that it was written by a 'foul dog, a ------- -----'. However, Bruno (fourteenth deposition) denied having criticized the breviary at all.[241]

3. Bruno habitually uttered blasphemies, and he foully gestured upwards and cursed against Heaven. Bruno (tenth

deposition) admitted that sometimes he had said blasphemies, but nothing major, and he denied having gestured against Heaven.[242]

4. Four accusers also said that Bruno entertained atheist ideas, did not believe in God, or had no religion. However, Bruno (fourth and ninth depositions) denied having abandoned the Catholic faith.[243]

5. Bruno allegedly held holy relics in contempt, saying it is useless to venerate them. Yet Bruno (in an unspecified deposition) denied having spoken against relics.[244]

6. Reportedly Bruno also disliked the veneration of sacred images, saying that it constitutes idolatry. However, in his twelfth Deposition he said he actually appreciated religious images.[245]

7. Some accusers said that Bruno practised divination, conjuration and judicial astrology, and said that magic is good and legal. Indeed, in his fifth and fifteenth depositions, Bruno said he studied astrology only to see whether it works, but that he did not believe that fate guides events in the world, only God's providence. He said he was curious about sciences because he was interested in medicine. In his tenth deposition Bruno also replied that magic is only illegal if it is used for evil purposes, because it is based on observing nature and mixing substances, purely physical operations.[246]

8. Next, Bruno allegedly condoned and performed sexual sins, saying the Church sins by prohibiting sex with women. However, Bruno (fourth deposition) replied that he agreed with the Church and acknowledged often that sexual acts are sins.[247]

9. He was also accused of apostasy, for having fled from the Dominican order. Bruno (fourth deposition) said he had not confessed sins for years, except twice, in Toulouse and

Paris. The confessor then said he could not absolve Bruno because he was an apostate priest. But nevertheless Bruno said he always asked God forgiveness for sins, and that he appreciated the Sacrament of Penance.[248]

10. Bruno had a previous criminal record with the Roman Inquisition in 1576. In his first deposition, in 1592, he said that in 1576 the Inquisition processed him merely for parting with religious images while keeping only a crucifix. Later, in his fifth deposition, Bruno said he originally didn't know why the Inquisition wanted to process him.[249]

11. Two accusers said that Bruno listened to heretical sermons in England, Geneva and Germany. Bruno replied that he only went to Protestant cities for his safety and that he only consorted with heretics out of curiosity, not by sharing beliefs.[250]

12. He read prohibited books by Luther, Calvin, Melanchthon and other heretics. Bruno admitted this, but not to adopt their doctrines, he said, just out of curiosity.[251]

13. Bruno rejected the Catholic restriction against eating meat on prohibited days. He admitted eating meat with heretics on such days.[252]

14. Allegedly Bruno said that if he were captured and forced to rejoin the Dominican order, he would burn down the monastery and escape. However, in his ninth deposition, he denied this allegation.[253]

II. Bruno's Alleged Heresies against the Bible and Catholic Dogma

15. Two accusers claimed that Bruno denied that the Church's dogmas are credible; saying that all about the Church was vanity, which could not be proved. However, in his thirteenth deposition, Bruno denied having criticized the Church.[254]

16. Those two accusers also complained that Bruno held erroneous opinions about Cain and Abel, saying that Cain was a good Pythagorean man for killing his carnivorous brother Abel. Yet Bruno replied that he was just joking in saying that Cain had a 'Pythagorean zeal'.[255]

17. Mocenigo and Celestino complained that Bruno held erroneous opinions about the Trinity, that he did not recognize three persons in God, saying the Trinity is an impossibility. In depositions, Bruno admitted that he had failed to understand what is meant by 'the Holy Spirit as a third person', except in the sense of the Pythagorean doctrine explained by Virgil: as soul of the universe. He said he didn't intend to doubt the Trinity, but speaking *philosophically* he had wondered how the Son (the Word) became incarnate; he said he didn't know how divinity became united with human nature.[256]

18. One accuser claimed that Bruno said that the cross that Christians placed above altars was originally a sign etched on the chest of the goddess Isis, venerated since antiquity, and that Christians stole it from the ancients, pretending that it had been used to crucify Jesus. When interrogated, Bruno did not entirely confirm this, but he said (tenth deposition) that the cross was an ancient symbol that pre-dated Christ: it was known in Egypt during the times of Moses, and was used on the chest of Serapide.[257]

19. Two accusers said that Bruno denied that people are punished for their sins. But in turn, Bruno (fourth deposition) didn't admit having said that, and he said he had 'always believed that good deeds are necessary'.[258]

20. Five accusers said that Bruno held erroneous opinions about Hell, saying there is no eternal punishment because God is not angry at anyone forever. However, in his ninth deposition, Bruno replied that he did not understand what these accusers had testified.[259]

21. Three of these accusers also claimed that Bruno said that at the end of the world even demons would be saved, because 'God saves men and beasts.' However, in his ninth deposition, Bruno denied this accusation. He said that demons would not be saved, and that infernal punishment lasts forever.[260]

22. Schoppe claimed that Bruno said that scriptures were just a dream.[261]

23. Schoppe also complained that Bruno wrote that human beings existed before Adam and Eve; only the Hebrews originated from Adam and Eve. Indeed, Bruno wrote this in his book *On Monads*.[262]

III. *Bruno's Alleged Heresies against the Prophets and Church Fathers*

24. Two accusers claimed that Bruno spoke ill of Moses, saying that Moses was a wise magician who only pretended to talk to God on Mount Sinai, and that Moses contrived his laws by magical arts, not from God, because they were tyrannical and bloodthirsty laws. Yet Bruno (tenth deposition) argued that Moses was an expert in all the Egyptian sciences, including magic, and that before Moses spoke with God, he could have used magic, which is not illegal since it is founded on the careful observation of nature and purely physical operations, not evil.[263]

25. Next, Bruno allegedly spoke ill of the prophets, saying that they were men who lied, did bad deeds and were rightly condemned. However, in his tenth deposition, Bruno denied having criticized the prophets.[264]

26. Mocenigo also accused Bruno of having said that the prophets and the apostles were magicians. However, Bruno (fourth deposition) denied that the apostles performed their miracles by magic, and he (tenth deposition) denied having criticized the prophets.[265]

27. Three accusers also said that Bruno disapproved of praying to the Saints, that he mockingly said that it was vain and ridiculous to invoke them because they cannot intercede for us. However, Bruno (in an unspecified deposition) denied having criticized the invocation of Saints.[266]

28. Some accusers also complained that Bruno spoke ill of the Catholic Church and its officials, for example, saying 'that the Church was governed by ignorant men and asses'. In contrast, Bruno repeatedly denied having criticized the Church (in his fourth, ninth and thirteenth depositions).[267]

29. Also Mocenigo claimed that Bruno, in his book *Cantus Circeus*, insulted the Pope by depicting him as a pig. But Bruno denied it in his sixteenth deposition.[268]

30. Furthermore, two accusers claimed that Bruno spoke ill of theologians, the Doctors of the Church, for example by saying that St Thomas Aquinas and others knew nothing compared to himself. To the contrary, Bruno (fourth deposition) replied that he didn't criticize theologians and that he especially admired Thomas.[269]

IV. Bruno's Alleged Heresies and Blasphemies against Christ and God

31. Mocenigo alone further claimed that Bruno denied the virginity of Mary, mother of Jesus Christ, by mockingly saying that it's impossible for a virgin to give birth. But again, Bruno defended himself, saying (in an unspecified deposition) that he didn't speak against Mary's virginity, and that she certainly conceived Jesus thanks to a miracle of the Holy Spirit.[270]

32. Another accuser said that Bruno denied that the Three Kings paid homage to baby Jesus. Yet Bruno (unspecified deposition) admitted only that he heard a sermon in which someone discussed whether the 'Three Kings' were not really kings.[271]

33. A couple of accusers also said that Bruno held erroneous opinions on Christ's divinity and incarnation. In contradistinction, Bruno (eighth deposition) replied that he had just wondered how Jesus became incarnate; and Bruno (tenth deposition) also denied having criticized doctrines about Christ.[272]

34. Four accusers testified that Bruno held erroneous opinions about the life and death of Jesus; for example, saying that Jesus was not crucified, but was hanged from two logs. To the contrary, in his tenth deposition, Bruno said that he always believed in Christ's passion and death. He said the four branches of the cross were not equally long, and he denied that Christ hung from a gallows.[273]

35. Two accusers claimed that Bruno had said that Jesus was a magician who forged his miracles with magic and necromancy. In his defence, Bruno (fourth deposition) said he did not deny Jesus's miracles; he denied that Jesus did miracles by magic.[274]

36. Next, reportedly Bruno said that Jesus had sinned mortally by refusing God's will, in particular when Jesus said: 'Father, if it be possible, let this cup pass from me' (Matthew 26:39). However, in his ninth deposition, Bruno denied having said that Christ sinned.[275]

37. Also, Bruno allegedly held erroneous opinions about transubstantiation: reportedly he said that the bread does not become the body of Christ, and he discussed this with heretics. Yet Bruno (fifth deposition) denied having doubted transubstantiation.[276]

38. Six accusers said that Bruno angrily and repeatedly voiced blasphemies against Jesus, especially in prison, saying, '---- of Christ', 'Christ -----, dog -----', 'Christ is a ------- dog', '------- -----, ----- of God', and so on. Consequently, in his tenth deposition Bruno only admitted that sometimes he had said Christ's name in anger, but said he never directly sinned against His name.[277]

39. Reportedly, Bruno also said blasphemies about God, such as 'God is a traitor, who doesn't govern the world well.' Bruno (tenth deposition) admitted that sometimes when angry he had improperly mentioned God, but that he never directly sinned against God's name; he denied saying that God is a traitor governing the universe.[278]

v. Bruno's Heretical Beliefs about Substance, Worlds and Souls

40. The consultors of the Inquisition complained that Bruno argued that substance cannot be created or destroyed, but only transforms, in accord with the teachings of Pythagoras. Bruno then admitted it, replying that there exists no true generation, but only transformation, because 'there is nothing new under the Sun.'[279]

41. The consultors also complained that Bruno denied Solomon's statement in Ecclesiastes 1:14, that the true being of individuals is vanity, because Bruno held that the underlying substance of all things is eternal, not illusory. Bruno had asserted this in his books. And then apparently Bruno insisted that the underlying substance of all matter is *not* vanity, but 'really is what it is'.[280]

42. Also, five accusers testified that Bruno believed that the universe is eternal, not created by God. Subsequently, Bruno (third deposition) argued that whether the world is eternal or temporal, the 'caused or produced world' depends on the First Cause, thus he did not reject the word 'Creation'.[281]

43. The consultors seemed to complain (second proposition) that Bruno believed that the universe is spatially infinite. Bruno replied that since God's power is infinite, it *must* produce infinite things. Accordingly, Bruno (third and fourteenth depositions) affirmed the infinity of empty space.[282]

44. Six accusers plus the consultors and the Inquisitors also complained that Bruno believed that many worlds

exist, infinitely many, and that all the stars are worlds; another necessary consequence of God's infinite power. Consequently, Bruno repeatedly *affirmed* (third, twelfth, fourteenth and seventeenth depositions) such Pythagorean beliefs: Earth is a star, stars are worlds, infinitely many worlds exist.[283]

45. Also, two accusers plus the consultors said that Bruno believed that the other worlds are similar to ours and are inhabited by rational beings. Indeed, in his fourteenth deposition, Bruno said that other worlds resemble Earth, with animals, plants and rational beings.[284]

46. The consultors of the Inquisition further complained that Bruno argued, against Aristotle and Catholic dogma, that the soul is not the form of the body, but that instead it is a spiritual substance that inhabits the body, 'as a captive in a prison', and controls it like a ship. Yes, Bruno had said such things, and he said that the false claim that the soul is the body's form is not in the Bible. In his fifth deposition, Bruno said that the human soul subsists outside the body, against Aristotle's claim that it is the body's form.[285]

47. Three accusers said that Bruno believed in the transmigration of souls, between human bodies, even into other distant worlds. Bruno (second and eleventh depositions) replied that the soul is immortal and 'speaking in a Catholic way' it goes to Paradise, Purgatory or Hell, but that 'reasoning philosophically', a soul *possibly* can go from one body to another, in accord with Pythagoras' opinion. Bruno insistently denied saying that *in fact* the soul transmigrates between bodies, but said instead, 'I posit this only as a possibility.'[286]

48. Two accusers said that Bruno also believed that human souls can transmigrate into animal bodies; and that the souls of some of his own friends might presently inhabit animal bodies. But Bruno (eleventh deposition) denied that human souls can inhabit animal bodies.[287]

49. One accuser claimed that Bruno himself said he remembered having lived previously in another body, and that his soul 'in another time in the world had been a swan', which made the other prisoners laugh. Another accuser twice testified that Bruno claimed to recall living in another body, whether human or animal. However, in his eleventh deposition, Bruno denied 'having been someone else in this world'.[288]

50. The consultors of the Inquisition twice complained that Bruno believed that the Earth has a soul, that Earth is a living and rational animal. Bruno then admitted that he believed that God created the soul of the world. He quoted the Bible, 'Producat terra animam viventem' (Genesis 1:24), which is often translated as 'Let the land produce living animals', but might be rendered as 'the Earth produces living souls', which Bruno interpreted as meaning that Earth produces animals and animates them with its worldly soul. He said Earth is a rational animal, as evinced by its regular motions.[289]

51. The consultors also complained that Bruno said that the Earth moves, which he said shows the Earth's rational mind. Indeed, Bruno insisted that the Earth spins and moves but the firmament does not, and he added that this doesn't undermine the authority of divine scriptures.[290]

52. The consultors complained that Bruno said that stars are angels: visible and animated rational bodies that convey the divine voice. And indeed, in his fourteenth deposition, Bruno quoted 'the Heavens declare the glory of God' (Psalms 19:1) and said that therefore the heavenly bodies are animated, rational, sensible beings, or angels, that interpret God's voice.[291]

53. The Inquisitors interrogated Bruno about his belief that the Holy Spirit is the soul of the universe, what he called 'the Pythagorean doctrine'. Yet Bruno *reasserted* this belief in his third, fourteenth and seventeenth depositions.[292]

> 54. Finally, the consultors also complained about Bruno's claims about how human souls originate. And again, Bruno admitted it: he replied that just as the earth in our bodies came from the world, so too our spirit comes from God and returns to God, while 'the universal spirit' always subsists.[293]

This list of accusations is remarkable. It shows that hearsay about Bruno was deeply offensive to Catholic officials. The accusers portrayed Bruno partly as a recklessly angry and embittered enemy of Catholicism, yet his replies show that he consistently denied any disagreements with Catholicism, to the best of his knowledge. At times he seems merely to have exercised intellectual curiosity.

I have grouped the accusations into five categories: departures from Catholic practices, heresies against the Bible and dogma, heresies against prophets and Church Fathers, blasphemies against Christ and God, and fifth, beliefs about substance, worlds and souls. It is striking that under interrogation Bruno directly denied *nearly all* charges in the first four categories, but he refused to recant most of the accusations listed in the fifth: beliefs about substance, worlds and souls. Of the fifteen accusations in that category, Bruno apparently denied only two: having said that human souls can be reborn in animal bodies, and he denied that previously he had lived as a swan. (Interestingly, Plato's *Republic* said that Orpheus had chosen to be reborn as a swan.) In his trial depositions and his replies to book censures, apparently Bruno boldly asserted or defended *thirteen* of those fifteen censured claims. *Fourteen* of the fifteen accusations were Pythagorean; again, without counting the one about stars being angels.

Someone might guess that by August 1599 Bruno recanted some beliefs about substance, worlds and souls, since he reportedly maintained only the Novatian heresy and the belief that souls resemble sailors. However, no direct evidence shows Bruno having denied such beliefs, as we do not know the six propositions that he recanted then, only the two that he did not.

At the time any Catholic person who was confronted by an Inquisitor was expected to deny or renounce any errors, blasphemies or heresies. If the accused admitted that such beliefs were mistaken, the Roman Inquisitors were often lenient, because they understood that their function was not merely punitive. For example, in the

Relief of Giordano Bruno's execution. Sculpted by Ettore Ferrari, on the base of Bruno's statue at the Campo de' Fiori.

fourth deposition in Venice, in 1592, the Inquisitors asked Bruno 'whether the miracles that were done by Christ and the apostles were apparent miracles and made by magical arts and not true'. Bruno replied: 'What is this thing? Who is it that has contrived such devilry? I have not said any such thing, nor did any such thing pass through my imagination. Oh God, what is this thing? I would rather sooner be dead than to have such a thing proposed to me.' Afterward, the Inquisitors in Rome and Venice did not ask Bruno about that accusation again.

Bruno's reactions to being accused of believing in many worlds were the opposite. It was one of the few doctrines that Bruno asserted as true. He repeatedly refused to recant this opinion. He refused at least four times, in four of the seventeen depositions: in the third (in 1592 in Venice), the twelfth (in 1593 in Rome), the fourteenth (also in 1593) and the seventeenth (in 1598). Bruno repeatedly insisted that the Earth is a star (what we call a heavenly body), the Sun is a star, radiant stars are orbited by planets such as the Earth, which are all worlds made of the same elements as the Earth. He insisted that 'innumerably many worlds' exist. And he was right.

IN ORDER TO find out how unusual Bruno's execution was, we should compare it to other executions in Rome. The Archbrotherhood of St John the Beheaded carried out the duty of accompanying all convicts to their deaths, and it kept records of such events. Consider, for example, the executions carried out from 1598 until 1604. The time frame is arbitrary: I considered counting how many persons were executed in Rome only during the rule of Governor Taverna: April 1599 until June 1604. However, since Bruno was executed soon after Taverna was appointed governor, I expanded the scope a bit more, from January 1598 until December 1604. During those seven years, 189 persons were executed in Rome.[294] Table 1 conveys the numbers and methods of execution.

By far the majority of executions in Rome were by hanging, while other convicts were executed by decapitation or bludgeoning. Most convicts were executed at the Piazza di Ponte, and then their corpses were exhibited at the Sant'Angelo bridge. The death records do not specify whether bodies were quartered only after death, or whether some individuals were still alive when their bodies were

Number	Method	Subtotal
136	Hanged	169
24	Hanged and quartered	
8	Hanged and burned	
1	Tortured with pincers, hanged and quartered	
5	Decapitated	8
2	Decapitated and quartered	
1	Decapitated and burned	
5	Bludgeoned and quartered	10
3	Tortured with pincers, bludgeoned and quartered	
2	Pincers, bludgeoned, decapitated and quartered	
2	Burned alive	2
	TOTAL EXECUTIONS	189

Table 1. Executions in Rome, from 1 January 1598 until 31 December 1604. Nearly all were executed for crimes, not for heresies.

torn to pieces. In any case, the rarest kind of execution was being 'broiled and burned alive'. It was feared as the most painful type of punishment, inflicted for the gravest offences.

The most surprising point about the records of executions, over those years, is that only two men were killed as 'heretics': Celestino and Bruno. Only the two of them were burned alive. Both were executed at the Campo de' Fiori. Both of them were described as 'heretics', both 'obstinate and impenitent'. Bruno had asserted to the Inquisitors that many suns exist, and apparently Celestino too.

Why did the Inquisition kill Bruno? In 1942 Angelo Mercati argued that the 'Summary' of Bruno's trial shows he was not prosecuted for scientific claims. In Bruno's writings, a topic such as the Earth's motion was not defended on the basis of experiments, mathematics or astronomical observations, but on philosophical arguments, for example that the Earth must orbit the Sun to partake of the seasons. Thus Bruno's trial was not essentially a conflict between science and religion. Instead Mercati concluded that the Inquisition condemned Bruno because of transgressions against Catholic orthodoxy.[295]

In the 1960s Frances Yates argued instead that what led to Bruno's execution was that the Inquisition considered him a Hermeticist and magician.[296] More recently, however, Maurice Finocchiaro has rightly argued that the problem with Yates's thesis is that 'there is little trace of Hermeticism and magic in the trial proceedings.' Among the many accusations against Bruno, Finocchiaro fairly notes that questions of Hermeticism or magic arise only a few times: that he practised magical arts and said the prophets, Moses, Jesus and the apostles did too. But such charges did not stick, since Bruno promptly and vigorously denied them.

Mercati too did not properly characterize Bruno's trial by saying it was about religious transgressions. Mercati overlooked the fact that some of the important censures against Bruno had a distinctly *philosophical* significance, instead of religious or scientific. Finocchiaro argues that in the 1590s the discipline of philosophy had autonomy, and that Bruno fairly questioned whether his (presumably philosophical) statements were really heretical. He asked *when* suchlike notions officially were declared heretical. By September 1599 Bruno had denied or retracted the censured theological propositions, so Finocchiaro infers that Bruno was unwilling to retract his *philosophical* beliefs. Then the Inquisition became intolerant towards his obstinacy.[297]

There were certainly substantial conflicts between Catholicism and Bruno's philosophy. So Finocchiaro rightly characterized the trial as involving a conflict between religion and philosophy. Yet the present analysis reveals a neglected dimension: many of Bruno's main philosophical and religious heresies were directly rooted in the ancient Pythagorean religion and its interpretations in the Renaissance. Even seemingly scientific notions, such as the Earth's motion, were linked to pagan ideas: that it has a soul. The idea that there are many worlds like Earth was connected to the Pythagorean belief that the beings living in such worlds can embody the souls of persons who once lived on Earth. The idea that the universe is infinite was connected to the Pythagorean denial that the world was created.

Again, why did the Romans kill Bruno? Since there were so many accusations and some documents are missing, some historians claim causal ignorance. For example, Thomas Mayer wrote: 'The simple fact is that we do not know why he was executed. Nevertheless, Francesco Beretta must be right that the articles could not have included any Copernican charges because they were all specified in the sentence as heretical.'[298] Such claims fail to recognize that charges of heresy did not need to correspond to pre-existing heresies. The Pope could designate something as heretical. Consultors and cardinals could also brand something as a heresy. This is what happened later with Galileo's belief in the Sun's immobility.

Also, some of Bruno's propositions were semi-Copernican (as with Digges), pertaining to cosmology, yet were deemed heretical. In particular, the notion of many worlds was categorized as heretical for centuries. Consultors and Inquisitors who mastered treatises on heresies knew that it was heretical, unlike recent historians. For example, Bellarmine knew Augustine's book *On Heresies* very well.[299] Bellarmine was also very familiar with works on heresies by Tertullian, Epiphanius, Irenaeus, Philaster and others, which he also cited often.

Schoppe complained that in Rome some people said that 'a Lutheran' had been burned, because they did not distinguish Lutherans from other heretics. Yet he explained that actually Lutherans and Calvinists were not really in any danger in Rome because the Pope had requested that they be treated with extraordinary civility and exhorted to investigate the truth. Certainly, the label 'Lutheran' was misused in Italy for many practices and beliefs, and it is also true that by the 1590s the Italian Inquisitions processed fewer Protestants than previously.[300]

Bruno had done something worse. In one paragraph, Schoppe complained that Bruno 'teaches these utterly horrific and most absurd things', which I here render as a list, in the same sequence, adding only numbers:

> (1) 'Worlds are innumerable',
> (2) 'souls go from body to body, and even migrate to another world',
> (3) 'one soul can shape two bodies',
> (4) 'magic is a good and licit thing',
> (5) 'the holy Spirit is nothing other than the soul of the world, and this is what Moses meant by writing that it nourished the waters',
> (6) 'the World exists eternally',
> (7) 'Moses effected his miracles by magic, in which he advanced more than the rest of the Egyptians, and that he forged his laws',
> (8) 'Sacred Scriptures are a dream',
> (9) 'the Devil will be saved',
> (10) 'only the Hebrews originated from Adam and Eve, other than those two God made everyone else on the previous day [of Creation]',
> (11) 'Christ is not God, but was a notorious magician who deceived men, and therefore he was rightly hanged, not crucified',
> (12) 'the prophets and Apostles were but men, magicians, and many were hanged.'[301]

Does Schoppe's list match the accusations that Bruno faced in the closed meetings with his Inquisitors? Yes, from Schoppe's list of twelve horrendous 'teachings', eleven – nearly all of them – had been articulated against Bruno during the proceedings in Venice and Rome, some several times.

Among these twelve 'horrors', five of the first six are Pythagorean. Others might be related to Pythagorean beliefs, for example, if we infer that the alleged critiques of the powers of Moses, Jesus and the apostles involved partly a belief that certain men, such as Pythagoras and Apollonius, had also performed similar feats, as Bruno said in some of his works. Anyhow, some of the latter claims in the list were hardly justified, since Bruno either denied them under interrogation or because apparently he said no such thing (as far as I know): that scriptures were a dream.

Schoppe was closely affiliated to one of the Inquisitors in Bruno's trial: Cardinal Lodovico Madruzzo. In 1598 Schoppe had initially met Father Bellarmine and Cardinal Baronio at Ferrara, on his way to Rome.[302] Once there, he promptly won the favour of Pope Clement VIII, thanks to Schoppe's writings and his emphatic abjuration of Protestantism. Schoppe's nemesis, Joseph Justus Scaliger, remarked that Schoppe had 'gone to Rome to lick the plates of the Cardinals'.[303] The Pope honoured Schoppe with the titles of Knight of St Peter and Apostolic Count of Clarvalla. In Rome Schoppe was honoured to live in the palace of Cardinal Madruzzi, who had been a papal nuncio to Germany, and was head of the Inquisitors who judged Bruno.[304] Thanks to the efforts of Baronio, Schoppe even received a pension from the Pope in May 1600.

One puzzle needs explanation. Schoppe said that, for Bruno, 'nourished the waters' meant the Holy Spirit is the Soul of the World. In Schoppe's account, the words in Latin are *fovisse aquas*, which may be translated literally as 'nourished the waters'. Historians have construed this as referring to Genesis 1:2, 'et spiritus Dei ferebatur super aquas' ('and the Spirit of God hovered over the waters').[305] Bruno had learned this from Ficino:

> Hence the saying [by Virgil]: *the Spirit nourishes within*. Likewise again and containing the soul power, insofar as with itself it is glued (so to speak), like oil that swims on and surrounds the other. Perhaps here this tends to 'the Spirit of God hovered over the waters', or infused in the waters.[306]

During his interrogations Bruno used other lines from the Bible and by Virgil, his 'Pythagorean poet', to interpret the Holy Spirit as the universal Soul.[307]

Another strange accusation was that 'the Devil will be saved.' Three witnesses had accused Bruno of saying demons would be saved, though he denied it in a deposition. A similar claim had been denounced by St Irenaeus (*c.* 180 CE). Irenaeus denounced men who believed in the transmigration of souls, and he rejected the idea that souls are imprisoned in bodies, and that eventually 'all souls are saved'.[308] Subsequently Origen argued that the Bible (Paul's letters) taught that everyone would be saved: even the Devil would ultimately gain redemption. Later Christians disagreed, saying instead that most of humanity is eternally condemned. Augustine, for example, argued that even unbaptized babies go to Hell.[309] Jerome complained that Origen said 'the Devil himself, after a certain time, will be as well off as the angel Gabriel.'[310] Thus, although Origen promulgated Christianity, long after his death some Christians declared him a heretic:

> If anyone says, or thinks, that the punishment of demons & of impious men is temporary, and will have ended at some time, or that a future restoration (*apocatastasis*) of demons or impious men will happen, he is anathema [a heretic]. Anathema to Origen, who is called Adamantius, who promulgated all this, along with his hateful and execrable doctrines & to every man who dares to think or assert this, or any part at any time forever and ever: In OUR LORD JESUS CHRIST, glory to him forever and ever, Amen.[311]

For this reason too Bruno was declared a heretic.

Schoppe summarized and exaggerated Bruno's crimes: 'In a word, whatever was affirmed by the pagan philosophers or by ancient and recent heretics, he defended it all.' Therefore Schoppe concluded that Bruno 'perished miserably by burning, departing, I think, so that he might tell them in the other worlds, imagined, the manner in which blasphemous and impious men are usually treated by the Romans'.[312]

This striking last sentence highlights the importance of Bruno's pagan heresy of the plurality of worlds. It appeared from the start, in Mocenigo's initial accusations against Bruno. Next, it was reiterated by five other witnesses against him. In sum, six witnesses accused Bruno of believing in many worlds on no fewer than thirteen separate instances, in ten depositions. In fact, it was the one accusation levelled more often than any other against Bruno throughout the protracted seven-and-a-half-year trial in Venice

Statue of Giordano Bruno where he was burned alive, at the Campo de' Fiori. Sculpted by Ettore Ferrari. The inscription reads: A BRUNO – IL SECOLO DA LUI DIVINATO – QUI DOVE IL ROGO ARSE, that is, 'To Bruno – The century predicted by him – Here where the pyre burned.' The statue was set in place in 1889, facing towards the Vatican.

and Rome. Moreover, the accusation against 'diverse worlds' was the single disciplinary admonition given to Bruno on his first meeting with Bellarmine. In the seventeenth deposition the Inquisitors noted that Bruno had 'relapsed' (*reincidit*) into this belief. This meant that he had remained obstinate in his heresy.[313] It was also included in the Summary of ten censures extracted from his books.

Schoppe pointed out the issue of innumerable worlds four times. His first letter describes Bruno's belief in the precise wording in which it was known as a heresy in Latin: 'Mundos esse innumerabiles.' In ancient times Cicero and Lactantius had disdained the claim 'mundos esse innumerabiles' as insane. Valerius Maximus and Seneca quoted the expression, attributing it to Democritus. Philaster declared that 'mundos esse infinitos et innumerabiles' was a heresy. Jerome too criticized Origen for asserting 'mundos esse innumerabiles'. Augustine described this heresy as 'Alia dicit esse innumerabiles mundos.' Likewise, Praedestinatus denounced the heresy of 'innumerabiles esse mundos'. Echoing Valerius Maximus, Ficino and Erasmus quoted the expression 'mundos esse innumerabiles'. Among various critics, Jacob Schegk 'refuted' this belief in

1550. Meanwhile, other writers often used other expressions, when writing about worlds but not writing about heresies.[314]

This was the first 'utterly horrific' doctrine listed by Schoppe, and again when he sarcastically said Bruno would go to such other worlds. Schoppe's letter to Wackher repeats this point. Moreover, the recently discovered document about another clergyman who was apparently burned for positing many suns seems to suggest that Bruno too was in that category: 'The Proceedings were burned, those in which the curate already noted had specified so many ☉ [suns] and for which he was burned, and soon will be burned an obstinate relapsed [heretic]', Bruno.

Is there any evidence that Schoppe's list of doctrines or accusations echoes the reasons why Bruno was executed? Yes. In his letter, Schoppe argued: 'But perhaps you might add: *the Lutherans neither teach nor believe such things, and therefore should be treated otherwise*. I agree with you, & therefore, precisely no Lutherans do we [Catholics] burn.' This means that if only the Lutherans 'taught or believed' what Bruno did, 'docere neque credere', they would be burned. It also means that Bruno was burned for his teachings and beliefs. And Schoppe listed the horrendous absurdities that Bruno 'teaches': 'quibus horrenda prorsus absurdissima docet'.

The convergence of evidence makes it preposterous to imagine that Schoppe interpolated whichever notions bothered him personally in Bruno's writings. He was echoing recurring accusations against Bruno. Schoppe's main point was to explain why Bruno was really burned. Importantly, he was an eyewitness on the day the Inquisitors condemned Bruno. Schoppe explained: 'Perhaps I too would believe the vulgar rumours that Bruno was burned for Lutheranism, but I was present at the Holy Office of the Inquisition when the sentence against him was pronounced, & so I know what heresy he professed.'[315] Schoppe specified the site of the condemnation: 'the Palace of the supreme Inquisitor', and indeed it was Cardinal Madruzzo's home, where Schoppe lived. His patron was the first to sign Bruno's condemnation on that very day.

Therefore I conclude that Bruno was burned for the heretical teachings that Schoppe listed, or for most of them. I cannot imagine that although Schoppe actually heard the heresies for which Bruno was condemned, and specified Bruno's 'horrific' teachings, and specified that anyone would be burned for such teachings, that yet

Bruno was burned for some *other* teachings that Schoppe chose not to specify. Bruno died for affirming that there exist many worlds, souls migrate into bodies, the soul of the world is the Holy Spirit and so forth.

For years, some historians, writers and astronomers have said that it's a myth that Giordano Bruno was burned for believing in many worlds.[316] But their claims did not take into account most of the present evidence. It shows the apparent myth was not a myth.

I have found more evidence showing that in 1600 the Church in Rome did condemn the thesis of the plurality of worlds. First, while the Inquisitors were deliberating Bruno's final fate, Cardinal Baronio was finishing a volume of his *Ecclesiastical Annals*. It was published in 1600 and it discusses, among many things, how Pope Zacharias reacted against Virgilius in the year 748. Baronio paraphrased Zacharias's critique:

> Regarding the perverse doctrine, which he [Virgilius] has spoken against the Lord & his own soul, namely that there is † another world, and other men beneath the earth, another sun & moon, if he is convicted of confessing this: the summoned Council will deprive him of the honour of being a priest of the Church. But we also communicate with the Duke mentioned above, regarding the aforementioned Virgilius we send a letter as we have presented this, & request a thorough investigation, [and] if an error is found, [we will] condemn it in canonical decrees.[317]

In the margin Baronio added a note corresponding to the dagger he placed next to the expression 'another world': 'To doubt the Antipodes is not heresy, but to posit many worlds is repugnant to divine scriptures, and therefore is proven to be a heresy.' Cardinal Baronio was librarian of the Vatican and held a privileged position as one of the closest clergymen to Clement VIII: he was the Pope's personal confessor.

Next, Johann van Heck was a Catholic physician, from the Netherlands, who lived in Rome at the time. In 1600 or 1601 Heck drafted a manuscript commentary on Pliny's *Natural History*. It should be remembered that Pliny had discussed, from the outset, the question of whether many worlds exist; Pliny had written that it was 'madness' to believe that there are innumerable stars and worlds. Hence Heck too discussed 'whether the world is infinite and [whether it is] one'.[318] He rejected the notion that the universe is infinite, and he attributed the thesis of the plurality of worlds to several writers: Democritus, followed by 'Archelaus, Metrodorus, Anaximenes, Diogenes, Leucippus, Epicurus and some others.' Heck did not specify it, but his list echoed an account by Stobaeus (*c.* 425 CE).[319] Heck quoted critical statements by St Albertus Magnus against the belief in many worlds. Heck added that this false belief was also asserted by the Manichaeans, 'truly heretics'. The *Kitâb al Fihrist* (*c.* 987), a history of Arabic literature, attributes to the Manichaeans a belief in the existence of 'five worlds' inhabited by devilish, lustful creatures.[320]

And here's the most significant point: Heck then wrote a timely remark that echoes the outcome of Bruno's trial some months prior: 'to posit a plurality of worlds is *totally condemned* by the Holy Roman Church.'[321] I've added the emphasis.

Furthermore, Heck also noted that the notion that the stars are animated 'contradicts the Catholic faith no less than [it contradicts] the Aristotelian philosophy'.[322] In 1603 Heck became one of the four founders of the scientific Academy of the Lynx in Rome, of which Galileo later became a member.

Finally, I discovered another piece of evidence regarding Bruno's condemnation. In 1626 a book was published by a 'Lucius Verus' that replied to critiques against Catholics by the prominent Calvinist politician Ludwig Camerarius. In response to a remark about Bruno, the pseudonymous 'Verus', an advocate for the Catholic emperor, wrote: 'for those who are interested, there are documents; & the records show, those which are available: Regarding the execution of Bruno of Nola it states: He was not a Lutheran, instead he was infected partly with Calvinism, even by Manichaeus, Borborites, Arian and Eutychian heresies, and he was convicted of those crimes.'[323] Rather than explaining all the possible connotations of these five kinds of heresies, we can answer a narrower question: did any of these heretical groups believe in the existence of many worlds? Yes. Epiphanius, Augustine and Albertus Magnus all complained that the Manichaeans believed that multiple worlds exist.[324] As for the 'Borborites', that was a name given to the Gnostics. Epiphanius accused the 'filthy' Gnostics of believing that, after death, a person's soul travels successively to multiple inhabited heavens, and can sometimes be reborn in 'this world' as animals.[325] It is significant that although books of heresies gave the names of hundreds of sects with various beliefs, the few groups that 'Lucius Verus' linked with Bruno include two that said many worlds exist.

There are other links between such sects and Bruno's heresies. Epiphanius complained that 'all the Gnostics and Manichaeans' believed in the transmigration of souls. Reportedly they thought that soul is dispersed in all things, and that the human soul is a fragment of God, imprisoned in the body.[326]

The Catholics who denounced, feared and hated the Christian reforms of Luther and Calvin likewise rejected Bruno's more eccentric efforts to reinterpret biblical scriptures and Catholic doctrines on the basis of pagan Pythagorean beliefs. Therefore, the Romans finally gagged him and burned him alive.

TWO

ALIENS ON THE MOON?

Giordano Bruno was not alone. Like him, Kepler and Galileo praised Copernicus and Pythagoras. In time, the less severe confrontations between Galileo and the Inquisition became more infamous than Bruno's deadly trial. For decades, historians have distinguished sharply between the proceedings against Bruno and Galileo. They have argued that Bruno was not prosecuted for the same transgressions as Galileo. It's true: Bruno's main crime was not that he claimed that the Earth moves. However, I will argue that Galileo's problems in 1616 were linked to the proceedings against Bruno.

This will not be a comprehensive portrayal of the Copernican Revolution, or of the Galileo affair.[1] But we will investigate the surprising links between Bruno and Galileo.

Prior to Bruno's works, some ancient sources told stories about life in other worlds. Lucian and Plutarch discussed the Moon, saying that it was inhabited by beings or souls. Iamblichus mentioned that some people believed that Pythagoras had lived on the Moon as a demon. St Augustine repudiated the idea of sacrificing to demons living on the Moon. Then Copernicus triggered the question: was it true that there exist worlds other than our Earth?

In 1597 a prominent professor of philosophy at Padua, Francesco Piccolomini, published a book discussing, among other questions, whether many worlds exist. He said this idea had been asserted by some Greek philosophers, including 'Democritus, Leucippus, Epicurus, Lucretius and others', who had posited in the immense vacuum atoms that randomly formed innumerable worlds.[2]

Piccolomini noted that Empedocles too asserted the plurality of worlds, and that 'Pythagoras said that the Moon is a Celestial Earth.'[3] Piccolomini discussed arguments for and against the plurality of worlds. For example, if such worlds are all similar then it's pointless that there be many. But if they are dissimilar then the universe is not really a universe or perfect. He also quoted authorities such as Aristotle and Thomas Aquinas, to conclude that our world is unique.

Piccolomini also discussed related questions. Is the world eternal? He discussed the opinion that the world has a soul, which he attributed to 'Orpheus and the most ancient theologians'. Interestingly, Piccolomini used brief expressions very similar to those of Bruno, whom he did not name: 'to understand their propositions, *Spirit nourishes within*: and moreover, *the spirit hovers over the waters*, and this spirit is the life of lives, and the Heaven of Heavens.'[4] This notion had been advanced by Orpheus, the Pythagoreans and the Stoics, Piccolomini said, and he quoted Virgil's words from the sixth book of the *Aeneid*, at greater length.[5] But this professor denied that the soul of the world is God.

Thus while Bruno defended such ideas against the Roman Inquisition, they were also criticized by Catholic professors of philosophy. Soon, the news of Bruno's death reached Protestant England. Robert Devereux, 2nd Earl of Essex, received this report:

> One named: Nolanus, Jordanus, politanus [from Naples], a notable, learned, and fantastical fellow, who with [Michel de Castelnau] Mauvissier the Ambassador was here in England, fell into the Inquisition's hands at Venice, and from thence sent to Rome, and there proceeded against by that holy order, disgraced, excommunicated and to avoid the spilling of blood, committed over to the secular power to be but burned.[6]

In England a prestigious physician, William Gilbert, voiced opinions very close to those of Bruno. In 1600 Gilbert published his book *On the Magnet, Magnetic Bodies, and the Great Earthly Magnet*. From experiments, Gilbert concluded that the Earth is really a great magnet. Like Copernicus and Bruno, he argued that the Earth moves. Citing Thales and Anaxagoras, Gilbert argued that magnets seem to be endowed with some kind of soul, because they have the power of moving iron. Since the Earth is magnetic and moves things and itself, Gilbert argued that it too has a soul, and that it is a living being, just as Bruno had argued, though Gilbert did not cite him in his book.

Gilbert credited ancient thinkers with the belief that the stars, Sun, Moon and the entire universe are alive and infused with a soul. He cited Thales, Heraclitus, Anaxagoras, Archelaus, Hermes, Orpheus, Pythagoras, Empedocles and Plato.[7] But Gilbert argued that life was not quite in the entire universe, only in the heavenly globes themselves, and to various degrees. He said: 'the whole world is animated, all the globes, all stars, and Earth', each having individual souls. He said God himself is a soul.[8] He said that living creatures arise and receive their life from Earth and the Sun. Like Bruno, Gilbert argued that Earth itself is a great animal:

> The Stoics attribute a soul to the Earth, hence they proclaim (while the learned laugh) that the Earth is an animal. This magnetic form is astral, whether it be a vigour or soul. The learned mourn & weep, since none of the foremost Peripatetics, nor even the popular philosophers, nor Joannes Costeus who laughs at this, were able to acknowledge and see the nature of this.[9]

Since the Earth was a soulful animal, it gave soul to some of its offspring, including magnets. It gave life to humans and animals by emitting a kind of breath (*spiritu*), and then souls lived in bodies almost as if in 'prisons', until they departed again.[10]

Since the world was a living animal, it also moved. In the Preface of Gilbert's book, Edward Wright argued that the words of Psalm 104:5 (103:5 in the Latin Vulgate), that God 'established the Earth on its foundations, it cannot be moved forever and ever', could not be used to prove that the Earth does not move. This was because the Earth could remain forever, 'resting on its centre', while moving in a circular motion, without wandering aimlessly.

In Spain, the Jesuit Juan de Pineda wrote a commentary on Job in which he complained about Zúñiga's support of 'the Pythagorean opinion of the Earth's motion'. Pineda declared: 'it is plainly false (some others call it delirious, worthless, temerarious, & dangerous to the faith; and taken from the hell of ancient philosophers by Copernicus & Caelio Calcagnino . . .).'[11] Living in England, however, Gilbert suffered no consequences for flirting with pagan heresies: on the contrary, in 1601 he became the personal physician of Queen Elizabeth.

Also in 1601 a book published in the Italian city of Reggio nell'Emilia listed and criticized the heretical beliefs of the contemporary 'Tower of Babel'. The author, the Dominican preacher Augustino Petreto, included the errors, confusions and blasphemies of ancient heretics alongside recent heretics, especially Luther and Calvin. It had been the duty of the Dominican Order, since 1216, to preach the Gospel and combat heresy. Father Petreto censured several heresies that Bruno defended, although he too did not mention Bruno's name. Instead Petreto credited such heresies and errors to whomever most infamously asserted them, or to broad groups like 'the philosophers'.

Alongside each heresy, Petreto briefly explained why it is false. For example, he denied the claim that 'the world is eternal', which he attributed to Aristotle.[12] And he blamed Porphyry for the belief that 'the soul is happy to flee from every body', but denied it by arguing that since the soul is the form of the body 'it is not repugnant to unity with the body.'[13] Petreto attributed the heresy that 'the Holy Spirit is the soul of the world' to Peter Abelard: Petreto countered that 'Since God is perfect, it is impossible that He is part of a composition.'[14] Against the claim 'They posit something of God as soul of the world,' he replied, 'But we assert that this is wholly false; because God does not come into compositions.'[15]

Kepler Announces Life in Other Worlds!

Other admirers of Copernicus and Pythagoras explored the idea that the world has a soul and many worlds exist. Consider especially the case of the young German astronomer Johannes Kepler. Like his parents, he followed the reformed Christian religion of Martin Luther, the 'new Porphyry', as Pope Leo x maligned him.

As a student at Tübingen, Kepler's teacher Michael Maestlin convinced him of the theory of Copernicus. Hence in 1593 Kepler began to draft a dissertation in which he mathematically analysed what the heavens would look like to observers standing on the

Moon. To present the subject, he sketched an introduction that described a voyage to the Moon. He read Lucian's 'True History'. Although Lucian had warned that it was fiction, Kepler was impressed and in some sense he did believe it to be true. Also in 1595 he read Plutarch's story of the Moon. Kepler's interest in lunar astronomy and geography led him to consider seriously whether there is life on the Moon. He conjectured that the Moon and the planets resemble the Earth. He wondered whether those other worlds were inhabited.

Kepler began to draft his own story about a dream, a lunar journey.[16] In his story an old mother reveals to her son that she could talk with the Moon and showed him the secret art of conjuring spirits. Then an alien daemon took the mother and boy on a dangerous voyage to the Moon. They travelled there to meet its inhabitants, who hid in caves to protect themselves from the Sun's heat, and to discuss their astronomy. They saw strange things and beings, including monstrous snakelike creatures that shun the Sun, and others who daily die by daylight and return to life at night. But Kepler did not publish this.

In 1594 the 22-year-old Kepler had become a teacher of mathematics in Graz, Styria (Austria). As an official mathematician of the district of Graz, he was expected to make astrological predictions. He did not really believe that astrology controls individual events. Still, in 1595 Kepler predicted that there would be a bitterly cold winter and the Turks would attack. Surprisingly, his predictions turned out to be correct, so he became a celebrity.[17] More importantly, Kepler worked on astronomy, fascinated by the Copernican theory. He struggled to answer questions such as why were there only six planets, including the Earth.

On 19 July 1595 Kepler was teaching a class when suddenly he had an idea. By interspersing geometric regular figures between the planetary orbits, he might be able to calculate the separations between the orbits. Kepler wrote: 'I believe that it was by divine ordinance that I obtained by chance that which previously I could not reach by any pains; I believe that so much the more readily because I had always prayed to God to let my plan succeed, if Copernicus had told the truth.'[18] Kepler calculated the distances between the orbits of the six planets by interspersing the five regular solids between them. It seemed clear evidence of God's wisdom.

Kepler believed that God 'revealed' to him this discovery, which had quietly waited since the dawn of Creation to be grasped by humans:

> It will never be possible for me to describe with words the enjoyment which I have drawn from my discovery. Now I no longer bemoaned the lost time; I no longer became weary at work; I shunned no calculation no matter how difficult. Days and nights I passed in calculating until I saw if the sentence formulated in words agreed with the orbits of Copernicus, or if the winds carried away my joy. In the event that I, as I believed, had correctly grasped the matter, I vowed to God the Omnipotent and All-merciful that at the first opportunity I would make public in print this wonderful example of His wisdom.[19]

If anything did not quite work in the numerical and geometric harmony of the universe, Kepler commented that he hoped Pythagoras might perhaps rise from the dead to help him – which did not happen, he wrote, 'unless perhaps his soul has transmigrated into me'.[20] But Kepler did not believe in numerical mysticism.[21]

In 1596 Kepler published his *Cosmographical Mystery*, endorsing the Copernican theory. He said Plato and the Pythagoreans ascribed great importance to the five regular solids in the order of the universe. Echoing 'Plutarch', Kepler stated that Pythagoras himself knew this.[22] Kepler wanted to include an introductory chapter explaining the consistency of Copernicus's theory with the Bible, but the senate of the University of Tübingen censored that part. Kepler acquiesced: 'we shall imitate the Pythagoreans also in their customs. If someone asks us for our opinion in private, then we wish to analyse our theory clearly for him. In public, though, we wish to be silent.'[23]

Kepler wrote that Pythagoras had been reborn as Copernicus. At the very start of his book, Kepler included a poem he wrote:

> What world, what cause, God, reason for creating,
> how God's numbers, which many regular magnitudes,
> that made six circuits, that each orb
> falls in intervals, that both Jupiter & Mars
> their orbs not primary, interspersed in gaps:

All this Pythagoras taught you with five figures,
thus this example teaches us, that we can be reborn,
two thousand mistakes, years, he becomes Copernicus,
hence you are called the best observer of the World.
But the seeds you discovered did not delay the crops.[24]

Thus Kepler linked the Creation of the cosmos with the transmigration of souls. He wrote nearly the same poem in a letter to Duke Frederick of Württemberg, early in 1596, but he replaced its last line with these: 'But you discovered Frogs, which is from seedful pleasure, with a herd of pigs posing among the trough of swine',[25] which show a resentment towards the unenlightened mob.

A year later, in 1597, Kepler sent a letter to a young and relatively unknown professor at the University of Padua, Galileo Galilei. In it Kepler tried to convince Galileo to publish on the Copernican theory and he referred to Pythagoras as one of their true masters: 'although you warn discreetly and secretly, by your own example, that one should retreat before common ignorance, one should not rashly incite or oppose the madness of ordinary learned men – in which respect you follow our true masters, Plato and Pythagoras.'

In 1600 Kepler was appointed imperial mathematician to Emperor Rudolf II. Although Rudolf had been raised in Spain, fortunately for Kepler he had withdrawn from Catholic observances and tolerated the Protestants.

One of Kepler's friends, Edmund Bruce, wrote him letters expressing eccentric opinions about astronomy. Bruce was an Englishman who studied and lived in Padua and participated in intellectual circles. Bruce knew Galileo and informed Kepler about him. For years Galileo had not replied to Kepler's letter of 1597.[26] In 1603, however, Bruce sent Kepler a frank letter voicing opinions that resemble those of Giordano Bruno:

> I have many doubts in astronomy about which you alone [can] make me more certain. For I think that there are infinite worlds. Each one of these worlds is finite, as if the Sun's centre is in the middle of the planets. And just as the Earth does not rest, so neither does the Sun. For it rotates most rapidly in its place around its axis, the motion of which the other planets follow. I think that the Earth is one among

> these [planets] but each [moves] more slowly the further it is from it [the Sun]. The stars are thus moved like the Sun, but they are not, as are the planets, driven around in a circle by the force of this body [the Sun], since each one of these [stars] is a Sun – and not part of this, our smaller world of planets.[27]

Bruce argued that the elements (earth, water, air, fire) exist not only on Earth but in outer space. Such ideas seemed 'probable' to Bruce, so he asked for Kepler's opinion. Mysteriously, Bruce didn't mention Bruno. Not much more is known about Bruce: 'Unfortunately after this letter, the brief trail left by Edmund Bruce dies out. Was he able to visit Kepler in Prague on his way back to England? Indeed, did he ever return?'[28]

According to Aristotle and ancient Greek astronomers, no change exists in the heavens. Yet a bright new star suddenly appeared in 1572, and vanished. Now, in 1604, another bright star appeared. For a year Kepler carefully observed it and published a book about it in 1606. He there said that heavenly matter is changeable, and he analysed his observations in terms of 'the Copernican hypothesis'. Kepler argued that, contrary to what theologians might claim, the new star was no miracle. Then, as he was about to opine about its origin, Kepler wrote:

> First I will say something: so that others may have material to discuss copiously. A certain Pythagorean brings to mind this Virgilian passage:
>
> *In the beginning, the sky and Lands, and regions of water,*
> *Shining globe of the Moon, and Titanic stars*
> *Spirit nourishes within; the totality, infused through*
> *the limbs,*
> *a Mind agitates the mass, & intermixes itself with*
> *the great body.*
>
> Thus from Aristotle: *Both plants and animals originate from earth and fluids* [*humore*]*: because there is a fluid in the earth, a fluid spirit, in its entirety indeed a* UNIVERSAL ANIMAL HEAT*:* by which Scaliger thought that these words lead us into the occult mysteries of nature.[29]

The next chapter compared the nova of 1604 to the nova of 1572. However, in another section of the same book Kepler explained that he believed in the soul of the world: 'If nothing were alive, it would vivify all matter, as the Earth; if nothing moved, it would attract everything to itself, some more than others.'[30] He argued that this soul was responsible for converting the sweat of women and dogs into fleas, dew into beetles, marshland into frogs, water into fish, soil into plants, corpses into worms, and so on, and ethereal regions into comets and stars. Kepler argued that a 'seminal reason' with some sort of moisture generates various species.

In describing such ideas, Kepler gave credit to Cornelius Gemma, a German physician, astronomer and astrologer. Gemma had described the new star of 1572 two days before Tycho Brahe. More importantly for our discussion, however, Gemma had argued that 'all that is called the world or universe' is a single animal endowed with an intelligent mind and one spirit.[31] He had also paraphrased a relevant quotation: 'In the beginning the sky & lands, regions of water, Spirit nourishes within, &c. *Aen.* 6.'[32] Moreover, in a work published in 1569, Gemma had argued that what some people called 'the soul of the world' was the same agency that Christians called 'the Holy Spirit', and that by it the heavens are moved and heat is diffused in the atmosphere, and so forth.[33] Again, these notions remind us of the heretic from Nola.

At some point, Kepler heard about Bruno's execution. Later, in 1607, he discussed the tragic event in a letter to a friend, the physician Johann Brengger. This letter is significant also because in it Kepler discussed his own belief in many worlds and alien life forms:

> You think that the orbs of the stars are perfectly unmixed and simple; but in my opinion they resemble our Earth. You, a philosopher, would remit the question to a philosopher; but Experience should speak, if she could be interrogated. [emendation] But Experience is silent, since nobody has been there; whence she neither affirms nor denies. I myself argue as you do, by induction from the Moon, which has many points of similarity with the Earth. And moreover I attribute moisture to the stars, and tracts which are rained on by evaporation, and living creatures to whom this is advantageous. For not only that the unfortunate Bruno, who was

> broiled on a wood fire at Rome, but my friend Tycho Brahe as well, held this opinion, that the stars have inhabitants. To this I agree more readily, that I hold, with Aristarchus, the motion of the Earth as well as of the planets.[34]

Brengger replied that he did not know that Bruno had been burned and said he felt pity for him. He asked whether Kepler was sure about this. Kepler then replied that he had learned about Bruno's execution from his friend Johann Wackher: 'I learned from Master Wackher that Bruno was burned in Rome, he endured his execution with firmness. He had maintained that all religions are vanity. Identifying God with the world, the circle, the point.'[35] Finally, Brengger wrote to Kepler: 'I can't stop being perplexed by the dementia of Bruno. What did he stand to win by enduring such tortures? If there were no God to punish the crime (as that was his opinion), could he not impudently play a bit in the comedy to thus save his own life?'

In 1609 Kepler published his *Astronomia nova*, vigorously defending the Earth's motion as fact. He argued that in theology authority carries most weight, but that in philosophy reason is more important. Following Copernicus, he criticized Lactantius for denying the Earth's roundness, and he also criticized St Augustine and the Inquisition: 'the Holy Office nowadays is pious, which, though allowing the Earth's smallness, denies its motion.'[36] Kepler argued that the truth is more pious, and that therefore he proved that the Earth 'is carried along among the stars'.

Meanwhile, a Jesuit theologian named Nicolaus Serarius wrote a commentary on the Book of Joshua, in which he took the opportunity to criticize Copernicus and the Pythagoreans for asserting the Earth's motion: 'such a hypothesis, if seriously held as true, I do not see how it can be devoid of heresy.' He said that the Bible always attributes rest to the Earth, and that where it says that the Sun and the Moon do not move it is by a miracle. This Jesuit then noted that 'All philosophers agree that this opinion [the Earth's mobility] should be rejected and condemned, except for a few such as Nicetas [Hicetas of Syracuse] and the Pythagoreans, relatives; all statements of the Holy Fathers [deny it], all Theologians in all [the monastic] schools.'[37]

At the same time, an innovative telescope enabled Galileo to seek visual proof of Copernicus's theory. In 1610 he published his

Message from the Stars.[38] He reported that the Milky Way is composed of stars, that Venus has phases, and that four bodies orbit Jupiter.[39] But above all, Galileo reported having seen mountains and valleys on the Moon. His findings raised sensational questions: is the Moon really another world? Is Jupiter another world?

Back in 1591, in his book *On the Immense and the Innumerable*, Bruno had insisted that 'There are innumerable distinct worlds, which we call stars.'[40] Now, on the title page of Galileo's book, Galileo boldly announced the discovery of 'INNUMERABLE STARS'.[41]

Bruno had also argued that multiple moons exist, not just one.[42] Now, similarly, Galileo reported that four bodies orbit Jupiter, just like our Moon around the Earth.[43] Furthermore, Bruno had proposed that the Moon and other heavenly bodies are surrounded by a kind of 'air', like the Earth.[44] Similarly Galileo now argued that 'vapours', or an atmosphere, surround the Moon and planets, like the Earth.[45]

In his 1584 book on *Worlds*, Giordano Bruno had written about 'the Moon (which is another Earth)', consisting of earth, water, air and fire, as well as seas, rocks, mountains and valleys.[46] Likewise, in 1585, Bruno referred to the Moon 'which we regard as another Earth'.[47] In his book Galileo now pondered whether his telescopic discoveries seemed 'to revive the old Pythagorean opinion that the Moon is like another Earth'.[48]

Like Bruno, Galileo did not attribute the notion of other worlds to Democritus, crediting instead the Pythagoreans, as 'Plutarch' did in his *Placita*. Galileo wrote that someone pursuing this view might well interpret the Moon's bright areas as land, and its dark areas as bodies of water. Moreover, Bruno had claimed that the Moon was allegedly the world closest to the Earth and similar to the Earth. Bruno had even referred to the Moon's *inhabitants*.[49]

Now, without mentioning Bruno at all, anywhere, Galileo announced that the Moon really does resemble the Earth. But this was a dangerous thesis for a Catholic. The Roman Inquisition had censured and condemned Bruno for obstinately insisting that other worlds exist.

Furthermore, Galileo declared that he would publish another book in which he would 'prove' that the Earth moves, by 'using countless arguments' from natural phenomena. He endorsed the Sun-centred theory that he attributed to the Pythagoreans.[50] But

Galileo Galilei discussing his telescopic discoveries. Etching by Stefano della Bella, 1656.

previously the consultors of the Inquisition had censured Bruno's Pythagorean belief that the Earth moves. Copernicus had claimed that the Earth moves, but Galileo's findings also seemed to confirm Bruno, who said not only that the Earth moves, but put forward other claims that Copernicus did not make: there are innumerable stars, multiple moons, and our Moon has mountains, valleys and an atmosphere.

IN MARCH 1610 Johann Wackher, counsellor to the Emperor, visited his friend Kepler to tell him, excitedly, that at Padua, using two lenses, Galileo had discovered four new planets. Kepler recalled:

> Wackher told me the story from his carriage in front of my house. Intense astonishment seized me while I listened to this strange account. I felt moved in my deepest being . . . [Wackher] was full of joy and feverish excitement, I with

> shame, at one moment we both laughed at our confusion, the next [moment] he continued his narrative and I listened intently – there was no end to it.[51]

Wackher was a Catholic convert, yet he had sympathized with Bruno and he believed in particular that the stars are suns. Hence Wackher imagined that Galileo had discovered planets that orbit the stars. But Kepler disagreed. He rejected Bruno's claim that stars are suns, and believed instead that the one Sun is the centre of the universe. Kepler also rejected the idea that the newly discovered planets might orbit the Sun, because he thought that there could be only six planets, which were separated by the five regular solids. Therefore Kepler imagined instead that the new heavenly bodies were actually moons orbiting some of the known planets.

Soon Kepler read Galileo's book. He then met with Martin Hasdale, who reported their conversation to Galileo in a letter. He said that Kepler told him that the book truly showed the divinity of Galileo's intelligence, but that it led Kepler to resent Galileo's country for 'not having made any mention of the authors who had discussed and given occasion to investigate what you have now found, namely Giordano Bruno, Copernicus, and himself [Kepler], professing to have discussed similar things . . .'.[52] Notice that Bruno was mentioned before Copernicus.

Then Kepler penned a public reply to Galileo to promptly remedy such omissions. Kepler finished his booklet, *Discussion with the Starry Messenger*, on 19 April 1610. It is very significant because it allows us to see how the leading expert in mathematical astronomy immediately responded to Galileo's findings. Did he mention Pythagoras? Did he relate Galileo's discoveries to Bruno's ideas? Interestingly, in this work Kepler mentioned Pythagoras six times, he mentioned Copernicus seven times, while he named Bruno more than either of them: eleven times. In Catholic countries it was considered inappropriate to refer to heretics, but Kepler spoke freely from Prague. He said that Bruno was 'the defender of infinity' and had often spoken with the voice of Apollo.

Echoing his own book of 1596, Kepler discussed 'the miracle of Pythagoras, Plato, Euclid', that the world could not have been constructed differently from how God made it, using the five regular bodies. Kepler said that 'Pythagoras interspersed the figures among the planets.'[53] But now Kepler discussed the discovery of 'new planets' that he had pondered, he said, 'for a long time from the speculations of the Cardinal of Cusa and Giordano Bruno'. He anticipated that since there were four previously unseen planets, 'what would prevent there being innumerably many others?'

Kepler acknowledged that Bruno had considered infinitely many worlds. Kepler even referred to his friend Edmund Bruce: 'recently, Bruno and Bruce – your friend, Galileo, and mine – envisioned infinite other worlds (or Earths, said Bruno) similar to ours.'[54] Kepler explained that according to Bruno the fiery stars had to be surrounded by watery Earths, as a law of nature. But Kepler rejected these ideas. It disturbed him to think that stars might be suns, surrounded by inhabited worlds. Still, he praised Galileo 'for improving the doctrines borrowed from Bruno'.[55] This statement is very important for emphasizing the apparent continuity between the works of Bruno and Galileo. That is, it was very apparent to readers at the time, even though many historians have neglected this point.[56]

Like Bruno, Kepler described the Sun as a living being, and 'as the origin of life and motion in the world'. Kepler said that Galileo's telescope might eventually reveal more hidden planets, 'that is, using Bruno's words, those being Suns, these Moons or Earths'.[57] But presently Kepler was relieved to find that Galileo's four new 'planets' did not orbit the stars or the Sun, but instead were satellites orbiting Jupiter. He noted that, according to Bruno, Jupiter was another Earth and he publicly commented to Galileo: 'If you could find planets around the fixed speakers [*discursitantes*: stars], then I am ready for chains and prison with Bruno's innumerabilities, or rather, being exiled in the infinite.'[58] He thanked Galileo for freeing him of that fear.

Still, Kepler was fascinated by the similarities between the Earth and the Moon. Galileo had asked whether his observations might seem 'to revive the ancient Pythagorean opinion that the Moon is like another Earth'. Kepler noted that Galileo's assertions might seem 'temerarious'. Yet Kepler brashly expanded them. He explained that 'following Pythagoras and Plutarch' he himself since 1593 had written about 'living creatures on the Moon'. From Galileo's descriptions of the Moon's topography, Kepler inferred that indeed certain beings live on the Moon! Kepler speculated that such inhabitants had large bodies and that they built barriers of clay to protect themselves from the insufferable heat from the Sun.[59] Kepler argued that it was not improbable that Jupiter too had inhabitants.

To give evidence that the Moon does have an atmosphere and moisture, Kepler cited a 1606 booklet (now lost) by his teacher Michael Maestlin, who had reported having observed the Moon during an eclipse in 1605 and seeing something remarkable. Maestlin said the Moon looked reddish but exhibited a dark spot: 'Call it a cloud, spreading over a great region, raining & laden with tempestuous showers; such as is not rarely seen from the largest mountain ridges to the lower valleys [on Earth].'[60] Kepler wrote that Galileo had confirmed some of what had been 'predicted' by his predecessors: Bruno, Bruce and Maestlin.

Soon after Kepler's *Discussion* was published, a young Bohemian astronomer, Martinus Horky, published a tract intended to refute the claims of Galileo. Horky had tried to see the phenomena with Galileo's telescope but had failed and seen nothing. Horky's critiques were furious and agitated. He especially assaulted Galileo's claim that four planets orbit Jupiter. He denied their existence as utterly ridiculous. He said that Democritus was laughable. Horky insisted that there exists only one Sun and one Moon, not many.[61] He accused Galileo of being a fraud. In June 1610 the General Inquisitor at Modena, Italy, approved Horky's essay for publication, noting that 'I find in it nothing contrary to the Catholic faith.'[62]

Kepler surmised that since Horky had failed to see anything with Galileo's telescope, he just had poor eyesight. Soon someone else published an essay critically discussing Horky's objections. John Wedderborn was a Scottish student in Padua who upheld Kepler's arguments for why Galileo's report was credible, even if one had not yet used a telescope. Wedderborn quoted Kepler's claim that 'If you could find planets around the fixed speakers [stars], then I am ready for chains and prison with Bruno's innumerabilities, or rather, being exiled in the infinite.' He distinguished Galileo from Bruno by saying that one could not derive the existence of the four new planets from Bruno's works: 'others have already said that new planets can be found: certainly Bruce and Bruno, old wives' tales, but only Galileo could help us discover new planets.'[63]

Galileo's discoveries did not merely support the notion that Earth is a planet orbiting the Sun. They immediately triggered associations to the controversial Pythagorean ideas of Bruno. Were the stars suns? Were they surrounded by countless other worlds? Were those worlds inhabited? Was the universe infinite? Does it have a soul? One way to demonstrate that the Earth *really* is a soulful being, a rational animal, was to prove a startling and counter-intuitive fact – *that it moves*.

Campanella Imprisoned and Tortured

In 1611 a prisoner of the Inquisition wrote to Galileo enthusiastically praising the discoveries in *Message from the Stars*. Tommaso Campanella was a questionable advocate: an apostate monk and astrologer who had been repeatedly accused by the Inquisition. Before discussing Campanella's views on Galileo, we should discuss his links to Pythagorean ideas and how he became a prisoner of the Inquisition.

At the age of fourteen, Campanella was so impressed by learning about Thomas Aquinas and Albertus Magnus that he joined the Dominican Order in Calabria, southern Italy. Campanella had been baptized with the names Giovanni Domenico, but he now took the name Tommaso in honour of Thomas Aquinas. He studied logic, philosophy and classics, but became sceptical about Aristotle's writings. On one occasion his professor of philosophy sent Campanella to debate with some Franciscans. Campanella performed so well 'that everybody was charmed with him, and cried out that the genius Telesio was transmigrated into him'.[64] Campanella had not heard of Bernardino Telesio before, but became curious to know more and became impressed by his philosophy, which said that truth should not be sought in authoritative books but by directly examining empirical facts.[65] Campanella developed irreverent philosophical views on the basis of recent and ancient thinkers. Soon the young Campanella debated an old Dominican professor, spontaneously and so successfully that the professor reported him to the Inquisition, as if Campanella had acquired his vast knowledge by magic.

Regardless, Campanella wrote his first book, *Philosophy as Demonstrated by the Senses*, which he published in 1591.[66] He systematically critiqued Aristotle's natural philosophy and cosmology, while defending Telesio's ideas. Above all, Campanella developed the idea that the Sun's heat is the primal and noble factor that gives life to all beings and connects them. Campanella transcribed a long passage from 1492 in which Ficino had commented on claims by Plotinus about the 'Soul of the World'.[67] Campanella approvingly quoted how Ficino described the World Soul:

> But it is a most subtle body, almost not a body, & almost more soul. Or almost not soul and almost more body. Its power is minimally earthly in nature, but more watery, plus more airy, yet more fiery stellar than most. In degrees from these measures, the quantities of stars have been produced from the elements. But it vivifies everywhere, it is the proximate cause of all generation and motion, of which he says: 'Spirit nourishes within'.
>
> By its nature all is luminous and warm, and humid and vivifying, from such endowments arise the higher qualities of the soul. Most of this was imbibed by Apollonius of Tyana, as testified by the Indian Iarchas in saying: 'No one should be surprised, O Apollonius, that you have gained the knowledge of divination, for holding so much ether in your soul.'[68]

Campanella was greatly impressed by the arguments of Plotinus and Plato, which Ficino construed as essentially Pythagorean. These same arguments, about the soul of the world and the Earth being an animal, had deeply influenced another prisoner of the Inquisition, Giordano Bruno. Bruno too construed them as Pythagorean, as he dangerously interpreted the Holy Spirit as the soul of the universe, as he explained to the Inquisitors, saying that this idea 'entirely conforms to the Pythagorean doctrine explained by Virgil in the sixth [book] of the *Aeneid*'.[69]

Unlike Bruno, Campanella did not identify the Holy Spirit with the soul of the world. Campanella argued that Varro misidentified God with the soul of the world. Instead, Campanella argued that God created things by means of heat, as a vivifying breath of warm air.[70] Telesio too had written a book about similar notions, titled *That the Universal Animal is Governed by a Unique Soul Substance*. Telesio had written it by 1565 but he did not publish it during his lifetime. His manuscript circulated in handwritten copies until it was published in 1590, two years after his death.[71]

Like Bruno, Campanella became disillusioned with the monastic approach to knowledge. He had studied philosophy, science and prophecy. He became fascinated with deciphering God's signs about the future. In 1592 the Inquisition in Naples examined Campanella because of his controversial ideas. Campanella then went to Rome, Florence and Padua, where, incidentally, he briefly met young Galileo in 1593. Someone, however, denounced Campanella to the Inquisition again on suspicion of heresies, so he was once more arrested in early 1594. The Inquisition transferred him to Rome, where he was imprisoned at the same time as Bruno, although we don't know if they met. The prisons of the Inquisition were on the south side of St Peter's Square.

The Inquisitors accused Campanella of supporting heretical doctrines. They said he had authored *The Three Impostors*, which argued that the fathers of the three monotheistic religions were impostors. Campanella defended himself, for one, by saying that that book was published thirty years before his birth.[72] The Inquisitors interrogated Campanella about four serious accusations. One was that he published 'a doctrine about the soul of the world, about living beings and objects, contrary to the fundamental dogmas taught by the Church'.[73] Unlike Bruno, Campanella recanted or denied it. The Inquisition and the Roman city government did not kill him, but still the Inquisitors subjected him to torture, declared him guilty of 'vehement suspicion of heresy', and forced him to make a public abjuration. In late 1597 they relocated him back to a Dominican convent in Calabria. In 1596 Telesio's book about the soulful, 'universal animal' was placed on the Index of Forbidden Books, 'until expunged'.

Meanwhile, by reading astrological and prophetic texts, Campanella became convinced by 1598 that certain dismal events and celestial signs would bring great upheaval at the end of the millennium. He became involved in a conspiracy to remove the Habsburg rulers in Calabria and re-establish the region (including the Pythagorean city of Croton) as an independent republic in opposition to the Spanish government in the Kingdom of Naples. But the plot was exposed. Campanella and others were arrested in 1599. He was put on trial in Calabria and Naples. Allegedly, Campanella sought not only to support a political revolt, but to contrive a new religion based on heresies and the recovery of ancient liberty. There were two separate and simultaneous investigations: one, headed by royal authorities, concerned the conspiracy for a rebellion; the other, an investigation for heresy, was headed by the Inquisition.[74]

It was alleged that Campanella had been influenced by the Devil.[75] Campanella later divulged some of the accusations against him: 'The first time, when they asked me: "How do you know that which you have never learned? Do you have a demon at your command?" I replied, "To learn that which I know, I have not used more oil than what you have drunken in wine."'[76] Witnesses testified that Campanella had said that Jesus Christ was not God's son and had not resurrected, that his body had just been stolen and hidden, as 'the same had been done with Moses, Pythagoras and others.'[77]

Allegedly Campanella had also said that the Christian religion was a political scheme; that Jesus had not established the sacraments; the Eucharist was 'an absurdity and it is craziness to believe in it'; Mary was not a virgin; heaven, Hell and demons do not exist; biblical miracles were imaginary, but Campanella himself 'could revive the dead and other extraordinary things'.[78] As if those were not enough heresies, allegedly he had also tried to prove that God does not exist and said that the Holy Trinity is an illusion, that Jesus had lovers, that sexual acts are licit, and that the tyrannical Pope was the Antichrist. Reportedly Campanella had further said that Christians are ignorant asses who were subdued by the apostles into believing by faith and against reason.[79]

Furthermore, Campanella later recalled that the Inquisitors accused him 'of having the opinions of Democritus', but he replied he had actually written books *against* Democritus. He also recalled:

> They accused me of being a heretic, I who have composed a dialogue against the heretics of our times. In sum, they accused me of rebellion and heresy, for having said that there are spots on the Sun, the Moon and the stars, against Aristotle who made the world eternal and incorruptible. For that it was that they threw me, like Jeremiah, into the lower lake, where there is neither air nor even light.[80]

Like Bruno, Campanella defended himself against the accusations by arguing that they were ridiculous and false. He also argued that his alleged participation in planning a revolt had consisted merely in making an astrological prophecy. Witnesses testified against him. He denied their allegations. So the Inquisitors tortured him by hanging him from a *corda*. He protected himself by acting

Tommaso Campanella, sculpted by Ettore Ferrari, on the base of Bruno's statue at the Campo de' Fiori.

insane. He babbled incoherencies. He yelled that he had been murdered but that the Pope would save him! When asked how long he had been imprisoned he said 'always, always'. To the threat of further torture he reportedly said, 'yes, yes, do it, do it.' And when asked what he expected from 'these lords', he replied, 'let me shit.'[81] Pope Clement VIII did not believe that Campanella was mad. At one point the Pope ordered that Campanella be tortured by both the *corda* and the *cavaletto*, a procedure whereby the prisoner was tied to a chair suspended from the walls for forty consecutive hours while kept awake.[82]

An account of another of the tortures endured by Campanella says that he was interrogated 'for thirty-five hours, held harshly and cruelly, all the veins and arteries around his rear end were broken by the inhumane device of torture, with such force that they emitted blood that could not be stopped'.[83] At some point Campanella wrote a poem to God about his sufferings, which I translate literally as prose:

> that the hard example of my lasting hell, may be seen without my consent. Six and six years, spent in painful punishment, the affliction of all the senses, my limbs tortured seven times, the blasphemies and the fables of fools, the Sun denied to the eyes, the nerves plied, the bones injured, the pulp lacerated, the roughness on which I sleep, the irons, the spilled

> blood, and the raw fear, and the food scant and foul; in hope worthy of Your lance and shield.[84]

If a prisoner endured torture without confessing, especially multiple tortures, then the Inquisitors usually surmised that he was not guilty of heretical intentions.[85] This could lower the accusations to the minor rank of unintended heresies. The prisoner could eventually be released. Also, having an 'injured mind' could prevent an ecclesiastical tribunal from issuing a death sentence 'because the accused's inability to perform penance would lead to a loss of his soul, with grave consequences for his judges'.[86]

Campanella's alleged insanity kept him alive, but his defence failed inasmuch as he remained imprisoned in Naples for years. Still, in 1602 Campanella began drafting a manuscript titled *City of the Sun*, in which he described a utopia based on astrology. Its inhabitants worshipped the Sun and shared wisdom like the Pythagoreans. Campanella managed to send some manuscripts out of prison, but his efforts were obstructed by a Decree of the Index in 1603, banning all of his works.[87] The same decree also banned all of Bruno's books. It prohibited such works for including 'false doctrines, heretical, erroneous, scandalous, corrupting', and 'against God'.[88]

In 1607 Gaspar Schoppe visited Campanella in prison in Naples. Whereas Schoppe despised the 'monstrous' Bruno, whom he never met, he had an opportunity to talk with Campanella. Schoppe and Campanella both shared an interest in producing a Christianized version of ancient Stoic moral wisdom.[89] Hence Schoppe became sympathetic towards Campanella's plight. Consequently, when Campanella worked on a manuscript against atheism, he dedicated its Preface to Schoppe. He there described the interrogations and tortures he had endured:

> I have been locked up in fifty prisons and subjected to the most atrocious torture seven times. The last time it lasted forty hours. I was strangled with a garrotte with very taut ropes that ripped my bones, suspended, my hands tied behind my back, over a sharp wooden stake that devoured me, the sixth part of my flesh, and drew out ten pounds of blood, after forty hours they thought me dead, so my ordeal ended; some of them [Inquisitors] injured me, and,

> to increase my pains, they shook the rope from which I was hanging; the others very softly praised my courage. Nothing has broken me and it could not tear from me a single word. After six months of illness, miraculously, I was finally cured; [but then] I was plunged into a pit. Fifteen times I have been put on trial.[90]

Despite Schoppe's sympathy, Campanella remained in prison.

Meanwhile, a philosophy professor in Florence criticized Copernicus and Galileo. His name was Ludovico delle Colombe. He circulated a dissertation that gave physical reasons why the Earth cannot move, by insisting that all heavy things must go to the centre of the world. He also denied the notion of a living Earth: 'it is an impossible thing that the Earth is of an angelic intelligence or substance that moves and spins.'[91] Delle Colombe also said that at first sight the telescope seemed to confirm 'the opinion of Pythagoras' that the Moon is 'another Earth', by apparently showing that it is irregular, with mountains and cultivated valleys.[92] But Delle Colombe argued that philosophy, reason and the scriptures showed that this was a 'deception of the senses'.[93] He insisted the Moon's surface is truly smooth. He said Galileo's claims were 'vanities', that everyone disparaged as 'temerarious'. He quoted the Jesuit theologian Pineda, 'it is worthless, temerarious and dangerous to the faith, and taken from the ancient philosophers.'[94]

Campanella had the opposite reaction. By 1611, still imprisoned after many years, he heard about Galileo's discoveries. He wrote effusively to Galileo:

> In astronomy Ptolemy and Copernicus had put us to shame, but you, enlightened man, not only have retrieved for us the glory of the Pythagoreans, stolen from the subtle Greeks, resurrecting their dogmas, but you extinguish all the glory of the world with your splendour. '*And I saw a new heaven and a new Earth*', as [John] the Apostle and Isaiah said, blinding us; you purged the eyes of men, and show us a new heaven and a new Earth on the Moon.[95]

Campanella thus quoted a phrase from the New Testament, about the Apocalypse, which I have italicized:

> Then Death and Hades were thrown into the lake of fire. The lake of fire is the second death. If anyone's name was not found written in the book of life, he was thrown into the lake of fire. *Then I saw a new heaven and a new Earth*, for the first heaven and the first Earth had passed away. (Revelation 20:14–21:1)

Campanella also referred to other notions by many writers. He cautioned Galileo that if 'theologians should murmur', some ancient theologians could serve to defend himself, including Augustine and Origen, so that the scriptures would not be used against mathematicians: 'You have Origen, who taught that the Earth is an animal and all the stars, and praises the Pythagorean dogmas and tested by the Scriptures.'[96] Campanella did not explain this remark further, but indeed in one work Origen had argued:

> but as our one body is provided with many members, and is held together by one soul, so I am of opinion that the whole world also ought to be regarded as some huge and immense animal, which is kept together by the power and reason of God as by one soul.[97]

Furthermore, Origen had similarly argued that stars and planets 'may be designated as living beings'. Because only bodies with souls can move, living things are always in motion. The stars move regularly so they must be rational beings. Furthermore, apparently stars were subject to sin (Job 25:5). Finally, because the stars receive orders from God, as the Bible says: 'I have given a commandment to all the stars' (Isaiah 45:12).[98] These had been the beliefs of Giordano Bruno.

Origen claimed to have discovered, from the scriptures, that heavenly bodies have life, reason and, probably, souls. But Origen did not attribute this to the Pythagoreans. He had criticized stories about Pythagoras, Euphorbus, Hermotimus and Apollonius. Origen said the Pythagoreans were wrong about transmigration and about souls' alleged descent from the vaults of heaven and from the planets.[99] So, in writing to Galileo, Campanella was linking Origen with ideas that seemed Pythagorean to him.

Campanella did not warn Galileo that, posthumously, the almost saintly Origen had been denounced for heresies, just like

Bruno. Against Origen, the Fifth Ecumenical Council of 553 decreed that anyone who says that the Sun, Moon or stars are animated is a heretic.[100] That Council condemned Origen for asserting the 'pre-existence of souls', or the transmigration and immortality of souls, which were Pythagorean notions. Therefore prisoner Campanella was not giving Galileo entirely useful advice. In addition Campanella cited titles of works that he had written, suggesting that Galileo read them, starting with his 'three books *On the Philosophy of the Pythagoreans*', now lost. And Campanella gave several other arguments to embolden Galileo. Still, Galileo did not pursue any claims about Earth, stars and planets being living, rational entities, unlike Bruno. Instead, he reported his telescopic findings in descriptive ways, though noting their broader significance. Then one of Bruno's Inquisitors became concerned about Galileo's claims.

Bellarmine and the Enemies of Bruno

Soon after Bruno's execution, Cardinal Bellarmine became known as 'truly the hammer of the heretics'.[101] This same Catholic compliment had been used for St Augustine, owing to his work *On Heresies*.[102] Bellarmine was admired for his encyclopaedic critiques of Protestant beliefs and for his extensive knowledge in theology. Pope Clement VIII praised him for being unmatched in learning in the Church of God.[103] Bellarmine laboured for decades to build and defend a Christian republic, a spiritual commonwealth that would transcend political boundaries, ruled by the Pope – what one historian rightly calls 'an empire of souls'.[104]

Bellarmine's views on Bruno are not known, because, as historians have noted, apparently his papers include no discussion of Bruno's trial.[105] However, I will illuminate this by identifying remarks that Bellarmine voiced in his theological writings. I'll also show his influence on his allies. There is a trail from Bruno's trial to the Inquisition's first confrontation against Galileo.

In 1603 all of Giordano Bruno's works were banned, placed on the Index of Forbidden Books, 'for religious reasons'. Still, some of his doctrines continued to be discussed, as we have seen. The following year a Catholic convert in the Netherlands, Justus Lipsius, published a book in which he tried to harmonize ancient Stoic ideas with Christianity. Lipsius was a correspondent of Cardinal

Bellarmine, who had defended Lipsius three times against critiques of one of his books in 1592, 1601 and 1611, even against other censors of the Index.[106]

In his new book of 1604 Lipsius briefly discussed Pythagorean notions that tied the Earth's motion to the belief that the Earth has a soul. He cited Diogenes for writing that Philolaus the Pythagorean said that the Earth moves. Lipsius noted that in the 'Life of Numa' Plutarch too said that about the Pythagoreans. Lipsius likewise mentioned Aristarchus, Hicetas and others, and linked their ideas to Plato's account in *Timaeus* that Earth and the Moon have souls. Then Lipsius complained: 'See these deliriums. What else to call them? Or what is this inquiry? Although (it is such a love of Paradoxes) the era of the famous mathematician has agitated this heresy, but is buried with him.'[107] I think he referred to Copernicus.

Importantly, Lipsius derided such beliefs as *heresies*. He also quoted Ovid's Pythagoras for saying that the Earth is a living animal, and enquiring whether 'primal matter' is permanent and 'nothing perishes'.[108] Lipsius also discussed 'Whether the World is one of many?' and whether it is eternal.[109] He did not refer to Bruno, but it was improper for Catholics to write about heretics. Yet Lipsius berated several of the offensive notions that Bruno advocated: Earth's motion, that it has a soul, because it's a living animal, plus the plurality of worlds, and the eternity of the material universe. Lipsius wrote, 'some in our times have the temerity to assert that there are *celestial Animals*, and even that the *World is animated.*'[110]

Incidentally, one of Lipsius's friends saw Bruno's execution: Gaspar Schoppe.[111] Schoppe's standing in the Catholic world had grown, thanks to his zealous attacks against the Protestants. He enjoyed titles and pensions, and by 1605 Cardinal Bellarmine personally granted him a licence to read heretical books.[112] Some zealots were allowed to study heretical works, to decry them and to educate Catholics in general.

In 1608 Ubbo Emmius, a friend of the Dutch astronomer Nicholas Mulerius, worried about opinions that are 'fantastical and absurd', namely that Earth moves and that 'the Moon is inhabited.' Emmius then complained:

> Good God, how far impiety goes! What could do more to overthrow all of our doctrine and religion! It contradicts the

> foundations. In fact, if this is true, as I sense that the author vehemently contends, [then] Moses is false, all of our Holy Scriptures are false.[113]

Also in 1608 one of Bellarmine's fellow Jesuits, Antonio Possevino, dismissed some of the notions that Bruno had defended. In a catalogue of definitions and truths of the Catholic faith, Possevino echoed and praised the critiques that Photius the Ecumenical Patriarch had raised against hypotheses allegedly voiced by Clement of Alexandria. Possevino said that it was 'blasphemy', 'monstrous', 'impious and ridiculous' to say that matter is eternal, that souls transmigrate from body to body, and that there existed many worlds before the first man, Adam.[114] Possevino commented that Clement's 'errors' probably came from the influence of 'Eastern heretics'. Possevino said that Photius had voiced the true spirit that could be recognized in Cardinal Bellarmine and other orthodox men truly invested in the works of the Church Fathers. Possevino was one of Bellarmine's longtime friends and supporters.[115]

While discussing works of St John Chrysostom, Possevino complained about 'certain heretics and philosophers who beyond the entire globe of stars have posited other starry heavens and other worlds'.[116] He did not name such heretics, but he cited arguments by John Chrysostom, Augustine and Thomas Aquinas to deny their claims.

Such worlds were part of Bruno's cosmology. Since God is infinite, His creation must be infinite too. Since God is everywhere, His Spirit infuses everything. Orthodox Catholics rejected these views. In a work on 'miraculous arts', a French jurist explained: 'False Gods have originated from the Devil's wickedness or pretentiousness and his pride, and from other fictions and empty reasonings.' Hence this jurist listed various false gods, including the 'obstinate reasonings' of philosophers such as 'Pythagoras of Samos and M. Varro that soul permeates the entire universe'.[117]

When Galileo published his discoveries, in 1610, Bruno's Pythagorean heresies were not yet forgotten. That same year the theologian Placido Padiglia published a long commentary on Psalm 50: 'Unto the end, a psalm of David.' In it, Padiglia mentioned a censured notion: 'not that we want to say, with Pythagoras, that God is a soul dispersed throughout all parts of the World, that by this perhaps we mean that which the [Solomon's book of] Wisdom says, "For the spirit of God fills the Earthly orb"'.[118] This is the same biblical quotation Bruno had voiced when interrogated by Roman Inquisitors to support his 'Pythagorean way' of making sense of the Trinity: to construe the Holy Spirit as the soul of the universe.[119] This was one of Bruno's 'horrendous and utter absurdities', according to Schoppe, along with the doctrine of many worlds. Padiglia did not mention Bruno, but he dedicated the book to someone who knew much about such matters: the 'most illustrious and most reverend' Cardinal Robert Bellarmine, the Protector of Padiglia's congregation. Accordingly, Bellarmine owned a copy of this book.[120]

Under interrogation, the Roman Inquisitors had accused Bruno of having said that stars are animated rational beings that speak. Bruno admitted it. He cited Psalms as evidence, 'the Heavens declare the glory of God', and that therefore heavenly bodies are rational beings or angels that interpret God's voice.[121] This issue arose in Padiglia's text, without mentioning Bruno. Padiglia translated *Caeli enarrant gloriam Dei* as 'the Heavens *write* the glory of God' (emphasis added).

Padiglia paraphrased John Chrysostom, saying that since 'the Heavens write the glory of God', then 'these very skies are books, on which is found recorded and collected all the secrets of God, and books fashioned to speak to all the World, and to teach.'[122] Padiglia paraphrased St Basil, saying 'all this World is a book open to the eyes of everyone, in which the secrets of the divine wisdom are described, and all other works of God, which we observe in this grand machine, are written in the many books of divine science abundantly, which we should read'.[123] Years later, Galileo used similar metaphors.

Yet Padiglia quoted more passages from scriptures and the Church Fathers to argue that none of this means that the Heavens actually speak. He explained, 'Chrysostom confirms it: *They have no voices, they possess no mouths, language is not theirs*; and we now know by science that Heaven is inanimate.' Padiglia did not mention that the beliefs he rejected, that the Heavens are animated and have voices, were known as Pythagorean beliefs, together with the idea of the soul of the world. In his illustration of 'the most ancient Doctrine of the Pythagoreans', Thomas Digges had described the 'innumerable' stars as divine, 'the very court of celestial angels devoid of grief'.[124]

Meanwhile, the Lutheran jurist who had received Gaspar Schoppe's letter about Bruno produced an annotated edition of Porphyry's *Life of Pythagoras*. Rittershausen referred to Porphyry by the name his parents gave him: Malchus, meaning 'King'. In his book Rittershausen discussed pagan notions: transmigration, that 'nothing in the world is really new', the 'error' that God has a body and soul, and 'the opinion of Epicurus' that things repeat 'innumerable times'.[125]

A decade had passed since Bruno was killed. Schoppe mentioned the event in a booklet published in 1611:

> Ten years ago I happened to see a memorable example of the obstinacy of hatred, in Giordano Bruno of Nola, in Rome. For he preferred omens and monstrosities, which he had learned from the ancient Epicureans and such kinds of philosophers and heretics, and which he advocated openly in books; first execrable insults and blasphemies about Christ and the Apostles, to recant, and he said that they were prestidigitators and artificers of Magic, which penance led to confess, miserably surrounded by faggots he preferred to be burned by luminous flames. To such a degree *is vindicated the good life itself*, as meant by [Publilius] Syrus the Mime: *Revenge on an enemy is to receive a second life*. For it was a great part of his [Bruno's] hatred for the Cardinal Inquisitors, whether it was from some intolerable, rough austerity and perversity.[126]

This account reiterates the 'monstrous' character of Bruno's beliefs, which echoed Epicurus and other philosophers. Writers such as 'Pseudo-Plutarch', Hermias, Augustine and Rupert of Deutz credited Epicurus with the theory that there are many worlds. 'Plutarch' and others also attributed such beliefs to Orpheus and the Pythagoreans. In Bruno's case, such heretical beliefs allegedly mingled with blasphemies about Christ and magic.

Meanwhile, Galileo voiced beliefs similar to Bruno's: countless stars exist that had not been seen, Earth moves, and the Moon is a mountainous world. In early 1611 an anonymous writer used Galileo's findings combined with Bruno's beliefs to ridicule the Jesuits. This was actually the famous English poet John Donne, who was also an anti-Catholic writer. In his book Donne mocked the founder of the Jesuits, Ignatius of Loyola, as possessed by the Devil and damned in Hell, where the sinful Jesuits arrived daily and willingly.[127] Ignatius advised the Devil, Lucifer, about which 'innovators' should be admitted into the most remote and exclusive chamber of Hell.

The first innovator trying to enter was Copernicus, because 'the Papists have extended the name and punishment of *Heresy* to almost everything.' According to Donne, since Ignatius did not even know Copernicus's name, this Jesuit was 'utterly ignorant in all great learning'. Hence Copernicus, to prove his evil, bragged that he had lifted Lucifer and Hell (in the Earth's centre) up into the heavens, against God's will. Donne's Copernicus dared to call himself 'almost a new Creator' and bragged that 'I was the soul of the Earth.' Since Origen the heretic inhabited the exclusive chamber of Hell, Ignatius wondered whether Copernicus too deserved entry. Ignatius asked whether by lifting the Earth Copernicus had threatened God as with the Tower of Babel, or whether he taught men 'that there is no Hell, or denied the punishment for sin?'. Donne's Ignatius pondered whether Copernicus belonged to 'a sect' of philosophers or deserved to enter the place reserved for the 'Antichristian Heroes'.[128]

Donne's Ignatius asked whether the Jesuits would issue a Decree by which the Earth's motion would become a heresy, so that Copernicus and his sectarians would earn their place in Hell. According to Donne, Galileo had discovered 'a *new World*, namely the *Moon star*, with its mountains, woods, and cities'.[129] He asked whether 'Galileo has done harm, in recently summoning other worlds, namely Stars', which are inhabited by 'peoples'.[130] And Lucifer himself came from a star.[131] Donne ridiculed the new astronomers for their arrogant vainglory, and their temerarious, disrespectful ideas.

Supposedly the founder of the Jesuits was so evil that he obstructed Copernicus, Paracelsus and Machiavelli from promptly entering into that exclusive chamber of Hell. Silly old Lucifer approved everything that Ignatius said. Donne's Machiavelli even described Ignatius as Lucifer's son, and he disdained Copernicus as 'futile and chymerical'. Donne further said that Ignatius was craftier than the Devil and worshipped the Devil as the true God. Donne added that the Pope too carried the image of the Devil.[132]

Ignatius was so evil that he deserved to be Lucifer's successor on the throne of the Empire of Hell. However, Lucifer decided that instead of sharing Hell, Ignatius should become the ruler of a second Hell. Lucifer would ask the Pope to get Galileo to somehow pull the Moon closer to the Earth so that all the Jesuits would move to the Moon! The 'Jesuits will beget a *Hell*', on the Moon, to be ruled by Ignatius. The Jesuits would then 'reconcile the *Lunatic* Church with the Church of Rome'. Lucifer told Ignatius: 'And similarly as you go from the Earth to the *Moon*, you will travel from it to other stars which are also thought to be Worlds, so your *Jesuits* will always easily travel, thus propagating many *Hells*, so that you may expand your empire.'[133]

Donne's Ignatius commented that Elizabeth of England was a 'Lunatic Queen' who was allied to 'the kingdoms of the rest of the stars, and all the planetary and firmament worlds'. He also mentioned that the Sun itself is a world, and that Ignatius had never thought about God. Allegedly, Pope Paul v then decided to canonize Ignatius, so that the 'spiritual butchers, and murderers of Kings' would have a saint to invoke.[134] Finally, too impatient to wait for his throne on the Moon, Ignatius seized the throne that was next to Lucifer, by pushing aside its former occupant, Pope Boniface.

Donne warned that having seen this Jesuit eject that Pope from a throne in Hell, perhaps the Jesuits would try to do the same thing in Rome.[135] Donne complained that they had already hired assassins to kill kings (in 1610 Henri iv of France had been assassinated). Donne ridiculed the 'sinful' cardinals in Rome. In particular, he denounced Cardinal Bellarmine for having evil and vow-breaking ambitions. He insulted Bellarmine as 'the Sword of the Roman Church'.[136] (Donne had previously attacked Bellarmine in a book of 1610, *Pseudo-Martyr*.)

Donne did not name Giordano Bruno, yet many of Bruno's actual and alleged beliefs appear in his book: the soul of the Earth, its motion, that the Moon is a star, that the stars are worlds, that the Sun, the Moon and the planets are worlds, that such worlds are inhabited, that there is no punishment for sins, and that souls can travel out of their bodies. Moreover, in a poem titled 'An Anatomy of the World' (1611) Donne complained about how the 'new philosophy calls all in doubt', by leading men to look for other worlds: 'And freely men confess that this world's spent, / When in the planets, and the firmament / They seek so many new.'[137] Thus, various kinds of evidence seem to suggest that Donne was influenced by some of Bruno's works.[138]

The Jesuits did not reply to Donne in print. He was safe in England. But not everyone was beyond their reach. In April 1611 Cardinal Bellarmine wrote a letter to the Jesuit mathematicians at the Collegio Romano asking them several questions about Galileo's alleged discoveries. He asked about the apparent 'multitude of fixed stars invisible to the naked eye'. He asked about Saturn, the phases of Venus, the Moon's surface, and the 'four moving stars' around Jupiter. Bellarmine asked whether these 'new inventions are well founded, or are just apparent and not true'.[139]

Five days later, four mathematicians of the Collegio Romano replied. They confirmed that very many stars exist that previously were not visible, that Venus exhibits phases, the Moon's surface *seems* to be uneven and that four stars orbit Jupiter.[140] Their letter shows that these Jesuits, astronomers skilled in mathematics, evaluated Galileo's claims fairly. They acknowledged most of his discoveries, speaking cautiously about their significance with respect to theology but without slavish allegiance to Aristotle.[141]

One of these Jesuits was the prominent astronomer Christopher Clavius. We should note two of his opinions, in a booklet of 1611. Clavius noted that some 'philosophers' claimed that the world is eternal, but that 'according to the Catholic faith, the world was created' from nothing. Clavius also complained that 'some Philosophers, among whom was Democritus, claimed that worlds are innumerable, with additional others, almost like spheres, or globes.' Instead Clavius insisted that Aristotle and 'our Theologians' taught that only one world exists, although God has the power to create infinitely many worlds.[142] Clavius did not name the infamous philosopher-heretic Bruno. Also he denied that the spots that Galileo had observed on the Moon were really mountains and valleys.

Still, in late 1611 Kepler published an addendum to his *Discussion with the Starry Messenger*. On the first page Kepler referred to Galileo's seemingly 'dangerous observations' and briefly raised the question of whether the new planets and stars pertained to what 'Cardinal of Cusa, and Bruno' had predicted: the infinite stars.[143]

Meanwhile, a professor at the Collegio Romano composed a booklet criticizing Galileo's claims. The philosopher Giulio Cesare

Lagalla finished it in late 1611 and published it in 1612. Lagalla argued that Galileo's flat, telescopic images were insufficient to show that the Moon had mountains and was therefore a world. He quoted Galileo's passage about 'if anyone wants to revive the ancient Pythagorean opinion that the Moon is like another Earth'.[144] Lagalla complained that the belief that the Moon is another world *led directly* to the false view that many worlds exist.

Lagalla knew Bellarmine, and he too lived in Rome at the time of Bruno's execution, when Lagalla's main patron was Cardinal Giulio Antonio Santori, Archbishop of Santa Severina and head Inquisitor in Bruno's trial. In his book Lagalla named several individuals who had wrongly claimed that worlds exist or that the Earth is a star: Orpheus, Thales, the Pythagoreans (according to Aristotle), 'Philolaus the Pythagorean' (according to Diogenes and 'Plutarch's' *Placita*), Democritus, Heraclides, Anaxagoras, Plutarch and 'Copernicus, the famous mathematician'.[145] Lagalla said that some of them wrongly thought the Sun is at the centre of the universe. And he paraphrased the ancient lines attributed to Orpheus: 'he posits another Earth above us called the Moon; how poetic, feigning that it has many mountains, many cities, many Dwellings.'[146] Lagalla insisted that such views are absurd, and grumbled that Kepler had been deceived into thinking that there are inhabitants on the Moon who build giant structures.

Lagalla noted that 'the Pythagoreans said that the centre is the prison of Jupiter', though Aristotle had not explained why. Therefore Lagalla tried to explain it. He said that since God moved the heavens in circles around the centre, then if Jupiter were near the centre it would be necessary to hold it in place, as if by a prison. Lagalla also argued that the universe is not infinite. If it were infinite it would be impossible to move it in a circle. He further argued that the Earth is not 'animated, and (what is worse), a rational soul, which is absurd and ridiculous'. He insisted that the claims of Orpheus, Philolaus, Heraclitus and Copernicus were 'false and impossible'.[147] Lagalla defended Aristotle's physics by echoing arguments to deny Earth's motion.

Next, Lagalla dedicated an entire chapter to discussing and refuting 'The Opinion of Democritus: that there are many Worlds'.[148] Lagalla griped that that opinion was voiced by 'Kepler in his *Discussion with the Starry Messenger*, and by a recent author who was found guilty with his laughing insanity'. Did he mean Bruno? Lagalla explained at length that Democritus believed that space is infinite and full of infinitely many atoms that are ungenerated and incorruptible, which interact by chance to produce many worlds that are infinite or innumerable. Lagalla asked whether anything prevents the Moon from being one of those terrestrial worlds, having its own mountains, valleys, seas, lakes and so on, given Galileo's telescopic observations. Such inferences would suggest, Lagalla said, that animals and beings live not only in the spheres of air and fire, but in the ethereal globes, beings that he called 'such lunar or solar or Jupiter's Daemons for each star'.[149] Such claims, said Lagalla, were absurd, baseless, empty, impossible and manifestly contradictory. These were all the senseless consequences of having supposed that the universe is infinite and full of atoms. Lagalla said that therefore he had shown the falsity of such assumptions. He rehearsed arguments by Aristotle and also by the great authority of the Jesuits, Thomas Aquinas.

In sum, Lagalla gave dozens of additional reasons against the opinion of many worlds. At one point he even mentioned Bruno by name: 'among whom was Bruno, the latest proponent of this proposition, who according to Elizabeth of England deserved to be called *Ἄπιςιος, καὶ Ἄςευῆ, καὶ Ἄθεὼς* – that is, *infidel and impious and atheist*'.[150] Thus Lagalla used foreign insults or English royal gossip against the notorious heretic. Without attributing specific arguments to Bruno, Lagalla proposed the opposite. For example, Bruno had argued that since God is omnipotent then He must have created many worlds, infinitely. Lagalla argued instead that although God has the power to make many worlds, He in fact did not, since the Gospel of John explicitly says, 'the World was made by Him', in the singular. Almost incessantly, Lagalla insisted that only one world exists. Finally, he ended his long chapter against the plurality of worlds, writing:

> it is not true that there are mountains on the Moon, or valleys, or Seas, as perhaps someone not sane of mind might think. If it were true, then necessarily the Moon would be a terrestrial globe, another World, and there would be other living beings, other men, who inhabit the lunar orb, as well as other inhabitants, that which hardly seemed absurd to

> the ancient Philosophers we mentioned above: Orpheus, Thales, Philolaus, as well as Plutarch, yet this proposition is exposed and refuted not only as absurd, but truly as false and impossible.[151]

And if this were not enough, Lagalla included more chapters further arguing about the nature of the Moon. He cited numerous ancient writers and works: Xenophanes, Anaxagoras, Anaximander, Plato's *Timaeus*, Virgil's *Aeneid*, Plotinus, Ptolemy, Plutarch's 'On the Face in the Orb of the Moon', Hipparchus, Pliny, Jerome, Averroes, Augustine, Bernardino Telesio and others, but Aristotle above all. And in another chapter he mentioned Bruno again: 'unless one thinks of the Moon as an Earth, and that vapours can rise up to it, mixing the ambient and surrounding air: which is not only around the Moon, but also around the other planets as believed by Giordano Bruno; but I've refuted this more than enough'.[152]

The point of Lagalla's expansive attack was that Galileo seemed utterly mistaken and temerarious in upending matters in astronomy, physics and even religion. Hence, Lagalla's tract answered the question that Cardinal Bellarmine had sent to the Collegio Romano: were the phenomena described by Galileo true or false? Lagalla answered: such phenomena are false and impossible!

And what did Bellarmine do? He was not an astronomer or mathematician, yet he could analyse claims about the world and the heavens from a more authoritative field: theology. I have said that we should expect that at some point in Bellarmine's extensive writings he could have noted some of his views of Bruno's heresies – and he did.

In a book of 1605 Bellarmine explicitly denied that the Earth or the heavenly bodies are divine, soulful or alive: 'God is not an inanimate thing; hence [God] is not the Earth, not Heaven, not the Sun, not the Moon, not the Stars: because all these things are inanimate, and inferior to the animate things.'[153]

Subsequently, Bellarmine denied one of Bruno's principal claims: that God is the soul of the world. Bellarmine's remarks appear in his commentaries on the Book of Psalms, first published in 1611 and again in 1612. While some philosophers used the Psalms to justify their eccentric beliefs, when Bellarmine analysed the Psalms he took the opportunity to deny eccentric philosophical beliefs. While discussing

Psalm 41:5, 'These things I remembered, and poured out my soul in me: for I shall go over into the place of the wonderful tabernacle, even to the house of God,' Cardinal Bellarmine commented:

> The soul does not move in the body, and entirely moves, rules and vivifies the body; and gives it sense; we know that in death the soul departs from it, the body gradually falls, and motion, sense, species, and whatever good it had; in a moment the body loses that which is the soul, it is a thing of universal God, but properly speaking, this does not mean that God is the soul of the world, what the Philosophers seem to have thought; but the likeness that souls seem to have: He remains immobile in Himself, the *word carries all* of His power, *and in Him we live, move, and are*.[154]

The last phrase is a quotation from Acts 17:28, which had earlier been quoted by Calcidius and Bishop Agostino Steuco in discussing the very same topic. Likewise, Bellarmine echoed the characteristic verb 'to vivify', which had been used by Novatian, Abelard, Ficino, Bruno, Campanella and Kepler – all to describe a universal spirit that animates living beings.

Bellarmine did not specify which philosophers thought that God is the soul of the world, in whatever sense, although, as we have seen, among them were Plotinus, Porphyry, Abelard, Ficino, Campanella and Bruno. All of them elaborated upon the notion in Plato's *Timaeus* and which the *Placita* attributed to the Pythagoreans. This was the notion that Ficino attributed to Apollonius the Pythagorean, and which Ficino pinpointed in Virgil's *Aeneid*. It was the controversial 'Pythagorean doctrine' for which Bruno too praised his 'Pythagorean poet', Virgil, even when confronted by Inquisitors. As we will see, Bellarmine trashed this notion again in 1615.

Catholic lists of heresies blamed Peter Abelard for the heresy that the soul of the world is the Holy Spirit. Bellarmine was a member of the Congregation of the Index, and in 1592 he prepared his own alphabetical list of 'Heresiarchs, that is, those who invented or fomented heresies, or were the leaders of heretics'. Heresiarchs were worse than heretics. And Bellarmine's list includes Abelard as a heresiarch – alongside Calvin, Luther, Melanchthon, Andreas Osiander and others.[155] Since then Bellarmine's role in Roman censorship had grown: he became the prefect of the Index in 1605 (serving in that capacity until 1621).[156]

Bellarmine further criticized other heretical views about the soul. In a book of 1613, about sins, he criticized six propositions by the 'pagan philosophers' about the origins of human souls. Two are relevant to our discussion because Bruno and the Pythagoreans advocated them. Bellarmine wrote: 'The First proposition is that human souls are fragments [*particulas*] of the substance of God, therefore are not properly created', but are somehow breathed [*inspirari*] by God.[157] Bellarmine objected that since the divine substance is really immutable and inviolable, whereas souls are not, then this proposition had been 'rightly condemned [*damnata*] as heretical'. Instead, he said, God makes human souls from nothing. Latin expressions such as *damnata* had a stronger meaning than our judicial sense of *condemned*, they could also mean the stronger, Catholic sense of *damned*.

Next, the 'second proposition' was the claim that souls inhabited heaven until 'God cast down souls into bodies as into prisons.'[158] Bellarmine complained that this 'error' entailed multiple 'utter absurdities'. Some statements in the Bible would mean their opposite. For example, by killing people with the Flood, God would be liberating their souls, while punishing Noah's by keeping him in his bodily prison. Also, God's benedictions such as 'Go forth and multiply,' in Genesis, would become 'maledictions': commands to imprison more souls from heaven.

Bellarmine did not mention the Pythagoreans. Yet both of these 'erroneous' propositions about souls had been attributed to the Pythagoreans by Cicero, St Hippolytus, Justin Martyr, Philostratus, Lactantius, St Jerome and Erasmus. And Bellarmine had read many of their works. Both of these propositions were consonant with the transmigration of souls. If the Pythagorean interpretations of scriptures were true then certain passages in the Bible would mean the opposite of what they say.

Furthermore, Bellarmine reasserted traditional interpretations of scriptures. So he denied the Earth's motion. In 1611 Bellarmine quoted Psalm 103:5 from the Latin Vulgate, that God 'established the Earth on its foundations, it cannot be moved forever and ever'. Bellarmine commented that God put Earth in the centre of the world, and that its 'weight rests on its eternal stability'.[159] He also quoted Psalm 118:8, that God 'established the Earth and permanently'. Bellarmine explained this phrase: 'it almost says that you [God] established the Earth immovable, and it always remains immobile.' He added: 'God ordained, on the contrary, that heaven or the Sun move always.'[160] These statements matter, because they show Bellarmine truly believed that the Earth does not move. So, its motion was *not* one of the questions he sent to the Collegio Romano. He only asked about the telescopic *phenomena* Galileo had described.

The Earth's motion had been censured by the consultors of the Inquisition in the works of Bruno. Strangely, historians hardly ever mention this key point when discussing Galileo.

Similarly, in 1611 the mathematical astronomer Nicholas Mulerius published a book on tables specifying positions of the Moon and the Sun, and warned about the theory of the Earth's motion: 'The Sacred Scriptures should have such a great authority among us, and our minds should be moved by such great reverence for it, that we would not dare to fall into the opinion of the Pythagoreans, which is openly contrary to Scripture.'[161]

Nicholas de Nancel was another writer who linked the Pythagoreans with heresies. De Nancel discussed 'various opinions' about the location and motion of the Earth and the existence of many worlds. He attributed to Pythagoras the belief that 'there are two Earths', and he credited Democritus, Epicurus and Origen with the belief that many worlds exist. De Nancel mentioned theorists who wrote about Earth's motion, including Philolaus, Heraclides, Ecphantus the Pythagorean, Seleucus and Copernicus. And then de Nancel said:

> And indeed long before (so that in religion, and also in the ancient opinions of philosophy, what heresies are spoken without fear, countless, met by new proponents who among the ignorant dare to sell the first authors) the sectarians of Pythagoras, who had that part of Italy known as Greater Greece, who as Cicero conveyed, said that the centre of the universe contains a fire, and that the Earth moves around it.[162]

De Nancel decried the 'Heresies of the ancient philosophers, and their innovators' but didn't elaborate the point. Meanwhile, Kepler praised Galileo's telescopic discoveries of the 'Pythagorean and Copernican world'.[163]

As we have seen, several writers attacked pagan cosmological notions. Lipsius, Possevino, Padiglia and Bellarmine obliquely criticized Bruno's beliefs, while asserting Earth's immobility. Schoppe demonized Bruno. Lipsius, Serarius, Mulerius and de Nancel criticized the Pythagoreans for advocating the theory of the Earth's motion. Horky and Lagalla ridiculed Galileo's claims alongside Bruno's. Meanwhile, Kepler and Wedderborn defended Galileo, but linked his claims with Bruno's.

In 1611 Galileo became a member of the exclusive Accademia dei Lincei, which had been founded in 1603 by Prince Federico Cesi with Johann van Heck and two other founding members. They had chosen the name 'Lincei' from the book *Natural Magic* by Giambattista della Porta, which celebrated the eyes of the lynx, which sharply examine things. They modelled their small Academy on the school of Pythagoras, stating that, 'if you sometimes hear of the Lynxes as Philosophers; this name lacks all pride, because it is not explicitly the Knowledgeable [men], but only Friends and Lovers of Pythagorean Knowledge, as they want to be called.'[164]

Galileo in Danger

In November 1612 Galileo heard that an old Dominican preacher, Niccolò Lorini, had bad-mouthed the Copernican theory. Father Lorini was a professor of Church history at the University of Florence. So Galileo sent him a letter, now lost. Lorini promptly retorted that 'the opinion of that Ipernicus, or whatever his name is, seems to oppose the Divine Scriptures.'[165] Then Galileo just shrugged him off as a 'clumsy speaker'.[166]

Undeterred by critiques, Galileo published more findings in March 1613: his *Letters on Sunspots*.[167] Again he argued: 'with absolute necessity we conclude, in agreement with the opinions of the Pythagoreans, and of Copernicus, that Venus revolves around the Sun; around which are moving all the other planets too, as the centre of their revolutions.'[168] Galileo chose to ignore the critiques of Horky and Lagalla, who had mixed Galileo's views with Bruno's, insisting that only one world exists.[169] However, Galileo wrote: 'positing inhabitants in Jupiter, in Venus, in Saturn and on the Moon

Galileo in 1613, when he was 49 years old. Engraving by Francesco Villamena.

is false, and damning, meaning by "inhabitants" animals like ours, and above all, men.'[170] He voiced agreement with 'Apelles' (the Jesuit Christopher Scheiner), who had recently said that 'the proposition that Jupiter, Venus, Saturn and the Moon have inhabitants is easily repudiated, since it is absurd.'[171] But then Galileo said he neither affirmed nor denied that there were some living beings or vegetation on the planets or the Moon.

In 1614 the German astronomer Simon Marius published a booklet describing Jupiter's moons. He claimed to have discovered them with a telescope before Galileo, back in 1609. He also claimed to be able to see the discs of some stars, meaning that the stars couldn't be as far away as Copernicus said. But most importantly for our discussion, Marius repeatedly described Jupiter as 'a World'.[172]

At the same time, George Locher (a student of Scheiner) criticized the 'Controversies and Novelties in Astronomy'. He criticized Copernicus and rejected as 'impossible' the ancient theory that the universe is infinite and that many worlds exist, which are born from combinations of atoms. Locher wrote: 'the true and Christian philosophy has repudiated this fiction already long ago; in order to establish its own subject-matter.'[173]

In December, during a Sunday sermon at the church of Santa Maria Novella in Florence, a young Dominican friar, Father Tommaso Caccini, complained about Galileo. Caccini lived with Lorini in the convent of San Marco. They sympathized with Ludovico delle Colombe, who had criticized Galileo. In the Bible, the Joshua miracle was that the Sun stood still. So in his sermon Caccini condemned as 'nearly heresy' the idea that the Earth moves. He disdained mathematicians as agents of the Devil, who should be exiled from Christendom. He was referring chiefly to astrologers, such as Galileo and Kepler, who were professional astronomers and mathematicians but were also astrologers, although they had some reservations against astrology.[174]

Now Galileo felt pressed to respond. On December 1613 one of his devoted students, Benedetto Castelli, had visited the court of the Grand Duchess Christina, where conversation turned to Galileo's discoveries. The Grand Duchess and Cosimo Boscaglia, a philosopher, argued that scriptures deny a moving Earth. Galileo wrote a letter to Castelli discussing religion and astronomy. Handwritten copies of Galileo's letter soon circulated in Florence. In it, Galileo analysed the use of biblical quotations in science. Several passages in the Bible seem to say that the Sun moves but the Earth does not. Galileo argued that such passages do not necessarily mean what they say.

Thus he violated Catholic rules on interpreting scriptures. Bruno had done the same indiscretion. They didn't have permission. Between 1545 and 1563 the Council of Trent had decreed that:

> in matters of faith and morals . . . no one, relying on his own judgment and distorting the Sacred Scriptures according to his own conceptions, shall dare to interpret them contrary to that sense which Holy Mother Church, to whom it belongs to judge their sense and meaning, has held and does hold, or even contrary to the unanimous agreement of the Fathers.

Regardless, Galileo now argued that the Bible's 'only' purpose is to persuade readers 'of those articles and propositions that are necessary for . . . salvation and surpass human reason'. He pleaded that God doesn't want us to abandon rationality. He said that in science, interpreters are not limited by 'the apparent meaning of words'. Galileo discussed 'problems that already have been debated by the greatest Philosophers for thousands of years, which are the stability of the Sun, and the mobility of the Earth: opinions that were held by Pythagoras and by all his sect', plus Heraclides, 'Philolaus, who was Plato's teacher', Plato himself, Aristarchus, Seleucus, 'Nicetas' and 'many others'.[175] He didn't name Bruno, who belongs in that list.

Galileo's letter to Castelli triggered complaints. The Bishop of Fiesole, a hill town near Florence, said Copernicus should be imprisoned, not knowing he had died. Lorini sent a copy of Galileo's letter to the Inquisition in early February 1615. Then Castelli met with Archbishop Francesco Bonciani at Pisa. The archbishop asked him about Galileo and then urged Castelli 'to abandon certain bizarre opinions, and in particular the motion of the Earth, telling me that this would be for my own good, and not to become my downfall [or perdition], that these opinions were dangerous, scandalous and temerarious, being in direct opposition to the Sacred Scripture'. Castelli alerted Galileo. He said the archbishop voiced arguments 'which summed up to this: that since all creatures were created in the service of man, it necessarily followed very clearly that the Earth could not move like the stars'. The archbishop said 'those opinions were hearsay and simply madness, and that this was the downfall' of Galileo, and that such beliefs 'deserved to be condemned'.[176]

Meanwhile, in Naples, Father Paolo Foscarini of the Carmelite Order published a defence of the 'New Pythagorean System of the World' in 1615.[177] He described 'the Pythagorean opinion, and of Copernicus', that the Earth moves. Foscarini admitted that it seemed to be 'one of the strangest, and most monstrous paradoxes', because it seemed 'repugnant' to scriptures, so everyone judged it as 'mere madness'.[178] It was denied by habit, not reason. New astronomical evidence supported it. He said it could be reconciled with biblical passages. First it had been defended *by Pythagoras himself*, Forscarini said, and then by 'many famous, and courageous men, such as Heraclides, and Ecphantus the Pythagorean, and all the Pythagorean school, Nicetas of Syracuse, Martianus Capella and many others'.

Foscarini submitted his argument to the judgement of the Church. But he quoted an irreverent line from the Epistles of Horace: 'I am not bound to swear as any master dictates.'[179] Horace was a Roman poet from around 20 BCE, known for the maxim 'dare to be wise.'

Father Foscarini accepted that the Church could not err in matters of faith and salvation. But he said it could err in practical and philosophical judgements. He said the Bible includes metaphors and common expressions: it refers to God as walking, having a face, eyes and anger. It refers to Death as eating, moving, having a voice and a shadow. Foscarini defended the 'monstrous and extravagant' idea that Hell, at Earth's centre, revolves in heaven around the Sun. He insisted that the planetary heavens are distinct from the spiritual heaven above everything. Since scholars have focused on how Foscarini's opinions pertained to Galileo, they haven't compared them to Bruno's doctrines.

Did Foscarini support any heresies that Bruno had defended? Yes. He said that thanks to the telescope we now know 'the Moon is mountainous', Venus, Saturn and Jupiter consist of multiple bodies, and many new stars exist. Galileo's telescope 'brought, and yielded new fixed Stars, and new Planets, *and new Worlds*'.[180] Foscarini argued 'the Stars are nothing other than so many Suns, so many Moons, but more distant.'[181] Furthermore, 'the Earth is nothing other than a Moon, and a Star.'[182] All these beliefs – many worlds, including the Moon, the stars are suns, Earth is a star – had been advocated by Bruno, not by Galileo. Foscarini didn't mention Bruno, but his booklet crept towards the views of the heretic.

Crediting Copernicus, Kepler and Galileo, now Foscarini concluded that Pythagoras was probably right and not in conflict with scriptures. But other churchmen disagreed. Cardinal Bellarmine intervened.

In a letter to Foscarini he said that Foscarini and Galilei proceeded prudently by speaking hypothetically, instead of speaking as if Earth's motion were absolute truth. However, Bellarmine seriously warned Foscarini that any claim that the Earth truly circles the unmoving Sun 'is a very dangerous thing, likely not only to irritate all scholastic philosophers and theologians, but also to harm the Holy Faith by rendering Holy Scripture false'.[183]

Bellarmine warned Foscarini that the Council of Trent had prohibited any interpretation of scriptures deviating from the Holy Fathers. He added that recent commentaries on Genesis, Psalms, Ecclesiastes and Joshua all agreed on the literal interpretation about the Sun's motion, while 'the Earth is very far from heaven and sits motionless as the centre of the world.' Bellarmine concluded that although the Earth's motion is not a subject of faith, it is a matter of faith on those speakers, the apostles, prophets and Catholic commentators, who wrote about it.

Meanwhile Father Caccini denounced Galileo to the Inquisition, but apparently the Inquisitors in Florence did nothing. Caccini's friend, Father Lorini, sent a confidential complaint to Cardinal Paolo Sfondrato (or Sfondrati), the head of the Roman Inquisition.[184] Lorini said the 'Galileists' were conceited and fixed in their opinions, including disrespectful errors against the Bible and its expositors.[185]

Soon a consultor of the Inquisition wrote a report on Galileo's letter to Castelli. Then Caccini gave a deposition against Galileo, arguing that the notion that the Sun is fixed at the centre of the universe was 'almost heretical' because it clashes with Psalm 18, Ecclesiastes 1, Isaiah 38 and Joshua 10.

The Inquisition summoned Father Ferdinando Ximenes, who likewise testified against Galileo and his followers, saying that Earth's motion and the immobile Sun were 'false and heretical'. Ximenes insisted the Bible says the Sun moves; God is a substance, 'and that it is nonsense to say that God is sensuous, laughs, and cries and that there is nothing but discrete quantity mixed with empty spaces'.[186]

The Inquisition also summoned Giannozzo Attavanti, who knew Galileo. Attavanti testified that Galileo did believe that the Earth moves and the Sun doesn't. Attavanti explained that Caccini apparently overheard but misunderstood a conversation Attavanti had with Ximenes about whether God is a substance or accident, whether God is sensuous, laughs or cries. Attavanti said these weren't Galileo's ideas. He said that Caccini had interrupted another conversation saying the Sun's immobility is 'a heretical proposition'.[187]

Meanwhile, Galileo's friends in Rome told him trouble was brewing. In late February his friend Giovanni Ciampoli relayed a conversation with Cardinal Maffeo Barberini, another friend. The cardinal was concerned that greater caution was needed in discussing scientific arguments in order not to trespass the limits of physics and mathematics. Barberini worried that if a bright person introduced new ideas not everyone would be dispassionate enough to portray them well:

Cardinal Maffeo Barberini in 1612. Eleven years later, he became the Pope. Medal etching by Guillaume Dupré.

> someone amplifies, another modifies; whatever came from the author's own mouth, so that it is transformed in spreading, such that he cannot recognize it as his own. And I [Ciampoli] know what he [Barberini] tells me: because your opinion regarding the phenomena of light and shadow in the bright and dark parts [of the Moon] posits an analogy between the terrestrial globe and the lunar; [then] somebody enlarges, and says that you place human inhabitants on the Moon; and someone else starts to dispute how these can be descended from Adam, or come from Noah's ark, with many other extravagances one never dreamed of. Surely one should attest often that one submits to the authority of those who have jurisdiction over the minds of people in the interpretation of Scripture, this is extremely necessary to remove this opportunity for malice.[188]

Cardinal Barberini told another of Galileo's friends that he had not heard anyone talking about the problem that concerned Galileo.[189] Barberini was a member of the Index. He wasn't an Inquisitor, so did he know that similar accusations had been raised against Giordano Bruno? He didn't necessarily know because of 'the extreme secrecy that bound the deliberations' of the Inquisition.[190]

Cardinal Bellarmine holding a crucifix. Portrait at the Church of St Ignatius in Rome.

In mid-1615 Galileo expanded his letter to Castelli to send it to the Grand Duchess Christina. Handwritten copies of both letters circulated in Florence, but the longer letter circulated only among Galileo's close allies.

That same year Bellarmine finished his book *On the Ascent of the Mind to God by a Ladder of Created Things*, in which he (incidentally) reiterated his rejection of the soul of the world:

> Since the soul is in the body, and governing, and moving [the body], it necessarily has the form of the body, and so it is conjoined with it, so that from the soul and body one man is made. God does not need, in order to make the soul, a soul of the world; neither is there made a composite substance from Himself and the world.[191]

Bellarmine was giving an argument he didn't include in his previous critique of the world soul in 1611. It was the same argument that the

Dominican friar Petreto had published in 1601 against the two heresies that either God or the Holy Spirit is the soul of the world.[192] It is fitting that Bellarmine used that argument, since he owned a copy of Petreto's tract.[193]

In the new book Bellarmine also said that God created just 'one Sun which was enough; making also one Moon'.[194] In an earlier book Bellarmine had already argued that only one Sun exists, just as only one God exists: 'It is very true too about the Sun, which has an individual nature, we say, this one alone is the true Sun, aside from it there is no other Sun.'[195]

Again, Bellarmine did not say that he was rejecting four of Bruno's heresies: the world has a soul, God infuses the substance of the world, the human soul is not the form of the body, and there are many suns and moons.

WITHOUT BEING summoned, Galileo went to Rome to defend himself. He met with several cardinals, foremost with Bellarmine. Was there continuity between Bruno's trial and these proceedings against Galileo? Was anyone present during both proceedings?

Years later Galileo recalled that he 'discussed the matter with some cardinals who oversaw the Holy Office at the time, especially with Cardinals Bellarmine, Aracoeli, San Eusebio, Bonsi and d'Ascoli'.[196] At first sight none of these names, other than Bellarmine, seem to match the participants in Bruno's proceedings. 'San Eusebio', however, was an honorary name of Ferdinando Taverna, who had served as Consultor in the Roman proceedings against Bruno from late 1593 to early 1595.[197] After he became governor of Rome in 1599 he presided over Bruno's final imprisonment and brutal execution; he became a cardinal in 1604.[198]

Someone else participated in Bruno's trial, someone with a higher ranking than Bellarmine and Taverna. Sfondrato had served as Cardinal Inquisitor in the Roman trial of Bruno for six years from 1594 until Bruno's condemnation in 1600.[199] A few months later he became a member of the Index. He was a close friend of Bellarmine.[200] By 1616 Sfondrato was the most senior Inquisitor and head of the Inquisition, under the Pope, and still a member of the Index. Sfondrato received the initial complaint against Galileo and decided it deserved attention.[201]

Bellarmine, Taverna and Sfondrato knew Bruno's heresies. They contributed to killing him. Cardinals Bellarmine and Taverna met with Galileo and asked him specifically to explain the claims of Copernicus.

Yet another authority had first-hand knowledge of Bruno's case, someone with a higher rank in the Church than even Sfondrato. Cardinal Camillo Borghese had participated in the Roman proceedings against Bruno from 1596 until 1600. He was also present, alongside Bellarmine and other Inquisitors, in March 1597 when they admonished Bruno to renounce just one particular delusion: his belief in many worlds. Later, in 1605, Borghese was promoted to the highest Catholic office, becoming Pope Paul v. Now, ten years later, he headed the investigation against Galileo.

Records of the Inquisition meetings clarify the sequence of events. On 7 February 1615 Lorini submitted his complaint to Sfondrato. On 25 February Bellarmine, Taverna and other officers of the Inquisition met to discuss Galileo's errors and misinterpretations of scriptures. On 19 March the Pope met with Sfondrato, Bellarmine, Taverna and four others to discuss Galileo's errors. On 2 April the Pope met with Sfondrato, Bellarmine, Taverna and others to discuss Caccini's deposition. On 25 November Sfondrato, Bellarmine, Taverna and others discussed Ximenes's deposition against Galileo.[202]

Thus in 1615 at least four official meetings of Inquisitors who had been at Bruno's trial discussed allegations against Galileo, twice in the presence of the Pope, who had also participated in Bruno's trial.

Now consider the accusations against Galileo:

(1) Galileo says the Earth moves.
(2) Galileo says the Sun and heavens don't move, and the Sun is at the centre of the universe.
(3) Galileo argues that certain ways of speaking in scriptures are inappropriate or literally false; for example, that when Joshua ordered the Sun to stop, the order was not really given to the Sun.
(4) Galileo argues that in disputes about natural processes, scripture holds the last place; and that in such disputes philosophical or astronomical arguments have more force than divine arguments.

(5) Galileo says that the expositors (Church Fathers) of the Bible are often wrong in their interpretations.
(6) Galileo argues that scriptures should not meddle in topics other than faith.
(7) Galileo is regarded with suspicion in matters of faith, partly because he corresponds with Paolo Sarpi, infamous in Venice for impieties and for interpreting scriptures against the interpretation of the Church Fathers.
(8) Galileo's disciples speak disrespectfully of the Church Fathers and Thomas Aquinas, rejecting Aristotle's philosophy.
(9) Galileo's disciples say God is not substance but accident; everything is discrete quantity and empty space; God is sensuous, laughs and cries; and saints' miracles were not real.[203]

Some of these accusations had been levelled previously against Bruno. These were the following: (1) Earth moves; (3) certain ways of speaking about scriptures are literally false; (5) some Church Fathers were wrong in their interpretations; (8) speaking disrespectfully of Church Fathers, Thomas Aquinas, and rejecting Aristotle's philosophy; and also, in a general way, (4) attributing undue importance to philosophy, and (9) having wrong opinions about God and the Saints.

MEANWHILE, TOMMASO Campanella sat in prison. But years later he published a book in which he claimed that, in early 1616, Cardinal Bonifazio Caetani in Rome contacted him to request that Campanella provide a theological evaluation of Galileo's astronomical assertions. This claim has been debated by scholars, such as Luigi Firpo, who doubt that a cardinal would contact a convicted monk to request theological opinions. Yet Campanella's claims do not seem like a fabrication; we just do not know. Anyhow, instead of a letter, Campanella wrote a treatise, which he allegedly sent to Cardinal Caetani: 'this inquiry, composed at your request: in which the Earth's motion, and the stability of the starry sphere, and the rationale of the Copernican system are examined following Sacred Scriptures'.[204] It was a strong defence of Galileo, arguing that his theory was based on the ideas of the Pythagoreans and Pythagoras, which allegedly stemmed from Moses. However, Campanella later said that Caetani received his article eight days after the Inquisitors 'had condemned the proposition of the Earth's daily motion as contrary to scriptures'. Too late. Therefore, Campanella immediately noted, his own approach was 'neither Copernican, nor of the Nolan', that is, of Bruno.[205]

On 24 February Pope Paul v had a meeting with the cardinals. They discussed Galileo, but then Cardinal Alessandro Orsini made a comment in favour of Galileo. According to the Tuscan ambassador,

> The Pope told him [Orsini] that it would be good for him [Orsini] to persuade him [Galileo] to leave this opinion. Orsini replied something, annoying the Pope, who cut off the argumentation and told him that he had remitted the business to the Holy Cardinals of the Holy Office; and Orsini having left, His Holiness called Bellarmine to him, and having discussed this matter, they concluded that this opinion of Galileo was erroneous and heretical.[206]

The Tuscan ambassador added that 'This point, this thing, today in the Court is shameful and abhorred.' He said everyone fled from contentious discussions about individual opinions, 'especially about astrological or philosophical things', especially because the Pope was hostile to them.

That same day, 24 February, eleven consultors of the Inquisition filed a unanimous official report concluding that the proposition that the Sun is immobile at the centre of the world is 'foolish and absurd in philosophy; and formally heretical, inasmuch as it expressly contradicts the sentences of Sacred Scriptures,' according to theologians and Church Fathers.[207]

About the proposition that the Earth is not the centre, but moves, all the consultors censured it too as foolish and absurd in philosophy, 'and regarding the Theological truth, it is at least erroneous in faith'.[208] We should recall that in 1597 other consultors of the Roman Inquisition had already censured the proposition about the Earth's motion when analysing the works of Giordano Bruno. It was so important to them that it was one of only ten censured propositions in Bruno's works; it was the fifth on their list.

Some hearsay accusations about Galileo were disregarded, lacking evidence. The Inquisition singled out for correction only two allegations: about the Earth and the Sun. But despite its consultors' internal report, the Inquisition itself did not issue a formal condemnation against anyone.[209] It was late February 1616, and Galileo was still in Rome.

Given the theologians' judgement 'against the propositions of the mathematician Galileo', that the Earth moves and the Sun sits still at the centre, Pope Paul v himself asked Cardinal Bellarmine to summon Galileo to 'warn him to abandon these opinions; and if he refuse to obey, the Father Commissary, in the presence of a notary and witnesses, is to issue him an injunction to abstain completely from teaching or defending this doctrine and opinion or from discussing it; and further, if he should not acquiesce, he is to be imprisoned'.[210]

Hence, Cardinal Bellarmine met with Galileo to warn him seriously that the two opinions were erroneous, and to abandon them immediately and completely. Also present were the Commissary of the Inquisition and two witnesses. Reportedly, 'Galileo acquiesced in this injunction and promised to obey.'[211]

Historian Peter Godman has analysed this meeting in the context of Bellarmine's extensive experience as censor and Inquisitor. The meeting was an 'admonition', the mildest form of discipline, and it was an unusually abbreviated censure, compared to others in Bellarmine's career. Its promptness indicates that Galileo was esteemed with favour. Other historians had long argued that Galileo's mild treatment was due to his social standing with friends in high places, such as Prince Cesi.[212] Godman says that such explanations focus only on Galileo's perspective, not his judges'. From Bellarmine's perspective it was common to treat secular scholars less severely than clergymen. Hence Galileo was spared with a mere admonishment for his 'error', whereas Father Foscarini's book was condemned. According to Godman, 'Discipline and authority were at stake and, if learned laymen might be spared, no quarter would be given to the rank and file of the *Ecclesia militans* who presumed to step out of line.'[213] This applied to deviant clergymen, such as Bruno and Campanella.

Did the prohibitions against Copernicus and Galileo have any relation to Bruno's heresy that many worlds exist, some inhabited? The extant depositions and Inquisition minutes do not specify this. However, on 28 February 1616, just two days after Galileo received Bellarmine's order to abandon the false opinions, Galileo wrote a letter in which he apologetically discussed only one topic: beings living on the Moon.

Galileo wrote to Duke Giacomo Muti, noting that 'a few days ago', when they had met in the presence of Muti's uncle, Cardinal Tiberio Muti, along with Alessandro Capoano, they had discussed the question of whether rational beings live on the Moon. Galileo wrote that Capoano tried to impugn Galileo's claim of mountains on the Moon. Capoano argued that since Earthly mountains exist for the benefit of plants and animals, which exist for the benefit of mankind, then it follows that mountains on the Moon would mean that 'other plants, and animals would exist in the benefit of some other intelligent and more perfect creature; a consequence that being utterly false, demonstrates that there exist no such mountains.' Remarkably, this is the problem that Archbishop Bonciani had denounced to Castelli. Moreover, this is also the problem that Cardinal Barberini apparently conveyed to Galileo's friend Ciampoli. And Barberini served alongside Cardinals Muti, Sfondrato and Bellarmine in the Index.

It is also significant that in December 1615 the Tuscan ambassador had voiced concern that Galileo had travelled to Rome, 'I know well that some *friars of Saint Dominic*, who have a great role *in the Holy Office*, and others, have a bad will against him [Galileo]; and this is not a country to come to dispute about the Moon, nor of the will, in the present century.'[214] Recall that Bruno had claimed that the Moon is another Earth, with mountains, plains and seas – and that it is inhabited.

Galileo now clarified his reply to Cardinal Muti's nephew, saying that the irregularities on the Moon's surface had been detected by the telescope, but that the alleged consequence was 'not only not necessary, but absolutely false and impossible, since I can demonstrate that in that globe not only can there not exist human beings in any way, but no animals or plants either'. This was because he didn't think the Moon is composed of soil and water, and also because the Sun doesn't affect the Moon like the Earth. The temperatures on the Moon vary too much and it seems to have no clouds. Galileo then immediately ended his letter:

> And this, as your Excellency can well remember, was all that was said that day, without entering into any philosophical

> discourse, nor that about that matter were any other words said. And with all humility I kiss your hand, and to the Lord God I beg that you be full of contentment.[215]

Philosophical discourse about inhabited worlds had been Bruno's speciality: the Index of Forbidden Books had banned Bruno's books, nine of which spoke about other worlds.

The evidence thus suggests that at least one archbishop (Bonciani), five cardinals (Bellarmine, Sfondrato, Taverna, Muti and Barberini), plus the Pope, were all concerned about the heretical belief in the existence of more than one world.

On 3 March 1616 the Inquisitors met with Pope Paul v to report Galileo's acquiescence. The meeting included Bellarmine and Taverna.[216] Two days later, on 5 March, the Index issued a Decree condemning several books for errors harmful to Christianity and prohibiting their printing, distribution, ownership and reading in any languages. The Decree states, 'This Holy Congregation has also learned about the spreading by many of the false Pythagorean doctrine, altogether contrary to the Holy Scripture, of the mobility of the Earth and the immobility of the Sun', as taught by Copernicus, Diego de Zúñiga and Paolo Foscarini. At the time Bellarmine was prefect of the Index.[217]

Whereas the Church had not censured Copernicus's book previously, it now censured it, partly because the cardinals had become more intolerant towards any deviations from the official interpretation of scriptures. The Catholics had lost half of Europe to the Protestant reformers, so the Catholic bureaucracy became increasingly authoritarian. The clergy increasingly interpreted the Bible very literally, to prevent modifications of its meaning.

The Index banned the works of Copernicus and Zúñiga until they could be corrected. It 'completely prohibited and condemned' Foscarini's scriptural defence of the 'New Pythagorean System of the World', a rejection more forceful and damning than the censures against Copernicus, Zúñiga and Galileo.

Consider now the expression 'the false Pythagorean doctrine, altogether contrary to the Holy Scripture', that the Earth moves and not the Sun. Historians and scientists have read this merely as an allusion to ancient mathematicians, scientists and astronomers. But were there religious connotations?

Suppose that the sentence said instead 'the false Lutheran doctrine, altogether contrary to the Holy Scripture', it would then be evident that there was a religious issue involved. But since the Pythagoreans had been misrepresented for centuries, the religious connotations became invisible to scientists and historians.

As had happened at Bruno's trial, most of the accusations against Galileo were dismissed, leaving few beliefs as deserving censure. Since Bruno believed in infinitely many suns, in his trial there was no issue of the Sun's centrality. But in the proceedings against Galileo this idea became described as 'Pythagorean'. Galileo himself had attributed it to 'Pythagoras, and all of his sect'.[218]

Copernicus and Kepler had merely credited the Pythagoreans, but Foscarini and Galileo took an extra step, they gifted the credit to Pythagoras himself. Others too, like Thomas Digges, had not distinguished between the Sun and the 'central fire' of the Pythagoreans, which Aristotle criticized. The claim that the central Sun is motionless upset the Inquisitors more than the opinion that the Earth moves. They labelled the former alone as 'formally heretical'.

I don't know all the Inquisitors and censors who in 1616 knew about the connection (evident in Bruno's trial) between the Earth's motion and various heretical Pythagorean beliefs. But at least it was known by Bellarmine, head of the Index, Sfondrato, head of the Inquisition, Taverna and the Pope. It was Sfondrato who signed the Decree of the Index as 'Bishop of Albano, Cardinal of St Cecilia'.

Bellarmine requires special attention. In addition to having served in the deadly proceedings against Bruno, Bellarmine had studied writings by many ancient Church Fathers relevant to the present discussion, including Tertullian, Lactantius, John Chrysostom and, of course, Augustine.[219] Bellarmine's life had been shaped by his militant struggles against heresies. In a work of 1616 – a significant date – he said that the 'various errors' of 'ancient heretics' would have been forgotten, 'had we not read them in the Catholic books of those who attacked them: Irenaeus, Philaster, Epiphanius, Augustine, Theodoret and the like'.[220] Among them, Philaster, Epiphanius, Augustine and Theodoret had denounced the pagan belief in many worlds.

Did Bellarmine notice their critiques of Pythagorean ideas? Yes. For example, in one book Bellarmine discussed the location of Hell, among other things, and the pagan opinion 'that the place of punishment for souls is this world, in which souls stay in bodies as if in a prison'. He said that this opinion had been refuted by Irenaeus since, according to scriptures, after death souls really 'descend into Hell'.[221] Similarly, we may well list some of the reputedly Pythagorean propositions that, as we have seen, Bellarmine explicitly denied or condemned:

(1) The world has a soul.
(2) God is the soul of the world.
(3) God is infused in the substance of the world.
(4) Human souls are fragments of the substance of God.
(5) Souls are punished by being imprisoned in bodies.
(6) The soul vivifies the body, and controls it like a sailor in a ship.
(7) Earth, Sun, Moon and stars are divine, animated beings, with souls.
(8) The Earth moves.
(9) The Sun does not move.
(10) There are many worlds.

Bellarmine didn't label all these beliefs as 'Pythagorean' because theologians usually referred to heresies by the names of *Christian* heretics, such as Simon Magus, Valentinus, Mani, Origen, Abelard and Novatian, rather than by the names of the reputedly earliest pagan proponents.

Bellarmine occasionally mentioned Pythagoras, without necessarily portraying him as evil, just mentioning him as a noble philosopher who taught much but did not write down his teachings.[222] Still, Bellamine did cite a complaint by Irenaeus against heretics who dared to set a portrait of Pythagoras alongside one of Jesus.[223] Bellarmine also quoted a damning version of an old story about Pythagoras:

> the judge is Saint Jerome. Thus in his book against Vigilantius, he said: 'Just as Euphorbus was said to be reborn as Pythagoras, so too the evil mind of Jovinian has reincarnated in this one, so in him, and in that one, we are forced to respond to the snare of the Devil.'[224]

Moreover, there is a sermon by Bellarmine, first published in 1615, in which he mentioned Pythagoras and Apollonius of Tyana as wayward, pagan philosophers:

> Look at the ancient philosophers, I beg you, what did their vain hope of wisdom and huge desire for knowledge compel them to do? Some forbade themselves from having any pleasures at all; others sank all their wealth into the sea; others approached the most remote regions with incredible labour and danger. There were some others who killed themselves in order to attend to philosophy more easily and freed of their bodies. Thus with much labour Pythagoras went to the soothsayers of Memphis [lower Egypt]. Thus with laborious troubles Plato travelled to Egypt and to the coast of Italy, which was called greater Greece, and with letters he almost chased the whole world, was captured by pirates and sold, and even obeyed a very cruel tyrant. Thus too, as noted by St Jerome in his letter to Paul, quoting Philostratus, Apollonius entered Persia, crossed the Caucasus, Albania, the Scythians, Massagetas, entered the most opulent kingdoms of India, and to the Brahmans, Parthians, Syrians, Phoenicians, Arabs, Palestinians; returning to Alexandria, he travelled to Ethiopia, and to the Gymnosophists and saw in the sand the most famous table of the Sun.
>
> Look finally at the fervour and diligence propagated by heretics in their madness and dreams. Daily they write heavy books, they labour assiduously to preach, always vigilant to capture souls by flattery, promises, lies, frauds, false miracles, corrupt books, and even weapons to try to persuade them of their terrible errors; they also often tormented and tortured themselves, and even burned themselves alive, but not as bravely as widely reported. For the punishment suffered by heretics (as St Cyprian wisely says) is not a crown of faith, but the penalty for treachery, and from that temporary burning they pass to the eternal fire.[225]

Not to be mentioned, Bruno of Nola was one of those heretics who had burned for treachery. In this sermon Bellarmine paraphrased lines from a letter from around 394 CE from St Jerome to Paulinus,

Bishop of Nola.[226] In it, Jerome urged Paulinus to study scriptures diligently. Jerome referred to the great labours of learning carried out by wise pagan philosophers and also by the apostle Paul.

According to Jerome, Apollonius could be described as 'whether a sorcerer, as the vulgar say, or as a philosopher, as the Pythagoreans say', citing his biography by Philostratus, as Bellarmine did later. The allusion to 'the most famous Sun table in the sand' requires some explanation. Philostratus wrote that Apollonius had seen an altar dedicated 'TO THE INDIAN SUN AND THE DELPHIAN APOLLO', but also that he met an Indian king who made libations to the Sun and had an immense round table resembling an altar.[227]

Prior to Bellarmine's sermon, Jerome's words about Pythagoras, Plato and Apollonius had been paraphrased by – of all people – Giordano Bruno! In 1588, in a farewell address to the Senate of the Academy of Wittenberg, Bruno thanked the people of Germany for having received him with kindness:

> Go now, whether Pythagoras, to the soothsayers of Memphis, or Archytas [the Pythagorean], to the shores of Italy, or Plato, in Sicily. Go now, or [Apollonius] Tynean, among the Persians, pass the Caucasus, the Scythians, the Messageti, enter the most opulent kingdoms of India and, across the great river Fiso, go to the Brahmans, travel among the Elamites, the Babylonians, the Chaldeans, the Medes, the Assyrians, the Parthians, the Syrians, the Phoenicians, the Arabs, the Palestinians, Alexandria, and go into Ethiopia to see the Gymnosophists and the most famous table of the Sun on the sand.

Bruno said he had found similar wonders in Germany and had been well received, even though being 'a stranger to the nation, exile, fugitive, laughing stock of fortune, small of body, poor of goods, without favour, viewed with hatred by the mob, then contemptible fools and those most ignoble who do not recognize nobility except where it shines of gold'.[228]

In their versions of this passage, Jerome and Bruno portrayed Pythagoras, Plato and Apollonius as role models, examples of earnest learning. In contradistinction, Cardinal Bellarmine used this same passage to *ridicule* the vanity of philosophers and the delusions of heretics as corrupted by lies, terrible errors and false miracles. Contrary to Jerome and Bruno, who praised travel as a way to gain knowledge, Bellarmine dismissed it as misguided wandering. Bruno himself had been a wanderer; born and educated in Nola, he had travelled to many cities and countries to seek knowledge: Naples, Venice, Padua, Rome, Genoa, Noli, Bergamo, Savona, Turin, Geneva, Lyon, Toulouse, Paris, Oxford, Wittenberg, Prague, Helmstedt, Frankfurt, Zurich and, finally, back to Venice and Padua, before being captured and imprisoned. Wanting to travel all over the world had been one of Mocenigo's original accusations against Bruno. Bellarmine knew this.

Also, by alluding to philosophers 'who killed themselves', Bellarmine's words seem reminiscent of Lactantius's claim that some Pythagoreans had misunderstood the immortality of the soul and had committed suicide.[229] Next, Bellarmine drew a contrast between Christians and pagan philosophers:

> And we Christians do not blush, who are not a little of the Earth, and greedy; not a little of smoke, and ambitious; but not of the most uncertain conceptions, as the pagan philosophers; not to propagate the errors of dreams, as the heretics; nor so much stipend to pay, as soldiers; but the eternal kingdom, the kingdom of heaven, blessed life, and immortal, the fellowship of saints, and angels equally, we want the likeness of God, do not blush, I say such a prize proposed, can we do nothing, or suffer for it?[230]

Here, the alleged vanities of pagan philosophers are just a step away from the errors of heretics. It reminds me of the claims by Tertullian, Lactantius and Jerome that 'philosophers are the patriarchs of heretics.' Indeed, Bellarmine knew and quoted these words by Tertullian, while discussing 'extremely serious' disputes of the origin of the soul, saying that philosophers like Plato and Origen had erroneous views about it.[231]

Origen had said the world has a soul, that the stars are rational beings, and that human souls are immortal, eternal because they pre-exist bodies, such that 'there is nothing new under the Sun.' St Jerome accused Pythagoras and Origen of teaching that souls fall from heaven, and may be clothed in various bodies in various worlds.[232] Bruno had advocated these heresies.

In 1611 and 1615 Bellarmine had criticized Bruno's views about the soul of the world, that the Earth moves and that God infuses everything. Bellarmine dedicated his book of 1611 to Pope Paul v, one of Bruno's judges. Bellarmine there asserted Earth's centrality in the universe. Again, this was contrary to Bruno and Galileo.

Bellarmine had studied an extensive theological literature that included writings by early Christian and numerous pagan authors, including Cicero and Porphyry.[233] Still, it is during the long proceedings against Bruno that we most clearly see the link between the Catholic denial of Earth's motion and the heretical beliefs of the Pythagoreans: the existence of other worlds, the soul of the world and the transmigration of souls.

SOON AFTER MEETING with Bellarmine, Galileo wrote to the Tuscan Secretary of State, saying that the Inquisition's deliberations were not directed at himself but against Copernicus. But that was false. Accusations and depositions had all been about Galileo, not about Copernicus.

Galileo said his enemies had failed to make the Church declare Earth's motion heretical and contrary to the faith. Instead the Church had only decreed 'that that opinion does not agree with Holy Scripture', and that only Foscarini's book would be completely prohibited. The works of Zúñiga and Copernicus were temporarily suspended. Galileo bragged that he wasn't mentioned: 'my own behaviour in this affair has been such that a saint would not have handled it either with greater reverence or with greater zeal toward the Holy Church.' He complained that his enemies were malicious gossipers, who 'have not refrained from any machination, calumny, and diabolical suggestion' to destroy his reputation.[234]

Galileo said the books by Zúñiga and Copernicus would be 'corrected', and that such corrections were already known. One sentence would be removed from Zúñiga's book. Ten lines would be removed from Copernicus, 'from the Preface to [Pope] Paul iii where he [Copernicus] mentions that he does not think such a doctrine is repugnant to Scripture; as I understand it, they could remove a word here and there, where two or three times he calls the Earth a star'.[235] Galileo said Cardinal Caetani was appointed to make the corrections.

We should note that the question of whether the Earth is a star pertains not only to the controversy of whether it moves, but to the more controversial question: if Earth is a star, then are stars worlds? In a deposition to the Venetian Inquisition in 1592, Bruno had said, 'I have indeed asserted infinite particular worlds similar to that of the Earth, which with Pythagoras I regard as a star.'[236] This statement was copied by the Roman Inquisitors in their Summary of 1598. It was also implied in the heretical statement 'Worlds are innumerable', in the initial and final accusations against Bruno, voiced by Mocenigo and Schoppe. It was also defended by Foscarini. But it had been pinpointed as a senseless absurdity by Giulio Cesare Lagalla of the Jesuit Collegio.

Despite Galileo's wilful impression that Bellarmine's warning was offered merely in kind courtesy, hearsay about it spread. Soon Galileo heard critical rumours. He felt the need to protect himself from slander. After all, the Inquisition had not accused him, had not put him on trial and his books were not censored. Therefore, he contacted Cardinal Bellarmine requesting support. In response, Bellarmine sent him a certificate, dated May 1616 and signed by Bellarmine, stating that Galileo was being slandered as if he had been ordered to abjure some opinion or doctrine, which he had not. Thus Bellarmine declared in writing that he had merely informed Galileo that the Pope and the Congregation of the Index had decreed that the doctrine of the Earth's motion and the Sun's immobility 'is contrary to Scripture and therefore cannot be defended or held'.[237] It was a generous understatement, because in fact Galileo had been the main source of the proceedings that led to the Pope's ruling.

Did Galileo hold any Pythagorean beliefs other than the Earth's motion? Yes – quietly, secretly, he did. In April 1615 Galileo confided to one of his supporters:

> It seems to me that in nature there is found a most spiritual substance, most tenuous and most rapid, which, spreading itself throughout the universe, penetrates into all without distinction, warming, vivifying and giving fecundity to all living creatures; and about this spirit of which the senses themselves show that the body of the Sun is its foremost reservoir, from which an immense light expands throughout the universe, accompanied by this spirit that heats and

> penetrates into all vegetative bodies, and gives them life and fecundity.[238]

This universal, vivifying spirit was not just light. Galileo paraphrased Genesis 1:2 to characterize the fertilizing spirit as 'nourishing the waters or incubating over the waters'.[239] That passage had been quoted by Giordano Bruno to suggest that the Holy Spirit is the soul of the world. Thus Galileo's words come alarmingly close to that heresy.

Galileo connected these beliefs to the Sun's centrality. He wrote that the Sun is located at the centre of the universe precisely 'because' there it can receive, focus and strengthen the 'fertilizing spirit' to project it outwards. He said that sunspots might well be 'nourishments' or 'excrements' of the Sun. Galileo further said, 'the vital spirit, sustains and vivifies all the limbs,' as if echoing Virgil. And Galileo noted, 'I could provide many testimonies from philosophers and serious writers, in favour of the marvellous force and energy of this spirit.' He did not quote them, but among them were Pythagoras (allegedly), Varro, Virgil, Apollonius (allegedly), Novatian, Trismegistus, Abelard (allegedly), Ficino, Bruno, Campanella and Kepler. Their claims had been denounced by clergymen such as St Augustine, Castro, Piccolomini, Petreto, Bruno's Inquisitors – and Bellarmine, in particular.

Did Galileo know these views were offensive? Yes.

He wrote to his confidant: 'I know and confess my excessive temerity in opening my mouth, being inexperienced in Sacred Scriptures', so he asked to be excused, saying that actually he did 'submit totally to the judgment of my superiors'. And lastly he begged, 'please do not let this [letter] reach the hands of anyone who, instead of with the sensitivity of the mother tongue, operates with the roughness and sharpness of a bestial fang, instead of polishing this, no, would lacerate and rip it apart entirely.'[240]

In his letter to the Grand Duchess Christina, Galileo mentioned the same belief. Galileo mentioned the soul of the world. Discussing the Sun, he wrote: 'I don't think that it is far from good philosophizing to say that he [the Sun], as maximum minister of Nature, and in a certain way as the soul and heart of the world, infuses the other bodies that surround him not only with light, but also with motion.' Galileo quoted Dionysius the Areopagite, stating that the

Sun 'gathers together all that is dispersed', and it 'renews, nourishes, protects, perfects, divides, marries, fosters, restores fecundity, increases, mutates, stengthens, delivers, moves and vitalizes all'.[241]

In the Renaissance many writers did not know that this 'Dionysius' was *not* the saint converted by St Paul; instead, he was a theologian around 500 CE who misattributed his writings. He tried to smuggle pagan beliefs into Christianity. He was the same 'Dionysius' whom Bruno praised as one of 'the most profound and divine theologians' who rightly worshipped God with silence: 'the negative theology of Pythagoras and Dionysius is much more renowned, above the demonstrative one of Aristotle and the scholastic doctors.'[242]

Despite any imprudence, Galileo managed to dodge the Inquisition. As for the young Father Foscarini, any inquisitorial punishment became impossible. The Inquisition arrested and jailed the printer of Foscarini's booklet, for printing it without a licence.[243] A few days later, on 10 June 1616, Foscarini suddenly died.

THREE

THE ENEMIES OF GALILEO

Soon after Galileo published his book on the Moon, the Dutch astronomer Nicholas Mulerius said the opinion of the Pythagoreans was 'openly contrary to Scripture'. Now, six months after Bellarmine admonished Galileo, Mulerius complained that the Copernican universe was so immense that it requires 'the existence of many suns. And this deserves to be called absurd and contrary to Christian piety.' Mulerius also referred to the new Pythagorean astronomers as 'a sect'.[1]

Other people too associated Pythagorean beliefs with heresies. The Jesuit priest Matteo Ricci carried out missionary activities in China. His manuscripts on the expedition were published five years after his death in 1610. Ricci said that some heretical beliefs of the Chinese resembled those of the ancient Greeks. He commented about one Chinese sect: 'It forges, with Democritus and others, many worlds, but mostly they seem to have borrowed the transmigration of souls from the doctrine of Pythagoras, and have added many other lies to it, to embellish the falsehood.'[2]

The Jesuit Antonio Rubio spent 22 years as a missionary in Mexico, teaching philosophy and theology, returning to Spain in 1599. Before he died in 1615 Rubio was writing a commentary on Aristotle's *On the Heavens*. Rubio spent four chapters discussing the question of 'Whether the world is one, or many'. He attributed the theory of worlds to a dozen philosophers including Democritus, while he counted Pythagoras among ten philosophers, plus Aristotle's followers, who said that the world is only one. Rubio did not mention that writers such as 'Plutarch', Hippolytus, Lucian,

Iamblichus, Theodoret, Bruno, Piccolomini, Kepler, de Nancel, Lagalla and Galileo had attributed the theory of more than one world to Pythagoras or the Pythagoreans. In addition, Pythagoras had been born more than a century before Democritus. In any case, Rubio duly cited several authorities who denied the existence of many worlds: Athanasius, St Ambrose, Albertus Magnus, Theodoret and Thomas Aquinas. Rubio noted that 'the multitude of worlds, and how large it is, is among the errors of the faith enumerated' by Augustine as the 77th heresy and by St Isidore. Rubio concluded: 'Now some heretics at the end, count among [their] other heresies to posit innumerable worlds.'[3] He did not name Bruno.

Galileo now abstained from publishing about Pythagorean doctrines. Yet Johannes Kepler did the contrary. Living in Protestant lands, Kepler hardly feared the Inquisition. Soon after Copernicus's book was prohibited, Kepler wrote, 'these people should be refuted not with arguments but with laughter.' So he ridiculed such censorship:

> They were able to castrate
> The bard lest he fornicate;
> He survived without any testicles.
> Alas, O Pythagoras,
> Whose thinking wore out iron chains;
> They spare you your life,
> But first they get rid of your brains.[4]

In 1618 Kepler published an *Epitome of Copernican Astronomy* aimed at general readers. In 1619 he published *Harmony of the World*, in which he further analysed planetary motions and voiced theological opinions. He wrote controversial ideas similar to those of Plato, the Pythagoreans, Plotinus, Porphyry, Origen, Novatian, Abelard, Ficino and Bruno. Kepler again insisted the Earth is an animate being and has a soul.[5] He said that the Earth's body has hairs (plants and trees) on its skin. It expels vapours, mucus and vomit. He also said the planets have souls, which produce a heavenly harmony. He argued that mathematics contains hidden divine knowledge, and that 'the Pythagorean philosophy disguises its teaching on divine matters with these, so to speak, veils.'

Kepler discussed numerology and the 'Golden Verses of Pythagoras', trying to decipher the mystery of 'the spiritual

Tetractys', the legendary sacred oath of Pythagoras.[6] Kepler conjectured that the ancient Pythagoreans knew the connection between the five regular solids and the order of the six planets, but had kept it secret.[7] Kepler also criticized Aristotle and Cicero for having denied that the heavens generate inaudible harmonies:

> These preconceived opinions are a considerable obstacle to readers who are striving towards the inner secrets of Nature, and could frighten off many who have great powers of judgment and are seekers after truth, to such an extent that they would disdain those Pythagorean pipedreams, scarcely recognized at arm's length, and throw away the book unread.[8]

Just like the ancient Pythagoreans (as criticized by Hippolytus), who worshipped a Creator who allegedly was 'the Great Geometrician and Calculator', Kepler claimed that geometry is God: 'Geometry, which before the origin of things was coeternal with the divine mind and is God himself (for what could there be in God which would not be God himself?), supplied God with patterns for the creation of the world.'[9]

Furthermore, Kepler refused to accept all the doctrines of his Lutheran religion. In particular, he refused to accept the Sacramental Union (that during the ceremony of the Eucharist the wine and the bread combine with the real blood and body of Jesus Christ). Therefore, the Lutherans excommunicated him in 1619, to his chagrin.

That same year the Index of Forbidden Books promptly banned Kepler's *Epitome* (1618), in which he not merely insisted that the Earth moves; he went beyond Copernicus by arguing that Earth moves because it has a soul. Kepler said that Earth's soul causes its bowels to emit heat, fire, exhalations and secretions. Its soul helps to move its rivers and seas.[10] He wrote that the Earth has blood, sweat, mucus, saliva and excrements.[11] Kepler said that God had initially set the Earth in motion, but subsequently, the Earth's soul kept it moving constantly.[12] Such claims, that the moving Earth is a living animal with a soul, had been defended by Bruno.

In 1619 the Decree of the Index, totally prohibiting Kepler's book, was signed by one of Bruno's judges, Bellarmine.[13] It was co-signed by Francisco Capiferreus, the Dominican who had co-signed the anti-Pythagorean Decree of 1616.

Soon an updated supplement to the Index of Prohibited Books was published; it included specific books by Copernicus, Zúñiga, Foscarini and Kepler, plus 'all books that teach the motion of the Earth and the immobility of the Sun'.[14]

Galileo had been careful not to associate his works with Giordano Bruno. Yet Kepler and others publicly linked the two. Kepler portrayed Bruno as Galileo's predecessor and praised them both. Still, Kepler rejected the infinity of worlds. He thought that stars are *not* suns surrounded by planets. Yet there were some astronomers who thought that the telescopic evidence supported the theory that infinitely many stars exist and planets are worlds. One was the French astronomer Jean Tarde, a canon of the Sarlat Cathedral in southwestern France. Tarde published a treatise analysing sunspots, arguing that they are not clouds in the Sun's atmosphere, as Galileo suggested, but small planets orbiting the Sun:

> Orpheus, Pythagoras, Anaximander, Leucippus and their followers, who imagined an infinity of worlds to exist, said that the Moon is a world, Lunar; Mercury, Venus and other planets, and the fixed stars, were also called worlds. And since the known stars did not exceed the number 1022, they thought that there were an infinity of others, and those were infinite worlds, those which in our century have been detected by glass lenses.[15]

Galileo's discovery of countless stars seemed to confirm Bruno.

The plurality of worlds was harmless speculation if it were presented dismissively as fiction. But presented as truth, it contradicted Christian dogma: that humans were unique in God's creation. Already theologians had to explain the puzzling destiny of humans born in the American continents. If natives had never heard the Word of God, through Moses and Jesus, would they then suffer the same damnation as European sinners who violated God's commandments? What about beings in other worlds?

In 1620 the Index censored eleven sentences in Copernicus's book. Catholics could now read it by making the required corrections. Hardly any history books actually state any of the censored sentences, so I must summarize them.[16] First, the Index censored the passage in the Preface where Copernicus criticized Lactantius for not knowing mathematics and being wrong about the Earth's shape. Second, they censored a sentence in book 1, Chapter Five, in which Copernicus said that the question of whether Earth rests in the centre of the world had not been settled absolutely. Third, they censored lines in Chapter Eight discussing Earth's motion as truth. (Incidentally, those lines are followed by words from Virgil's *Aeneid* that illustrate relative motion: 'as we sail from the harbour, the lands and cities recede.') Next, an objectionable line in Chapter Eight said that it is 'absurd' to think that a container can move without moving its contents, plus, another sentence said that it is 'more probable' that Earth moves rather than not. Next, they censored the first sentence of Chapter Nine, which said 'nothing prohibits the Earth's mobility', and that it could well be regarded 'as one of the wandering stars'. They also censored assertions in Chapter Ten that said 'we are not ashamed to admit' that Earth moves with the wandering stars around the immobile Sun, that Earth's motion can be verified, and that it is the divine work of God. They censored the title of Chapter Eleven, which 'demonstrated' or proved Earth's triple motion (instead, this demonstration now had to be viewed as hypothetical). Finally, in book 4 they censored the title of Chapter Twenty, on 'The magnitude of these three stars: Sun, Moon and Earth', because *none* of these bodies should be considered stars.[17]

This list shows that what bothered the censors was not merely the Earth's motion and the Sun's immobility, but the claim that the Earth, Moon and the Sun are stars. Whereas they censored the Sun's immobility only in one passage, they censored the claim that the Earth is a star three times.

Unorthodox claims about the Earth's location and motion pertained directly to the question of whether Earth is a star.[18] This involved the question: are the stars other Earths? I point this out to show the continuity with Bruno's condemnation and heresies. He had affirmed: 'there are infinitely many individual worlds like our Earth. I regard it, like Pythagoras, as a star, and the Moon, the planets and the stars are similar to it.' In 1630, as we will see, the most important cardinal Inquisitor became annoyed at Galileo *precisely* for implying that Earth is a star.

Meanwhile, other controversies continued. In 1620 Kepler published an addition to his censored *Epitome*. He elaborated his eccentric religious views. Building upon his previous claim that

Earth has a soul, Kepler noted that he did not deny that planets have individual souls, yet he suggested that 'the soul of the world' flowing outwards from the Sun, with light and heat, penetrates all bodies with power to animate them.[19] The Sun's spinning motion was not caused merely by 'the Creator's omnipotence', but was maintained by its 'motor soul', the 'superintendent' of worldly motions.[20] He argued that sunspots are exhalations from the Sun's innermost bowels, and that sunlight too is evidence of the Sun's soul. He said the soulful Sun is the source of the motions of all things, and the source of life of all living things.[21]

That year, the perennial prisoner Campanella published *On the Sense of Things and Magic*. He envisioned a universal animation, a great concord by which even the living Earth could perceive things. Like Bruno and Kepler, Campanella connected the Earth's motion to the theory that Earth is alive. He explained:

> Since the heavens have sense, the Earth also feels. Thus Philolaus and Copernicus, who would give it movement, even more they give it sense. Pythagoras posited that Earth is truly a corpulent animal, whose fur and hairs are plants and trees, bones are stones, the animals on it are like the lice on our head, and he says that the heat in its centre is like the heart in ours. And what Ovid seems to think, but this is disputed elsewhere.[22]

Campanella criticized Aristotle, while praising Pythagoras, Trismegistus and others. Campanella cited Varro and Virgil's *Aeneid* to discuss how the great spirit nourishes animals, he argued, by heating air within their heart and spreading it within their organs.[23] He said that air is the common spirit, which interacts with our human spirits, also aerial.

Regarding those 'who posit the soul of the world', Campanella replied that each natural thing suffices for its own actions, because God's instruments can accomplish anything without any assistance from a world soul.[24] Remember that Bellarmine had voiced the same argument. Later in his book, however, Campanella described the world's divine intelligence, and said that if it is not a primal wisdom then it is 'the soul of the world', or the immortal mind of a most noble Being.[25]

Like others, Campanella mentioned Bruno's doctrines without embracing them and without mentioning him. Regarding the doctrine of 'infinite worlds', Campanella attributed it to Democritus and Epicurus. He said that God's power is not exhausted with this one globe, because God could well create infinitely many diverse worlds. He said that if this is true it could only be revealed by God. He then subsumed his speculations to higher authorities: 'However, the body of Jesus, we confess, is above all the heavens, where the Legislators place happiness, and the divine Being of Pythagoras.'[26]

Campanella discussed infinite worlds not as a consequence of God's infinite power, as Bruno did, but as a plausible intuition: 'Behold, when a man meditates, he extends his thoughts above the Sun, and then higher, and then wanders beyond the sky, and finally conceives the infinite outer worlds, as the Epicureans themselves realized, giving the truth.'[27]

Campanella also denied that there was a demonstrative argument to show that God sends each soul into a body to be 'captive in a prison', or as 'a portable tomb'. Still, Campanella said:

> But when I consider the spirit of the Sun implanted in the form of the body, and living in that prison, and enjoying it, just as I have seen very many rowers willingly live in a galley, and others for a long time in prison, they almost have forgotten the good knowledge of freedom.[28]

Campanella rejected the transmigration theory he attributed to Pythagoras, that a human soul can inhabit any animal body: 'I say that the human soul is not commensurable with all bodies, because fleas, lice, sea urchins, oysters, sponges are utterly different from us; it is not possible for our soul to vivify them.' Campanella denied the Pythagorean religion in general, but he allowed some exceptions:

> The pronouncements of Pythagoras contain a kind of religion, but no certainty. And in the Church of Christ and God's knowledge, this opinion is considered a heresy by the law of nature, only God with justice can make it happen sometimes, as Daemons have often entered into the bodies of humans and animals.[29]

Campanella said that transmigration was condemned by theologians and even in Origen, if indeed he had affirmed it. He also discussed what 'Origen skilfully wrote', that stars are the seats of evil spirits and demons.[30] Campanella's book was both praised and criticized.

Also in 1620 a treatise, *On the Nature of Demons*, mentioned and denounced Pythagorean ideas. The author noted that Pythagoreans, Persian Magi and Apollonius spread the fraud of the transmigration of souls. He criticized the shameless Pythagoras for teaching silence and the Tetractys, 'as the instinct of demons shakes off the knowledge of the true God from the chests of men, taking the effigy of Jupiter'.[31]

Likewise, a professor at the Ambrosian College in Milan, Antonio Rusca, published *On Hell and the State of Demons before the End of the World*. One chapter focused entirely on Pythagoras.[32] Rusca tried to settle questions about Hell, its inhabitants and torments. He cited Diogenes' story that Pythagoras descended to Hell and saw the souls of Hesiod and Homer being tortured. Rusca rejected Diogenes' and Iamblichus's claims as fictitious terrors. He discussed the Pythagorean belief in immortality, and commented that Pythagoras visited Hell because he had nothing in life but education. He denied that Pythagoras was Euphorbus during the Trojan War, and later Hermotimus. Rusca denied transmigration and various subterranean punishments as figments. He criticized Plotinus' 'heresy' that souls are reborn in human or animal bodies depending on how they lived. In other chapters, Rusca wrote mostly about demons, satanic spirits and Lucifer's sins. He also provided maps of Limbo, Purgatory and Hell in the depths of the Earth.

In 1621 Cardinal Bellarmine died, nearly 79 years old. He never wrote a public account of what transpired between him and Galileo, nor between him and Bruno. People knew that Bruno was imprisoned and executed for heresies, but the particulars were not widely known. In 1621 Gaspar Schoppe's eyewitness account of Bruno's death was published. Finally readers had a chance to learn some of the particulars. Schoppe itemized Bruno's 'horrific' teachings worthy of execution, starting with: 'Worlds are innumerable, souls go from body to body, and even migrate to another world, one soul can shape two bodies, magic is good and licit, the Holy Spirit is nothing other than the soul of the world,' and so on.[33] His account could alert ignorant readers that opinions about many worlds, the world soul and transmigration were heretical.

Some historians used to doubt whether Schoppe's claims echoed the Inquisition's accusations. Such doubts became unjustified once the 'Summary' of Bruno's trial became available. Moreover, Marta Fattori discovered a document from 1621 in which a censor of the Index approved and praised Schoppe's account as 'fruitful and useful'. Thus Fattori found that in 1621 the Index actually upheld the Inquisition's sentence of 1600.[34] Hilary Gatti agreed and said, 'It is clear that in the minds of the Roman Curia, Bruno's trial and execution and the budding Galileo affair were closely connected.'[35]

It was improper to write about heretics, so Catholic astronomers hardly mentioned Bruno. In Protestant lands, however, astronomers could. Denmark had been Lutheran since the 1530s, so in a discussion about celestial discoveries the Danish astronomer Christen Sørensen Longomontanus remarked: 'But first we respond to the Pythagoreans, whom others in our age have defended, and principally the Italian Bruno of Nola.'[36] Longomontanus named Bruno as the top advocate of the Pythagoreans.

Also in 1622, Giulio Lagalla denied the transmigration of souls. He argued that such 'Pythagorean fables' were false, absurd, impossible and condemned.[37] Following Aristotle, Lagalla insisted that the soul is really the form of the body. Lagalla denied that the human soul is like 'a sailor in a ship'. He *repeatedly* emphasized this point.[38] Bruno had defended this belief. It was one of the heretical propositions that Bellarmine had demanded that Bruno recant in 1599. But Bruno refused. Later Bellarmine rejected it again as a wrong way to speak about the union of Christ's divinity with His humanity.[39]

Also in 1622 one of Rusca's colleagues at the Ambrosian College critiqued Pythagoras. Franciscus Collius discussed whether ancient pagan philosophers received salvation or eternal damnation, and began by citing Philaster's book against heresies.[40] Collius dedicated eleven chapters to discussing Pythagoras. He cited claims about Pythagoras by many writers, including Porphyry, Iamblichus, Philostratus, Diogenes, 'Plutarch', Pliny, Plotinus and Ficino. He quoted many critics, including Justin Martyr, Tertullian, Jerome, Lactantius, John Chrysostom, Epiphanius, Theodoret, Pico della Mirandola, Ludovico Ricchieri and, especially, Augustine. Collius denied claims about Pythagoras as inconsistent and absurd.

Divination based on astrology was 'perpetually condemned' by the Church.[41] Successful predictions by Pythagoras had been caused by demons; claims about transmigration and that Pythagoras had lived in Hell came from demons too, and his alleged miracles and magic were caused by demons.[42]

I noted earlier the important fact that Pope Leo x denounced Martin Luther as 'a new Porphyry', blinded by the Devil. The implication is that Porphyry, advocate of Pythagoras, was one of the greatest enemies of Christianity. To confirm that the Pythagoreans were as despised as the Protestants we would expect to find that some writers had actually compared the main reformers to Pythagoras himself. And indeed such expressions exist. Nicolas Romaeus, a Jesuit, published a book denouncing the ideas of John Calvin. As Romaeus criticized the Calvinists, he twice referred to their misleading leader as 'their Pythagoras'.[43] Romaeus also criticized Luther's disciples, including Melanchthon and Osiander, and other reformers, such as Zwingli. He complained that the 'filth of the Heretics that now floods the Christian world' followed the shameless claims of Luther, 'their Pythagoras'.[44] Thus Romaeus ridiculed Luther as 'the devil's close friend' and 'the architect of all the heresies of our time'.

Also in 1622 the Index banned a book, *On Pythagorean Symbols*.[45] The record states that it was banned 'for religious reasons'. The author was a German Franciscan monk concerned with the 'calamities and transformations' of governments. He discussed the significance of 'signs' such as comets, eclipses, earthquakes and floods. Since our present interest is not political astrology, but links among Pythagorean works, I should point out that this book briefly referred to 'Alexander the Great, King of Macedonia, who like the Philosopher Anaxagoras, said that there are many worlds'.[46] The book also discussed 'the Earth's motion', but only in respect to earthquakes, which the author called 'premonitions of fate and future evil'.[47] In December 1622 the Decree of the Index prohibiting *On Pythagorean Symbols* was signed by one Cardinal Maffeo Barberini.[48] The next year he became the new Pope. It was also signed by Capiferreus, who had co-signed the two Decrees against the 'Pythagorean doctrine' of Copernicus, Foscarini, Zúñiga and Kepler.

Campanella Defends Galileo from Prison

From his youth Campanella wrote poems. Most are undated, but they convey the feelings he experienced for years. For example, these lines convey the age of darkness he endured:

It befits our age to wear black robes.
  Originally white, afterward various, today Moorish,
  nocturnal, river, infernal, traitorous,
  with ignorance and horrid fright and sick.
We shun with shame all cheerful colours,
  that mourn our end and living under tyranny,
  the chains, knotted ropes, the metal and the snare
  of sombre heroes, and of the suffering soul intact.[49]

Copying Emperor Charles v, who dressed entirely in black, many aristocrats and clergymen had dressed in black since the 1530s.

Even though Campanella remained imprisoned in Naples in 1622, he still managed, finally, to publish in Frankfurt his long-delayed *Defence of Galileo*. Campanella's goal was to answer 'whether the philosophical view advocated by Galileo is in agreement with, or opposed to, the Sacred Scriptures'.[50] Campanella began by summarizing arguments against Galileo. Among them, Campanella echoed claims that some of Galileo's opinions apparently contradict the Church Fathers by teaching that the Earth moves and is not the centre of the universe, and that the Sun and stars do not move.[51]

He cited multiple ancient and religious authorities to analyse whether the theory of Earth's motion was heretical. He mentioned Cicero, Ovid, Pliny, Diogenes, Eusebius, Lactantius, John Chrysostom, Augustine, Copernicus, Bruno and Kepler. For example, he remarked that according to John Chrysostom 'it is heretical and contrary to Scriptures and the Church to affirm many heavens and orbs.'[52]

Campanella made a vague reference to Giordano Bruno, without saying his name. He stated the claim that multiple previously unknown Suns and planets exist: 'This opinion was defended by a person from Nola and by others who, being heretics, we cannot mention by name.' This remark confirms that heretics other than Bruno believed 'many suns' exist. But Campanella said they had not been condemned specifically for such opinions. He gave multiple examples of prominent men who held some such views, including Cardinal Cusa, Johannes Kepler and William Gilbert.[53]

Campanella argued that the question of whether Earth is the centre of the universe is not a Catholic dogma, but it seemed closely related to questions about the location of Hell: whether Hell is at the centre of the universe, and whether Hell is inside the Earth. Campanella quoted the apostle Paul, 'Christ descended to the lower regions of the Earth' (Ephesians 4:9), and also Psalm 15:10, which speaks of Christ descending to the lower regions: 'You will not abandon my soul in Hell.' Campanella inferred that therefore Hell probably is located in the lower regions of Earth, 'unless we assume that there are other Earths'.[54] Apparently Hell was within the Earth, but the location of Earth itself remained unclear. Campanella concluded: 'Therefore we do not know whether the Earth is in the centre of the world', which he said helps Galileo's case.

It is unclear how much Campanella, in prison, knew about the proceedings against Galileo. Yet his *Defence* includes the interesting claim that Catholic censors specifically opposed the implication, in Galileo's writings, that other worlds exist. Campanella explained:

> If someone were to hold that the infernal darkness, which Christ called 'exterior', is located outside this world, as Origen conjectured in his commentary on Matthew and as Chrysostom thought in his commentary on the Letter to the Romans, then there are other worlds outside of our world. The censors condemned this idea in Galileo because they had not carefully examined the Scriptures and the books of the holy Fathers.[55]

This is a remarkable claim! Recall how promptly, after Bellarmine admonished him, Galileo wrote to Cardinal Muti's nephew to clarify his position on the existence of other worlds.

Campanella exaggerated words by John Chrysostom – who once wrote that *we can only know* that Earth is cold, dry and dark – to mean specifically that we don't know whether it moves or is at the centre of all. Citing other Catholic authorities who did not specify that Earth is central, stationary, Campanella interpreted them to mean that perhaps it wasn't. Furthermore, he argued that if Hell is really at Earth's centre, as Catholics believed, then it necessarily follows that Earth moves. He said fire is *always* in motion.[56] So he argued:

> If indeed it is true that Hell is located in the centre of the our Earth where the damned are tormented by fire, as Gregory[57] and others seem to think, then it must be that the Earth moves. For as Aristotle reports, Pythagoras, who located the place of punishment in the centre of the Earth and who taught that fire is the cause of motion, concluded that the Earth moves and is animated, as was also believed by Ovid in Book xv of his *Metamorphoses*, by Origen in his commentary on Ezekiel, by Alexander of Aphrodisias, and by Plato . . . Hence Galileo's view is not inconsistent with St Gregory but with Aristotelianism.[58]

Like Thomas Aquinas, Campanella construed Aristotle's claim about the Pythagoreans and the centre of the world as being about Pythagoras himself.

Campanella painstakingly interpreted passages in scriptures to show that they are not incompatible with the notion that the Earth moves, for example, 'A generation passes, and a generation comes, but the Earth stands forever' (Ecclesiastes 1:4). Campanella argued that this passage does not mean that Earth stands unmoving, but instead, that it *remains* in a constant state of motion. He attributed this view to recent astronomers and to 'Pythagoras and all his followers, Heraclitus, Aristarchus, Philolaus and so on, whose opinion St Thomas did not condemn as heretical'.[59] Campanella said that therefore it was permissible for Galileo too to interpret it in this way. Likewise, the Joshua miracle, when interpreted as the Earth stopping, was not a denial but merely an explanation.

Campanella argued that the 'firmament' in Genesis does not move, and is the same as the starry heavens. Therefore, 'it follows that the Earth moves in a circle, like a ship, and the stars appear to move in a circle, like an island or a tower on the shore.'[60] He reviewed different theories about this firmament, the heavens. Campanella agreed with 'the view of Empedocles and other Pythagoreans' that it is composed of the same four elements as Earth.[61]

Campanella explained that years earlier he had not appreciated the theories of the Pythagoreans. Thinking the heavens were made of fire, he had 'tried to refute all the arguments of Copernicus and Pythagoras'. But following the discoveries by Brahe and Galileo, Campanella increasingly doubted that stars and planets are different from Earthly things. Still, he would 'suspend my own judgment on these matters', to willingly 'obey the commands of the Church and the judgment of others who know more than I do'.[62]

He concluded that 'the theory of Copernicus and Galileo' was probable, and that nothing obviously harmful seemed to ensue.[63] Nevertheless, he acknowledged: 'But it is up to the Church to judge whether Galileo should be permitted to write about and debate these matters.'

Another argument against Galileo's theory, said Campanella, was that it implied the plurality of worlds, which seemingly revived an old heresy:

> Further, from Galileo's opinion it follows that there are many worlds and earths and seas, as Mohammed said, and that there are human beings living there, if the four elements of our world also exist in the stars. For if every star is composed of all four elements, then clearly each will itself also be a world. But this seems contrary to the scriptures, which speak only of one world and one human race. I will omit the point that to say that Christ also died on other stars to save those inhabitants is to revive the heresy, which some have maintained, that at one time Christ was crucified a second time in the other hemisphere of the Earth to save the humans living there as he has saved our part of the world. It would also be necessary to agree with the heretic Paracelsus that there are other humans who live in the air and in the waters and under the Earth, who enjoy beatitude, even though it is doubtful whether they are included in the redemption. The Jesuit Martin Delrio has written against this in his *Disquisitiones magicae*.[64]

If rational beings inhabit distant worlds, then how would they receive Christ's redemption? Campanella noted that some writers had denied that people might live in distant places on Earth because they would need to know about Christ, and therefore 'Christ would have had to be crucified twice, once here and once there.'[65]

This was a common Christian reason to deny that many worlds exist. Another book published in 1623 argued that 'only one world was created, not many worlds, because the Sacred history is silent about the plurality of worlds', and also because 'In this way we assert that one divine person was incarnated, not many, because the Evangelical narration does not affirm that many were incarnated.'[66] So, one world, one Christ.

Yet Campanella believed that Galileo really discovered new worlds! He said that just as Galileo had revealed mountains on the Moon, the Bible agrees with him because Genesis and Deuteronomy 'speak of fruits and mountains and hills in the heavenly bodies'.[67] Campanella didn't quote these passages. What they say is:

> The God of your father will be your helper, and the Almighty will bless you with the blessings of heaven above, with the blessings of the abyss that lies beneath, with the blessings of the breasts and of the womb. (Genesis 49:25)
>
> Likewise, to Joseph he said: 'His land shall be from the blessing of the Lord, from the fruits of heaven, and the dew, and the abyss which lies below, from the fruits brought forth by the Sun and the Moon, from the top of the ancient mountains, from the fruits of the eternal hills, and from the fruits of the Earth and from its plenitude. (Deuteronomy 33:13–16)

The first passage doesn't say what Campanella claimed. The second says 'fruits from heaven', which Campanella interpreted literally, but it does not specify whether the mountains and hills are in the heavenly bodies.

Campanella mistakenly said that nowhere in the Church canons does any decree deny that many worlds exist.[68] His mistake illustrates the ignorance of those untrained in the prosecution of heresies. Still, he knew that Thomas Aquinas criticized Democritus for such beliefs, so Campanella argued that the critique referred not to the idea of many worlds in itself, but only to the belief that many worlds might exist *without any ordering*, as if by chance.[69] As in *Sense of Things*, he again argued that God certainly can make many worlds, and we cannot limit His powers.

Campanella claimed that living beings might exist in the stars, even if they are somewhat different from us. He said that Galileo at least denied that there are *humans* in the stars, and that Kepler's published comments about beings on the Moon were just playful, humorous.[70] But Campanella added that 'even if the theory of the plurality of worlds is false', then it would not undermine Galileo's discovery of 'a plurality of systems' that are ordered uniformly. He then named Bruno as having argued the same thing, the only instance in which he dared to name Bruno in this book.

While defending Galileo, Campanella attacked Aristotle's astronomy as 'completely false', incompatible with Galileo's telescopic discoveries. He insulted Aristotle's fans: 'potbellied theologians' who were 'fools', 'insane and most ignorant'. He denied their faith: 'They are not true Christians who would chain us to Aristotle or Ptolemy.'[71] Campanella claimed that it is 'heretical to say that theology is based on Aristotelianism', or that theology essentially needs such teachings. He complained that trusting Aristotle above Christ was 'wickedness and foolish blindness'.[72]

Campanella said that Copernicus rightly corrected old errors in astronomy, and 'returned to the teachings of the ancient Pythagoreans, which provide a better account of the appearances'. Campanella claimed that saints Leo, Anthony, Bernard, John Chrysostom and others had rightly said that 'the world is the book of God', which should be studied diligently.[73] And Galileo truly knew 'something new about the world, the book of God'.[74] Campanella proclaimed: 'Scripture, which is the book of God, does not contradict the other book of God, that is, nature.'[75]

Regarding Copernicus: 'I think that this philosophical theory should not be condemned.' Campanella said that if Catholics attacked physicists and astronomers then the Protestants would more vigorously ridicule Catholics, for assaulting both scriptures and nature.[76] And he said that Cardinal Bellarmine was aware of this; but Bellarmine had died in 1621.

Campanella praised Galileo, saying, '[you] rise up to the celestial houses; you bring the distant stars closer to our eyes, and you subject the heavens to your genius'. He said 'falsity cannot be found in Galileo' because he proceeded carefully from observations. Campanella tried to prove that scriptures 'agree more with Galileo than with his adversaries'.[77] He characterized 'the opinion of Galileo' as four claims: Earth moves, the Sun does not; the Sun is at the centre of the universe; stars are composed of the four elements, including water.[78] He said these views were not original to Galileo, but were very ancient. Campanella credited Pythagoras.

Ancient stories said that Pythagoras visited Egypt; and Moses had lived in Egypt. Therefore Campanella speculated that Pythagoras probably based his philosophy on the teachings of Moses. Thus Campanella conjectured that the theory of Copernicus or Galileo originated from Moses himself, and that 'Pythagoras, who was of Jewish stock although he was born in a Greek city', brought these ideas 'to Greece and to Italy, and taught it at Croton in Calabria'.[79] He further explained:

> It is likely that Pythagoras was born on Samos to a Jewish family . . . According to Diogenes Laertius, Plutarch, Aristotle and Galen, Pythagoras was the first to promulgate to the gentiles this marvellous philosophy that the Earth moves, that there are many systems in the heavens, that the Sun is at the centre, that the Moon is another Earth, and that all four elements, and not just water, exist in the stars. Therefore it seems that Pythagoras derived these teachings from Moses, for he could not have had such wisdom without a previous revelation.[80]

Campanella added that 'Pliny rightly says that Pythagoras was the wisest of philosophers.'[81] Allegedly Pythagoras received his knowledge from God.

Campanella did not mention some opinions he had previously attributed to Pythagoras: that the world has a soul, and that Earth is a living animal. Bruno had been judged for these doctrines, and others.

Thus Campanella tried to rehabilitate Pythagorean ideas by linking them to Moses. He said that the truth had been buried, but 'Our ancestors, however, are not to be blamed, for they had not yet discovered the new earths, and celestial worlds, and new phenomena, and the agreement of Scriptures with philosophy.'[82] Campanella promoted the Pythagorean views as a key to understanding the Bible.

Ending his book, Campanella advised that Galileo's writings should not be prohibited. He said that censuring Galileo would undermine the credibility of Catholics. Finally, Campanella submitted himself to the censure of the Church.

Despite his attempt to legitimize Galileo and the Pythagorean theory, publication of Campanella's *Defence of Galileo* was prohibited in 1623 and he was forced to disavow the edition published in Frankfurt.

Meanwhile, in *City of the Sun*, Campanella described his ideal city as follows:

> The inhabitants of the City of the Sun do not fear death, because they all believe that the soul is immortal, and that when it has left the body it is associated with other spirits, wicked or good, according to the merits of this present life. Although they are partly followers of Brahma and Pythagoras, they do not believe in the transmigration of souls, except in some cases by a distinct decree of God . . . They have but one book, which they call Wisdom, and in it all the sciences are written with conciseness and marvellous fluency of expression. This they read to the people after the custom of the Pythagoreans.

Did Campanella actually believe in transmigration? He had made the same claim in his *Sense of Things*: that souls are not reborn, except in special cases decreed by God. He there said it was heretical to believe in transmigration. Here again, his views resemble those of Bruno, but were understated.

Campanella envisioned an alternative to the Catholic Roman government. The City of the Sun would have a central temple of the Sun, with a model of the universe, a globe on an altar and stars painted on the ceiling. Seven ceiling lamps would correspond to seven planets. Around this temple, there would be circular walls, dividing the city into seven districts where people would live, corresponding to the seven planets. The inhabitants of the city, called Solarians, would worship the Sun and share knowledge from astrology and talismans. Scientific arts, including the telescope, would be superior to the best achievements of Greeks, Jews and Romans. Campanella wrote: 'I learn more from the anatomy of an ant than from all the books that have been written since the beginning of time.'

Galileo Defends the Pythagorean Doctrines Again

In 1623 Galileo published *The Assayer*. He argued that philosophy is written in a grand book, the universe, which cannot be understood at all without mathematics.[83] Galileo seemed to say that mathematicians such as himself or Kepler had a special skill for understanding nature, the work of God. This kind of claim had been criticized by Hippolytus, St Augustine and others.

That same year, a friend of Galileo, Cardinal Maffeo Barberini, became Pope Urban VIII. Hence Galileo was kindly invited to papal audiences on six occasions. He also gained confidence that the Copernican account might win favour. Thus far, the Index had temporarily suspended the books by Copernicus and Zúñiga, while it had prohibited books by Bruno, Foscarini, Kepler and Campanella. In 1624, and thereafter, the Index included a prohibition of all works teaching Earth's motion and the Sun's immobility.[84]

Sometime before 7 June 1624 Cardinal Zollern spoke with the new Pope about the controversial theory of Copernicus. Zollern told Galileo what he said to the Pope: that 'the heretics all have the opinion' of Copernicus (presumably Zollern meant Protestants such as Kepler), but 'one should be very circumspect in reaching any determination.'[85] Cardinal Zollern quoted the Pope's reply: 'that the Holy Church had not condemned it [the opinion of Copernicus] nor would it be condemned as heretical, but only as temerarious, but that there was no need to fear that anyone would ever prove it as necessarily true'.[86]

Importantly, the Pope's remark clearly shows that he did not view Copernicus's theory as terrible, nor even as particularly dangerous. Years later, a theologian for the Pope quoted the Pope's comments to 'a most learned man', whom historians construe to be Galileo. The Pope argued that even if Earth's motion could account for all astronomical observations, God still had the power to move the orbs and stars in any way while generating the same effects: 'we should not restrain the divine power and knowledge in this way. Hearing this, the most learned man became quiet.'[87] No matter what argument purported the Earth's motion, it could never be proven true, because God could have made the world in any way He wanted. Human reason lacked the certainty to impose necessity on God – to constrain His omnipotence. Besides, there were worse doctrines around.

In 1624 Marin Mersenne, a theologian in Paris, published a book denouncing deists, atheists and freethinkers, including Giordano Bruno. Father Mersenne said that Bruno was a 'miscreant', an impious 'brigand' who pretended his deceptions were miracles.[88] Mersenne said that Bruno could have been forgiven if he had restricted himself to discussing only questions about geometry, atoms and divisibility. But Bruno had also discussed 'the infinity of starry worlds, and other things', and since he had 'attacked the Christian truth, it is reasonable to decry him as one of the most evil men that the Earth has ever produced'. Therefore, Mersenne scorned Bruno as 'an Atheist, who has been burned in Italy for his impieties'.[89]

To reassure readers that 'you do not think that I speak without knowing', Mersenne suggested that they read the last chapter of Bruno's book on *The Minimum*, where 'you will easily realize that he [Bruno] favoured the transmigration of souls from one body to another, and that it seems that he did not invent a new way of philosophizing, but just to secretly combat the Christian Religion.' Mersenne argued that if one tried to decipher what Bruno really meant by writing about the sphere and the centre, about birth, life and death, one would realize that Bruno's intention was not to talk about 'a reasonable immortality of the soul, but about that [transmigration] which he attributed to the soul of animals, and plants, and to all individuals, who are down here'.[90]

In the words of historian Miguel Granada, Mersenne sensed that Bruno's beliefs about the universal soul and transmigration were 'extremely dangerous' and 'radically incompatible' with Christianity.[91] Mersenne cited one reason why Bruno was killed. Regarding Bruno's beliefs about human souls and the world soul, Mersenne reported that Bruno's 'evil book' *On the Cause, the Principle, and the One* (1584) includes 'the dialogues for which he was burned in Rome, as some individuals have informed me'.[92] Recall that in this book Bruno agreed with the Pythagoreans and Virgil for arguing that human souls originate from the world soul, while the universal 'Mind agitates the mass and intermixes itself with the great body'.

Mersenne further argued that the wicked Bruno wrongly tried to prove that God has no freedom at all, 'by convincing himself of infinite worlds' – as if God had no freedom to create otherwise.

Mersenne argued that God could create the world in any way. (This was essentially the same argument the Pope conveyed to Galileo and Cardinal Zollern that Earth *does not necessarily* orbit the Sun.) Mersenne also complained that Bruno's impiety led him to question the true religion and to disrespect the miracles of Christ. He said that Bruno 'thinks that one can elevate oneself in the air, like St Paul, who was raised up to the third Heaven'. Mersenne was referring to 'I know a man in Christ who fourteen years ago was caught up to the third heaven. Whether it was in the body or out of the body I do not know – God knows' (2 Corinthians 12:2).

Father Mersenne discussed other impertinent 'liars', including Campanella. Mersenne insisted that Bruno's 'many Earths' do not exist.[93] Did Mersenne know that belief in many worlds was a heresy? In a previous book, published in 1623, Mersenne made a remark that answers this question. He included an article 'Proving that the World is Unique'. Having acknowledged that St Jerome criticized Origen's belief in many worlds, Mersenne yet claimed that scriptures did not specify its falsehood. He argued that the world is one and he wrote, 'anyone who claims the contrary is not a heretic, but is at most temerarious.' Mersenne claimed that 'no Ecumenical Council' had defined that it was a matter of faith to not believe in many worlds.[94] Like Campanella, Mersenne just didn't know that belief in many worlds was officially a heresy. Still, Mersenne recognized that the existence of many worlds led to absurd issues, such as whether Christ had embodied the shape of the alien beings in such worlds and died for them too.[95]

In 1625 Mersenne published another book in which he discussed Bruno, criticizing again his belief that the world is infinite. He also criticized another 'error' of Bruno, that the universe is eternal. Mersenne commented, 'one well knows that when one embraces a principle that is false, afterward there follow many absurdities and impertinences.'[96]

Editions of John Donne's satire against the Jesuits, *Ignatius His Conclave*, were published anonymously in Latin and English in 1611. By 1621 Donne had become a prominent Anglican priest and Dean of St Paul's Cathedral in London.[97] Both versions were reprinted, in 1625 and 1626, respectively, and again people could read about how Galileo's discoveries seemed to support extravagant anti-Christian notions: the Moon is an inhabited world; the Sun, stars and planets are inhabited worlds; and the moving Earth has a soul. These notions had been defended by Bruno. Likewise, in a sermon to the nobility in the Earl of Exeter's chapel in St John's, Clerkenwell, Donne preached: 'Did God satisfy Himself with this visible and discernible world; with all on Earth, and all between that, and Him? . . . Let every star in the firmament, be (so some take them to be) a several world, was all this enough?'[98] Donne returned to the question of many or 'infinite worlds' in several other sermons.[99]

Meanwhile, some important writers used the memory of Bruno's death to bicker about religion. In 1625 the Calvinist politician Ludwig Camerarius wrote:

> When the Holy Office of the Inquisition sentences Lutherans to the death, how simply they declare that the punishment be done without a profusion of blood. Hence burn them alive, and without any shedding of blood. Oh what leniency worthy of Catholic men! As Schoppe tells us, in his Letter about the execution of Giordano Bruno of Nola, who was burned publicly on 9 February of the year 1600 at the Campo dei Fiori in Rome, in front of the Theatre of Pompey. The Lutherans should give many thanks to the Inquisitors, that they use their blood to feed the flames, instead of extracting it with iron.[100]

As a Protestant, Camerarius complained that Schoppe was a 'truly wicked and cruel' Catholic who believed that 'all Protestants or Calvinists are Heretics, proscribed by the empire to the ultimate punishment.' Camerarius was the chief counsellor of Elector Frederick v of the Palatinate (and briefly King of Bohemia), a Calvinist who plotted to recapture lands and titles that the Catholic emperor Ferdinand II had taken from him.

Camerarius's use of Bruno's execution to criticize Catholics elicited a reply. The following year a 'Lucius Verus' published a book replying to claims by Camerarius. In response to the point about Bruno, the pseudonymous author wrote:

> for those who are interested, there are documents; & the records show, those which are available: Regarding the execution of Bruno of Nola it states: He was not a Lutheran,

> instead he was infected partly with Calvinism, even by Manichaeus, Borborites, Arian and Eutychian heresies, and he was convicted of those crimes, and therefore nobody except Mystes [Camerarius] can be an advocate of that wickedness. He said many blasphemies against God, but this happens to the impious, as the Roman proverb says:
>
> *Chi contro à dio gitta pietra, in capo li ritorna.*
>
> That whoever throws a stone against God, it falls back on his head. Clearly one cannot defeat the Sun.[101]

Meanwhile, after 27 years of imprisonment, the unfortunate Campanella was finally released from the Neapolitan prison in 1626, thanks to Pope Urban VIII. Campanella was transferred to Rome, to be held by the Roman Inquisition, but with accommodations. The following year Campanella began to write long commentaries on poems written by the Pope. Campanella praised the discoveries of Columbus and Galileo as ranking equally with scriptures as sources of knowledge. He used the Pope's words to ramble freely: 'Pythagoras, Trismegistus and Plato have argued regarding the immortality of the soul, given the disagreement between souls and bodies, that the soul is not in its own home, but almost in a prison.'[102] Bruno had made similar claims.

By 1629 the Roman Inquisition liberated Campanella. By flattering the Pope and fitting him into expansive, encyclopaedic views, Campanella gained the Pope's favour: incredibly, he became an advisor on astrology for Urban. Campanella also tried to convince the Pope to include magic in papal policy. At the time, incidentally, various books on magic discussed the powers of Pythagoras.[103] However, the strangest magic trick, if we may call it that, was how Campanella the heretic managed to go from bloody torture in dungeons to being directly in the service of the new Pope. Campanella's long sufferings and bold, erudite writings had turned him into a European celebrity. At the time the papal court included an unusual mixture of men, with various nationalities and social backgrounds. One observer remarked, 'as shown by our daily experience, at the Roman court there is nobody of so low a condition that he may not climb to a great position at some point.'[104]

Galileo had Urban's friendship. Yet some of his supporters began to vanish. In August 1630 Prince Cesi, the leader of the Lincei and his main protector, died. The Accademia dei Lincei fell into disarray and Galileo could not print his next book with them. Meanwhile, Kepler, though ill, planned to finally publish his expanded 'dream' about the Moon, but he too died in late 1630. It was published later (in 1634), with a dedication in which Kepler's son recalled that his father fell into a fatal sleep, but hopefully 'His soul flew above the lunar into the ethereal region.'[105]

Old Galileo prepared to publish about the Pythagorean doctrine on which he had kept silent for years. As is well known, some Inquisitors soon became annoyed by Galileo's book. To ascertain whether their concerns were connected with heresies from Bruno's trial, we must inquire whether, by 1630, Inquisitors or censors of the Vatican were aware of the pagan connotations of Pythagorean doctrines. There is a significant work that sheds light on this question.

The Pope's nephew, Cardinal Francesco Barberini, was a top Inquisitor. He had employed a classicist, a German Catholic convert named Lucas Holste, to live in his household. Barberini owned an important private library in Rome, where Holste worked with its collections of Greek and Latin texts. In 1630 Holste finished an extensive, scholarly compilation of the works of Porphyry, in Greek and Latin, with commentaries. It began with Porphyry's *Life of Pythagoras*, followed by another brief biography of Pythagoras, some other works by Porphyry and, finally, Holste's commentaries and dissertation.

Christians had destroyed some of Porphyry's works, yet Holste collected extant treatises for careful examination. In addition to analysing Porphyry's life and beliefs, Holste discussed how certain 'heresies originated from philosophy'.[106] This had been proclaimed by Tertullian, Lactantius, St Jerome and Cardinal Bellarmine.

As a young man Holste had been fascinated by Platonist philosophy, so he laboured to promote it. It drew him closer to Catholicism: 'I turned completely to those Latin and Greek texts that deal with this contemplative and mystic theology, which excites the soul of God.'[107] Therefore, he better understood the ancient Church Fathers. He studied works such as Proclus' notes on Plato's theology, and Iamblichus' 'On the Pythagorean Way of Life', despite its 'strong occult tendencies'.

Holste showed extensive familiarity with many ancient works, including manuscripts in the Vatican library, which discussed Pythagoras, Porphyry, Apollonius and other Pythagoreans

– including works by Philostratus, Proclus, Plutarch, the *Placita*, Alexander Polyhistor, Aristotle, Herodotus, Cicero, Pliny, Diogenes, Macrobius and Hierocles. He cited works by Church Fathers who had criticized Pythagorean beliefs, including Tertullian, Lactantius, Epiphanius, Eusebius, Jerome and Augustine.

Holste summarized the Pythagorean theory of transmigration: the human soul is captive inside the body as if chained in a prison, subsequently it resides in other bodies, and afterwards is purged to kindred stars, returning to a celestial choir of souls. The soul is a particle of God, and torn from God, it transmigrates into five bodies, then to the Moon, the Sun and later to the beatific regions.[108] Holste said, 'these deliriums were abundantly refuted by Epiphanius.'[109] He also cited Augustine's critique of Porphyry's account of transmigration, contrasting it to the Christian doctrine of bodily resurrection. He also discussed the Pythagorean abstinence from eating flesh, partly in regard to Porphyry's tract on it, while denying that human souls are reborn as animals.

Furthermore, Holste mentioned the fake miracles of Pythagoras, 'which if true, no one can easily excuse him for the crime of magic'. He alluded to 'the incantations, and other magical tricks' of the sect of Pythagoras, as well as their 'symbolic and mystical method of teaching'. He noted that Tertullian denounced philosophers as 'the patriarchs of heretics'.[110]

Nonetheless, Holste defended Porphyry's abilities as a philosopher. Likewise, St Augustine had described Porphyry as 'the most learned of the philosophers, though the bitterest enemy of the Christians'. Holste explained that Eusebius and Augustine accused Porphyry of superstitious crimes. Echoing Augustine, Holste criticized Porphyry's willingness to hear pagan oracles and demons, which Porphyry misconstrued as gods. Eusebius had 'refuted the blasphemies of Porphyry'.[111] Holste also wrote two chapters about Porphyry's censored *Against the Christians*.[112]

Holste dedicated his book to his patron, 'the illustrious and most reverend' Cardinal Barberini, and he noted having used the Cardinal's library.[113] Holste submitted his manuscript to the Master of the Sacred Palace, Father Niccolò Riccardi, for approval for publication. Riccardi gave the work his imprimatur. It was published in 1630, by the Vatican's press. On its title page, the book features the emblem of the Barberini family: three bees.[114]

Bust of Cardinal Francesco Barberini, 1682, under his family's emblem with the three bees, located at St Peter's Basilica.

Meanwhile, Galileo finished his *Dialogue on the Two Chief Systems of the World: Ptolemaic and Copernican*, which argued persuasively in favour of the Copernican or 'Pythagorean' theory. Cardinal Barberini's opinion on Holste's works on Porphyry and Pythagoras is not known, yet we know something about his opinion on the moving Earth. In early February 1630 Galileo's friend Benedetto Castelli met with Barberini and others and they discussed Galileo's forthcoming work. Castelli argued that Galileo did not require that Earth truly moves, but that Galileo just argued that *if* it moves then the tides would necessarily follow. Castelli recalled that the Cardinal 'showed himself to be very averse' to the matter. This was immensely important because Cardinal Barberini was the head of the Inquisition. He then spoke with Castelli privately and complained: 'if the Earth really has motion, it seems necessary that it be a star, a thing that then seems too contrary to the theological truth.'[115]

This striking sentence explicitly denies a controversial claim about the Earth that Giordano Bruno had defended to the Venetian and Roman Inquisitors: 'I regard it, like Pythagoras, as a star.'[116] It was one of the claims censored repeatedly, in 1620, in Copernicus's book: the Earth is a star.

Castelli informed Galileo about the conversation and warned him to be careful to prove that the Earth is *not* a star, by arguing that the Moon is the Moon, Mars is Mars, and so on. Castelli also spoke with the censor at the Vatican, who had expressed positive opinions on Galileo and his previous works. The censor, the same Father Riccardi, now had to review Galileo's manuscript, require changes or approve it for publication.

Master Riccardi was known as 'the Monster'. He began work but did not finish it. Galileo hoped to get the book approved and printed in Rome, but the process was interrupted by the plague. Impatient, Galileo asked if he could obtain permission from the censors in Florence to print it there. Riccardi had reviewed and edited parts of the manuscript. He expected Galileo to return to Rome to work on it, but he agreed to let the rest be reviewed in Florence instead. Riccardi wrote that the book would not be problematic if the author followed instructions faithfully. Galileo, however, somehow obtained permission to print in Florence, without any censor there actually reviewing the text.[117]

Galileo sent the book to press in mid-1631, before receiving Father Monster's corrections and draft Preface. Riccardi's objections arrived – 'absolute truth should never be conceded to this opinion, but only the hypothetical, and without Scripture' – and he specified a requirement for publication: 'the author must add reasons from divine omnipotence dictated to him by His Holiness, which must quiet the intellect, even if it were impossible to get away from the Pythagorean doctrine.'[118] The Pope had personally instructed Galileo to say that even if evidence seems to necessitate the Earth's motion to explain some phenomenon, like the tides, there was really no necessity because omnipotent God could well proceed otherwise. Therefore Galileo added the Pope's opinion that one cannot impose necessity on God, because God could create the world in any way He pleased. Strangely, Galileo put these words into the mouth of the one moronic character in his dialogue, Simplicio.

While Galileo was trying to publish, two books were published in late 1631 arguing the opposite. One was titled *Anti-Aristarchus, or Earth-orb Immobile. In which the Decree Issued by the Holy*

Congregation of Cardinals in 1616 against the Pythagorean-Copernicans is Defended. Its author was Libert Froidmont, a professor of theology at Leuven. He began by attributing to Pythagoras and his followers the theory of Earth's mobility and the Sun's immobility and centrality.[119]

Froidmont complained that 'the Pythagoreans . . . also known as the Copernicans', were a 'particularly alien sect that has invaded the Catholic faith'. He named some of them: Kepler, Michael Maestlin, William Gilbert, Foscarini and Galileo.

He rejected heliocentrism and other ideas. He dismissed Kepler's belief that Earth is a living animal with a soul. He quoted the old denunciation by Pope Zacharias against the notion that aside from our world there exists another Sun, another Moon and another Earth, also inhabited.[120] He said that Pope Zacharias rightly condemned such notions and almost condemned Bishop Virgilius for advocating them.

Froidmont cited a century-old account by Johannes Aventinus, a Bavarian historian and philologist. Aventinus had briefly mentioned the dispute between Zacharias and Virgilius. Aventinus commented that 'in our age that [topic] is investigated not with arguments, but is known by experience, that men surround the Earth on all sides, and conversely stand on their feet, what the Greeks called the Antipodes. This is now acceptable, what Virgilius called other worlds, other men beneath the Earth, finally another Sun, and another moon are claimed.'[121] Froidmont noted that the works of Aventinus were on the Index of censured books. Froidmont concluded that beliefs in another Sun, another Moon and another inhabited Earth were 'heretical, or are nearly so'.[122]

Yet Froidmont hesitated to condemn the heliocentric theory as heretical, pending the overt judgement of Pope Urban VIII. Until then, Froidmont remarked, 'I dare not yet condemn Copernicus of open heresy.' But Froidmont wrote, 'The Copernican opinion is nevertheless temerarious, at the very minimum, and at least one foot has entered the threshold of heresy, unless the Holy See sees otherwise.'[123] Still, Froidmont quoted the opinion of Justus Lipsius, who in 1604 called it 'delirium', a 'love of paradoxes', 'which arouses heresy'. Lipsius had commented: 'Nowadays there are some who view the Copernicans severely and openly call them heretics.'[124]

If Galileo's readers in 1632 were unaware of the condemnation of heliocentrism in 1616, Froidmont's book would undo that. The other book putting the opposite case was *The Famous and Ancient Problem of the Earth's Motion*, published in Paris in December 1631. The author was Jean-Baptiste Morin, a physician and professor of mathematics.

Morin somehow knew of Galileo's manuscript and had seen Froidmont's brand new book. The title page of Morin's book includes a telling quotation: 'The Earth rests forever; the Sun rises and sets' (Ecclesiastes 1:4–5). Morin dedicated his book to Cardinal Richelieu, alerting him that it was 'temerarius' to assert the seemingly impossible claim that Earth moves. It was 'repugnant' to the scriptures and Catholicism. As usual, Morin attributed the 'nefarious dogma' to Pythagoras and Aristarchus, saying that Ptolemy and Aristotle refuted it.[125]

Morin commented, 'the same thing has happened with this dogma as with heresy in Religion', the truth had been divided and subdivided by imaginary considerations. He said that the Holy Spirit had foreseen the vain ingenuity with which some writers would argue about the Earth. Morin quoted the Bible to prove that Earth does not move, for example, that God 'established the Earth on its foundations [*stabilitatem*], it shall not be moved forever and ever' (Psalm 103:5).[126]

Morin complained that the opinion of Earth's motion was a fiction invented 'against the pristine faith', scriptures, and that its proponents were 'temerarious and completely obstinate'.[127] He criticized Copernicus and Kepler for such ideas, along with 'a manuscript that was produced in Italy by a certain clever man', who argued that the tides were caused by Earth's motion.[128] Morin did not name Galileo, but he gave physical arguments to refute him. The evidence seemed to support instead the system of Tycho Brahe: the Earth is immobile at the centre, while the Sun moves around it, and the planets orbit the Sun.

Furthermore, Morin dismissed Campanella's defence of Galileo as 'frivolous'. Morin argued that 'it is temerarious to assume, contrary to the rules of St Augustine's book on the Christian doctrine, where he said it is wrong to pervert the literal meaning of scriptures, when not incompatible with the faith, because: Heretics open the way' to errors and deceptions.[129]

Morin said that the claim of 'Kepler, Galileo, Campanella and others' that the Moon is like the Earth, with mountains, valleys, continents and seas, was fictitious.[130] Furthermore, among the various reasons to deny the Earth's motion, Morin rejected at length Kepler's 'absurd' claim that the world has a soul and animal faculties. Morin denied that the Earth lives and has veins within it. He said God does not flow as a kind of soul into the Earth.[131]

Morin concluded his book, saying, 'supreme Popes, without waiting for evidence of natural reasons; the proposition of the Earth's motion should be condemned and prohibited, urgently in these days of ingenious temerity.'[132] The theologian censors at Paris promptly approved Morin's book: 'in the entire work nothing is said against the teachings of Catholic Apostolic Roman Faith.'[133]

Regrettably, Galileo did not read the books by Froidmont and Morin before his own book was printed.[134] Its printing was finished on 21 February 1632. Like Froidmont, Galileo or his censors decided to begin the *Dialogue on the Two Chief Systems of the World* by discussing the Decree of 1616. Hence the Preface begins with the Catholic censorship of the Pythagorean doctrine:

> Some years ago there was published in Rome a salutary edict which, to prevent the dangerous scandals of the present age, imposed opportune silence upon the Pythagorean opinion of the Earth's motion. There were some who rashly asserted that the decree was the offspring of extremely ill-informed passion, and not of judicious examination.[135]

Galileo claimed that he now wrote not by audacity, but as a 'sincere witness of the truth'.

But instead of speaking directly, Galileo posited three characters in conversation, who occasionally referred to 'the Academician', that is, Galileo. (Similarly, in some of Bruno's books, Bruno too had used dialogues and referred to himself only indirectly, as 'the Nolan'.[136]) The Pope and the censors had instructed Galileo to speak of the Earth's motion only as a hypothesis. Yet Galileo's *Dialogue* vigorously defended the claim that the Earth moves.

In person, the Pope had been receptive to Galileo's hypothetical discussions about astronomy. But the Pope's tolerance towards eccentric ideas suddenly became subverted. It happened because of a confrontation of March 1632 that had nothing to do with astronomy.

Recently the Pope had tried to maintain Rome's independence from the Habsburg rulers in Spain. To do so, he had been allied with France, but fearing a universal Spanish monarchy, France suddenly allied itself with an invading Swedish army in 1631. The Pope, however, could not support that new alliance, because the invading army was Protestant. Pope Urban then sacrificed the Vatican's independence by giving allegiance to Spain and the Emperor. Soon the Spanish ambassador at the Inquisition began to accuse the Pope of protecting heresies in Rome.[137] Then an ugly confrontation happened on 8 March 1632: the ambassador of Spain, Cardinal Borgia, directly challenged the Pope at a council on the state of the Church.[138] Borgia first stood and surrounded himself with all the Spanish and pro-Spanish cardinals, to shield him as he read the statement. Borgia said that Madrid wanted the Pope to oppose heresy vigorously. This instruction was so offensive that the Pope's brother lunged at Borgia, but only managed to grab his arm.

Soon some Spaniards threatened to depose the Pope for protecting heresies.[139] To defend himself, the Pope agreed to enact stricter persecution of heresies, by giving more influence to Spanish dignitaries. This change of ideology meant that eccentric individuals whom the Pope had sponsored would no longer receive the same tolerance at the Vatican. For example, the former heretic and prisoner of the Inquisition, Tommaso Campanella, would have to leave Rome: 'The liberal atmosphere that had characterized the early years of Urban's pontificate disappeared.'[140]

Galileo's alleged aim in his book was to show that Italians like himself had thoughtfully contemplated arguments for and against Earth's motion. He sought to show that any experiments carried out on Earth's surface would not prove its motion or rest because they would produce the same results. Still, he would defend the Copernican theory by showing that it matches celestial phenomena, and that ocean tides might be explained as being caused by Earth's motions.

Galileo mentioned the Pythagoreans only briefly to illustrate certain points. For example, he wrote that Pythagoras himself believed that the Earth moves.[141] Galileo's First Dialogue referred to the Pythagoreans as follows: 'I know that they did not expose wonderful things to the ridicule and contempt of the populace, to damage, as sacrilege to publish the most recondite properties of numbers.'[142] Speaking as one of his characters, Galileo wrote: 'That the Pythagoreans held the science of numbers in high esteem, and that Plato himself admired the human understanding and believed it to partake of divinity simply because it understood the nature of numbers, I know very well; nor am I far from being of the same opinion.' In response, the character of Simplicio commented: 'I do not want to join the number of those who are too curious about the Pythagorean mysteries.'

At another point Simplicio denied that there is generation or mutation on the Moon. He complained that the notion of men living on the Moon was 'impious', or wicked. The other two characters replied that even if the Moon is not inhabited by beings similar to those on Earth, nevertheless, there might well exist some strange beings on the Moon: 'things, that adorn its operations, and moving, and living; and perhaps in a very different way from ours, seeing, and admiring the grandeur and beauty of the World, and of its Maker, and Ruler, and continually singing His glory'.[143] Historian David Wootton has rightly commented on Galileo's 'remarkable' claims: 'The Moon may be uninhabited, but you do not have to force the text to find in it Bruno's heresy: around other stars there may be other planets, other worlds . . .'.[144]

Furthermore, note that Galileo's *Dialogue* mentioned that the Sun is a star, and the stars are suns.[145] It also referred to his telescopic discovery of 'innumerable stars'.[146] Galileo was defending opinions regarded as erroneous or heretical: the Earth's motion, the Sun's immobility, the Sun is a star and the existence of other inhabited worlds.

In April 1632 the Pope became annoyed by Giovanni Ciampoli, one of Galileo's closest supporters, a papal courtier and fellow Lincei.[147] Ciampoli had supported Galileo in 1616. He had been a source of useful hearsay for years and he had secured a provisional permission from Riccardi so that Galileo might publish his *Dialogue* in Rome. Yet somehow Ciampoli upset Urban. Some said it was because Ciampoli had dared to try to improve (and hence critique) some of Urban's poems. Others said that one night Ciampoli had secretly met with Cardinal Borgia. Pope Urban arranged to have Ciampioli exiled from Rome, under the dubious honour of appointing him governor of an insignificant village. Galileo was losing allies in Rome.

At the same time Galileo asked a friend to take eight copies of his book to the Vatican. They arrived in late May 1632 and he instructed that the first copy be delivered to Cardinal Barberini, the others to Riccardi, Campanella, other acquaintances, and to a consultor of the Inquisition, Ludovico Serristori.[148]

The book promptly upset some readers. Even the Pope became alarmed. In July the Pope instructed Riccardi to contact the Inquisitor in Florence to stop the publisher from distributing the book and confiscate all copies. Riccardi complained that 'there are many things that are not liked' about the book.[149]

In Galileo's *Dialogue* there were few references to the Pythagoreans. I don't want to imply that a main reason why the work seemed offensive was because it praised them. In particular, it did not discuss their religious beliefs. Nonetheless, it was offensive because it embodied the same brazen heresy that Bruno advocated, that in certain matters philosophers have more authority than Catholic theologians. Moreover, the main thesis of Galileo's work, the Earth's motion, was one of the Pythagorean claims that had been censured during Bruno's trial and again by the Inquisition in 1616.

On 5 August 1632 Tommaso Campanella sent Galileo an enthusiastic letter. This congratulatory letter is significant because it shows an educated reader's immediate impression on reading Galileo's book. Campanella praised the 'novelty of ancient truths, of new worlds, new stars, new systems, new nations'. He spoke of a new century beginning, and said that Galileo's ideas 'were of the ancient Pythagoreans and Democritics'.[150] This letter's importance has been emphasized by historian Pietro Redondi, who argued that it shows how an informed reader quickly understood Galileo's work and, allegedly, that it reveals the relevance of the atomism of Democritus, which was disdained as heretical.[151] Historians have carefully analysed this argument, since an earlier manuscript report on Galileo's book of 1623 did discuss atomism. However, historians concluded that despite the significance of atomism in that earlier book, there is insufficient evidence that it was an issue when Inquisitors read Galileo's book of 1632.

Nevertheless, Campanella's letter shows how Galileo's claims were construed to be those of 'the ancient Pythagoreans' – including explicit reference to the heretical notion of new worlds. Although Redondi conjectured that Campanella's mention of the 'Democritics' implied the atomistic theory, Redondi did not point out that the name Democritus had *another* relevant connotation. Campanella did not mention atomism, but he did enthusiastically emphasize 'new worlds, new stars, new systems', and Democritus was known as a proponent of many worlds – according to Cicero, Varro and Valerius Maximus. Hippolytus had ridiculed Democritus for claiming that there are infinitely many worlds, some inhabited, some with Suns and Moons.[152] Philaster, Ambrose and Thomas Aquinas denied that more than one world exists, and criticized Democritus for asserting infinitely many.[153] Ficino too credited Democritus with the claim that 'worlds are innumerable.'[154] Hence Matteo Ricci, Nicholas de Nancel, Clavius, Campanella and others had credited that doctrine to Democritus.

How did the Pope react to Galileo's book? Urban had been Galileo's friendly supporter. Now he saw that Galileo had misused his advice. Galileo personally offended him by placing Urban's measured outlook in the voice of an idiotic character, Simplicio. Galileo also violated his commitment to treat the Earth's motion as a mere hypothesis.

In August 1632 the Pope convened a special committee to decide whether Galileo should be brought to the Inquisition. From the Inquisition's archives, the committee received a document that, Riccardi said, 'is alone sufficient to ruin Mr Galileo completely'.[155] This document from 1616 forbade Galileo 'to hold, teach or defend in any way whatever, orally or in writing', the opinion of Earth's motion, or else be prosecuted by the Inquisition. Galileo had revealed no such injunction to the Pope or Riccardi. Now the Pope had evidence that Galileo had been deceitful.

On 4 September 1632 the Pope met with the Tuscan ambassador, Francesco Niccolini, and complained that Galileo's book was 'pernicious', 'troublesome and dangerous'. The Pope became agitated and bitterly enraged. Niccolini reported:

> His Holiness exploded in great anger, and all of a sudden told me that our Galilei had also dared to enter where he should not have, and into the most severe and dangerous matters that could be stirred up at this time . . . matters,

> which involve great harm to religion and more awful than were ever devised.

The Pope denounced it as 'the most perverse subject matter that one could ever handle'.[156]

But why? Galileo had personally offended the Pope, but yet that affront can hardly be construed as the most perverse subject ever. On 13 November 1632 Niccolini met with the Pope again, to plead for Galileo. But the Pope 'again said that one is dealing with an awful doctrine'.[157] Years later the Pope reiterated that Galileo had defended 'an opinion so very false and so very erroneous', which had generated 'a universal scandal to Christianity with a doctrine that had been damned'.[158] So it was not just personal.

In 1624 the Pope had told Cardinal Zollern that the opinion of Earth's motion was *not* heretical, it was merely temerarious, and it was *not* something one should worry about. Hence it seems that what upset the Pope in 1633 was not merely that Galileo had defended the hypothesis of Earth's motion, hardly a major danger to Catholicism, but Galileo had '*also*' done something much worse: he had engaged 'the most severe and dangerous matters . . . which involve great harm to religion and more awful than ever devised'. Another remark reiterates that it was not just *one* subject-matter: the Pope complained that Galileo 'had been ill-advised to give out his opinions'.[159]

Ambassador Niccolini explained in a letter that the Pope's concern was *not* about science: 'the Pope believes that this involves many dangers for the Faith, not that we are dealing here with mathematical matters, but about the Holy Scripture, about religion and about the Faith.'[160] What were those many dangers?

In March 1633 the Pope met again with Niccolini, saying that Galileo would be interrogated. Then the Pope became 'incensed' and 'infuriated', insisting that 'one should not impose necessity on God.'[161] And in April the Pope reiterated the gravity of Galileo's crime: 'it pains His Holiness that he [Galilei] has entered into this matter, which he [the Pope] still considers to be extremely grave and of great consequence for the [Catholic] religion.'[162]

What was so dangerous to Catholicism about Galileo's work? Historians have advanced various conjectures. For example, John L. Heilbron argues that maybe there were other considerations indirectly connected to the *Dialogue*: maybe Galileo's support of atomism (in 1623) seemed incompatible with transubstantiation, or perhaps there were implications about astrology, or even the problem of salvation and grace. According to Heilbron, 'Although Galileo's trial was not about atomism, astrology, freewill, salvation, grace, or divine attributes, many or all of them probably were in Urban's mind when he castigated Galileo's doctrine as "the worst [menace] ever perceived".'[163]

We do not know what Pope Urban had in mind, but we can understand his words if they referred to the central matter of Galileo's *Dialogue* and the Decree of 1616. If the Earth's motion were not merely hypothetical but actually true, then it would mean that some passages in the Bible were *literally false*. Cardinal Bellarmine had warned Foscarini that the doctrine of Earth's motion and Sun's immobility 'is a very dangerous thing' because it would irritate all theologians and 'harm the Holy Faith by rendering the Holy Scripture false'.[164] Galileo was proposing a *Pythagorean* way to interpret scriptures, as if the Pythagorean notions were true. Galileo's work would then seem to undermine not just Catholic authority, but Christianity – in favour of what? A blend of Christian and ancient Pythagorean ideas? The intrusion of an alien pagan sect into Catholicism? That is precisely how Froidmont had recently described it.

Members of the Inquisition were disturbed by Galileo's *Dialogue*, partly because it offended the Pope but especially because it defied the personal injunctions of 1616 and the public Decree: his book taught and vigorously defended the 'false Pythagorean doctrine' as if it were true. Moreover, it was inadmissible to argue that a proposition was 'probably' true after the Church had declared it contrary to scriptures.[165]

In February 1633, having failed to get sympathy from the Pope, the Tuscan ambassador approached the nephew Cardinal Francesco Barberini, to request support for Galileo. Cardinal Barberini was the head of the Inquisition. He replied that he esteemed Galileo and wished him well, 'but that this matter was very delicate, because it could introduce some fantastical dogma into the world'.[166] What was that outlandish belief, '*dogma fantastico*'?

Incidentally, in the passage in the *Dialogue* in which Galileo defended the idea of beings living on the Moon, Galileo began: 'I have many times been fantasizing [*fantasticando*] . . .'.[167] Coincidentally, in England Bruno had been described as a 'fantasticall' man who was burned alive.[168] At the time, in England and Italy such words were used to refer to falsehood, monstrosity, false worship, heresies, ridiculous inventions, absurd fables, deranged conceit, foolishness, satanic opinions, deceptive magic, insanity and awful lies.[169]

Those who appreciated Galileo's revival of ancient ideas were distressed by the Inquisition. In March 1633 Lucas Holste complained to his French friend Nicolas-Claude Fabri de Peiresc about the ignorance of cardinals at the Index, and he became so annoyed by what they said that he decided not to attend their meetings anymore.[170] Holste wrote to his friend:

> It would take me a long time to review the cause of the hatred against the fine old man [Galileo]. Surely nobody saw without indignation that the men chosen to judge about Galilei's book and about the entire Pythagorean or Copernican proposition are plainly dissonant, while it is principally about the authority of the Church, which will suffer vehemently from a less correct judgment. I earnestly advise them to realize that the first authors were great mathematicians, and who were more excellent than others in investigating the truth; and those who in our century have brought their doctrine back into the light, to their credit their erudition has approached that of the ancients.[171]

Holste's advice to the clergymen raises the question: who were these 'first authors'? Since he had just mentioned Pythagorean knowledge and 'the ancients', it seems that Holste referred to ancient philosophers, such as Pythagoras and Democritus. It also seems that the cardinals did not view such men as 'mathematicians', which moved Holste to describe them thus. Still, despite Holste's tolerance for eccentric ideas, he like others lacked the courage to endanger himself by openly supporting someone who seemed guilty of heresy.

On 12 April 1633 the Inquisitors first interrogated Galileo. I need not summarize his depositions since they have been widely published. Aside from discussing the publication of his *Dialogue*, he described at length his impressions about the events of 1616. However, I want to note a trace of important topics left unspoken.

Galileo said that there was something from 1616 that he preferred not to voice, except to whisper it to the highest authority. He told the Inquisitors:

> one morning the Lord Cardinal Bellarmine ordered that I be summoned, and he told me a certain particular that I want to say to the ear of His Holiness prior to anyone else; but the conclusion was that he told me that the opinion of Copernicus could not be held or defended, for being contrary to Holy Scripture.[172]

But the Pope would not meet with Galileo, so we don't know what unspeakable thing Bellarmine voiced in 1616, and why Galileo did not say it in 1633.

To defend him, Campanella hoped to be appointed to the committee that would evaluate Galileo's case. But no. Riccardi explained to Niccolini it was impossible to appoint Campanella, because Campanella had 'composed a somewhat similar work, which was prohibited, and cannot defend him while being himself guilty'.[173] Still, Riccardi 'felt obligated' to defend Galileo himself. So instead of Campanella he recommended a Jesuit (namely Melchior Inchofer) 'who is his [Riccardi's] confidant, and guarantees that he proceeds with proper intentions', along with the Pope's theologian, Agostino Oreggi, who 'truly had good will'.[174]

The special committee, however, concluded that Galileo had defended the censured opinions in his *Dialogue*. The case fell to the Inquisition. In April 1633 three consultants analysed Galileo's book: Agostino Oreggi, Melchior Inchofer and Zaccaria Pasqualigo. They had to determine whether Galileo held, defended or taught that Earth moves and not the Sun. Oreggi concluded that Galileo indeed held and defended those censured views. Pasqualigo, a young professor of theology, reported likewise that Galileo transgressed the injunction against him because his book did 'teach and defend' it, and therefore also Galileo 'is strongly suspected of holding such an opinion'.[175]

Finally, the longest reports against Galileo were submitted by the Jesuit theologian Melchior Inchofer, who was Riccardi's close friend.[176] Inchofer analysed Galileo's *Dialogue* and reported to the Inquisition: 'I am of the opinion not only that Galileo teaches and defends the view of Pythagoras and Copernicus but also . . . that he is vehemently suspected of firmly adhering to it, and indeed holds it.'[177] Inchofer complained not only about many arguments 'ad nauseam' that Galileo wrote in favour of such opinions, but about how Galileo argued: with arrogance, insolence and combative bitterness.

Inchofer listed many instances where Galileo argued in an absolute, non-hypothetical and non-mathematical way about Earth's motion and the Sun's immobility. Since we are focused also on other Pythagorean beliefs, we may note that Inchofer mentioned another objectionable belief sometimes attributed to the Pythagoreans: the eternity of the world. This was one of the heretical propositions that Inquisitors brought up against Bruno. Inchofer didn't mention Bruno and he didn't say that Galileo made such a claim. Inchofer used it as an example of a belief to avoid: 'Philosophers also inquire whether the world could have existed from eternity; yet no Christian says that it has existed from eternity.'[178] Inchofer also complained that Galileo 'declares war on everybody and regards as dwarfs all who are not Pythagorean or Copernican'.[179]

I wonder whether members of the Inquisition were aware of the heretical connotations of the term 'Pythagorean'. It is particularly striking that Lucas Holste authored an expansive work on Porphyry and Pythagorean ideas under the sponsorship of Cardinal Barberini. I have found no comparable book published by anyone anywhere in the preceding decades. It seems impossible to imagine that the consultants and Inquisitors, though having expertise on theology and heresies, were nonetheless indifferent to Pythagorean heresies about whether souls pre-exist bodies, transmigration, pagan gods, other worlds, the soul of the world or its eternity. Still, the available evidence does not show that all the consultants and Inquisitors were overtly moved by such concerns in 1633. However, the strongest critic among them – Melchior Inchofer – certainly was, as we will see.

It is also noteworthy that one Inquisitor from the previous investigative proceedings against Galileo, in 1616, was present in Galileo's trial in 1633: Bishop Felice Centino, known as Cardinal d'Ascoli.[180] Historian Annibale Fantoli argues that it is conceivable that d'Ascoli played a significant role in the final opposition to Galileo.[181]

During the trial Cardinal Barberini commanded considerable power among the Inquisitors, yet he expressed some sympathy for Galileo.[182] Contrary to writers who construe Barberini as one of Galileo's allies, however, I view him as duplicitous. Recall how annoyed he was in 1630 when he inferred that Galileo argued *that Earth is a star*. And in 1633 he told Niccolini that the matter was 'very delicate, because it could introduce some fantastical dogma into the world'. Barberini did not sign the final condemnation, so some writers speculate that this shows some disapproval. However, apparently he could not sign it because he was meeting with the Pope that day. Moreover, Barberini's name was included among the authors of that final condemnation against Galileo. Barberini and Urban felt some sympathy for Galileo, but their allegiance to Catholic dogma was far greater than any sympathy for an irreverent friend. Remember that in the times of Bruno and Campanella, Inquisitors could be courteous, kind, patient, sympathetic and accommodating – while at the same time they enabled obstinate suspects to be brutally tortured and burned alive.

Likewise, I view Father Riccardi as duplicitous. It was his sacred obligation to defend scriptures. He chose his Jesuit confidant Inchofer because of his 'proper intentions', although earlier, in mid-1632, Riccardi had anticipated that 'the Jesuits will persecute him [Galileo] most acrimoniously.'[183]

In May or June the Inquisitors submitted a final report to the Pope. It reviewed the accusations and proceedings against Galileo from 1615 to 1633. It included not just the accusations about Galileo's belief in Earth's motion, but also the most spurious early accusations: that he said God is an accident, God laughs and cries, and his denial of the miracles of the Saints.[184] Similarly, at the end of Bruno's trial, the Inquisitors cited some of the earliest and spurious accusations against him: that Bruno allegedly had doubted transubstantiation and Mary's virginity – whereas actually Bruno wrote no such things, he denied such accusations, and therefore such accusations were not the focus of the proceedings.

On 21 June 1633 the Inquisitors ordered Galileo one last time to admit whether he had *ever* believed that the Earth moves around the Sun. Galileo cautiously denied it, saying that before 1616 he merely thought the theory was 'disputable'. Unconvinced, the Inquisitors insisted that he 'tell the truth, as otherwise one would have recourse to torture'.[185] Galileo denied it again – he lied – endangering himself.

To condemn Galileo, the Inquisition took him to the Church of Santa Maria Sopra Minerva, between the Pantheon 'Rotunde' and the Jesuit Church of St Ignatius.

The next day, old Galileo was humiliated by being forced to wear sackcloth, the penitential garb of guilt, and publicly taken to the Dominican convent of Santa Maria sopra Minerva, in the centre of Rome, to face the cardinals and other members of the Inquisition.[186] They declared Galileo guilty of 'vehement suspicion of heresy', a grave and punishable offence.[187] It is noteworthy that the first cardinal to sign Galileo's condemnation was Cardinal d'Ascoli, the one who had been present in the proceedings of 1616 and the only one who heard any comments by Sfondrato, Taverna, Borghese and Bellarmine – who had all judged both Galileo and Bruno. They knew that Bruno's belief in the moving Earth pertained directly to his heresies about the many worlds and the soul of the world. Yet no such issues show up in the extant trial transcripts of 1633.

At this point, if Galileo had refused to abjure the 'suspicion of heresy' then it would become a *proven* heresy, and 'the inevitable penalty in this case was to be burned at the stake.'[188] To avoid being burned alive, the Inquisitors now compelled Galileo to kneel and recant his claims.

He then said, 'with a sincere heart and unfeigned faith I abjure, curse and detest the above-mentioned errors and heresies, and in general each and every other error, heresy and sect contrary to the Holy Church.' I suspect that Galileo was renouncing not just all sects in general, but one in particular: the Pythagorean sect. In late

Inside the Church of Santa Maria Sopra Minerva, where Galileo was condemned for 'vehement suspicion of heresy' in 1633.

1631 the Catholic theologian Froidmont had accused Galileo of being a member of this awful pagan sect.

Galileo was condemned to imprisonment for a period to be specified. He was required to repeat the seven penitential psalms once a week for three years. His sentence was commuted to live permanently under house arrest.[189]

Froidmont soon heard about Galileo's trial and humiliation. He immediately drafted a pamphlet to celebrate the victory of the Catholic faith and of his *Anti-Aristarchus* of 1631. Echoing the ancient account by Plutarch about the pagan Pythagorean temple of the central fire, Froidmont titled his new pamphlet *Vesta, or Anti-Aristarchus Vindicated*.[190] It was approved for publication in September 1633 and published in early 1634. Froidmont proudly declared, 'this year in Rome the Most Eminent Cardinals have judged and condemned the errors of Pythagoras and Copernicus, and all the subjects [members] of the Apostolic See are barred from this doctrine.' And in the margin of one page, he succinctly summed up Galileo's final judgement: 'Galileo was forced to abjure his Pythagoreanism.'[191]

Inchofer Against the New Pythagoreans

Historians have puzzled over the motivations of the individuals who led the case for Galileo's conviction. The most damning testimony against Galileo was provided by Melchior Inchofer. But why? Riccardi and Francesco Barberini had been friendly towards Galileo. Why did they choose Inchofer?

As early as 1623 Inchofer had been in contact with Francesco Barberini, requesting a copy of a letter that allegedly had been written by the Virgin Mary.[192] Historian Thomas Cerbu reports that members of Barberini's entourage were interested in the letter as early as 1622. In 1629 Inchofer published a book arguing that the letter was genuine, but an archbishop disagreed and so Inchofer was summoned by the Inquisition. Inchofer, however, quickly convinced them that his book could be readily modified. He then argued that the letter 'seemed to be' authored by the Virgin Mary. Most importantly, Inchofer obtained the privilege of having his revised book published in 1631 by Ludovico Grignani, the printer who 'specialized in works by authors attached to the Barberini papal court'.[193] Also, since one of Barberini's attendants had published a work on the letter, Cerbu remarks that it is plausible that 'Francesco Barberini both sanctioned and funded' the publication of Inchofer's book.[194] Inchofer became friends with the Vatican librarian, who explained that Inchofer 'ingratiated himself with many Cardinals of the Holy Congregation of the Index'.[195]

Cerbu conjectures that Riccardi and Barberini chose Inchofer to analyse Galileo's *Dialogue* 'to displace the scrutiny of Galileo away from scientific controversy'.[196] Presumably, since Inchofer would focus on theology instead of astronomy, then his criticisms would not be as severe, since Galileo did not discuss theology openly in his *Dialogue*. If that were expected to soften the evaluation, then it would be surprising to receive Inchofer's utterly negative assessment.

Inchofer himself elaborated his concerns in a book he drafted during the trial and promptly published.[197] In 2006 Richard J. Blackwell fairly noted that Inchofer's lengthy tract had been 'largely ignored by Galileo scholars', so he published an English translation and analysis. Inchofer analysed the theological case against heliocentrism. Blackwell characterized its importance: 'This tract was written while the trial was being conducted, and thus in a special way it gives a direct insight into at least one participant's view of the religious rationale behind the trial.'[198] No other publications by Inquisitors or consultants are as extensive and explanatory about their concerns. Instead of focusing on scientific issues, the theological case against the heliocentric theory turned out to be very strong.

In the summer of 1633 Inchofer published *A Summary Treatise concerning the Motion or Rest of the Earth and the Sun, in which it is briefly shown What Is, and What Is Not, to be Held as Certain according to the Teachings of the Sacred Scriptures and the Holy Fathers.* It justified the Catholic opposition to the heliocentric theory, 'to rally everyone as soldiers of religion'. Among various objections, Inchofer complained 'since the Pythagoreans have gradually come to oppose the faith, it must be shown that the truth is found in the scriptures, and as our major authors knew, is opposed to them'.

The new Pythagorean sect argued that parts of the Bible were literally false, though lacking any authority to do so, which was intolerable to the clergymen. Inchofer required that 'the Copernican theory and its related Pythagorean philosophy should not be taught at all.'[199]

Inchofer enumerated seven arguments that had been proposed to supposedly show that the Bible is consonant with the claims that Earth moves, not the Sun. I summarize all seven arguments, to note Inchofer's worries about incorrect interpretations of biblical passages.

Seven Arguments Criticized by Melchior Inchofer

First Argument
The passages of scripture on the stability or fixity of the Earth refer actually 'to the situation and ordering in which the Earth remains in stable and perpetual motion'.

Second Argument
'The starry heaven is called a Firmament, only because it is stable and unmoving, therefore the Earth really moves.'

Third Argument
'Scriptures speak metaphorically and in the common ordinary way in many places, especially in Mathematics and in

Theoretical Truths; therefore the Earth's stasis and Sun's motion are said in the ordinary sense, from which it does not follow that this happens in Physics and Mathematics.'

Fourth Argument
'That the Firmament rests and the stars in it, is not contrary to Scriptures, therefore by necessary consequence the motion of the Earth is not contrary to Scriptures.'

Fifth Argument
'Hell is in the Centre of the Earth, and in it is a fire tormenting the damned; therefore it is absolutely necessary that Earth moves . . . Because fire is a cause of motion, therefore Pythagoras, who as Aristotle reports put the place of punishment in the Centre, felt that Earth is animated and endowed with motion.'

Sixth Argument
According to Genesis I, there are waters in heaven above the firmament and beneath it. 'Therefore the Earth's Water is not contained only in the solidity of the Earth, and consequently the natural place of the Earth is not the centre, but possibly, outside it and carried in circular motion in a Great Orb.'

Seventh Argument
'This opinion of the Earth's motion, and quiescence of the Sun at the Centre is most ancient, not first by Pythagoras, but had been conveyed by Moses from whom Pythagoras received it as a kind of Jew.'

Inchofer rejected these seven arguments. Since it was inappropriate to discuss heretics, he did not mention Galileo or Campanella, or much less Bruno. Still, he criticized many claims Campanella had voiced: that the scriptures refer to the starry heavens as the 'firmament' because they are immobile, that Copernicus's book had been approved by Pope Paul III, and that recent astronomers (allegedly) had found it impossible to establish astronomical tables correctly without using Copernicus's book.[200] Inchofer also criticized Bruno's beliefs, without naming him either.

Since Inchofer considered the Earth's motion to be an ancient Pythagorean idea, and he described its advocates as 'Pythagoreans', then all of these seven arguments were Pythagorean, both to him and to his readers. But among these arguments, consider the two that explicitly mentioned additional Pythagorean notions: the fifth and the seventh.

The fifth argument states that since the fires of Hell are in the Earth's centre, therefore Pythagoras argued that Earth is *animated* and moves. Whereas Aristotle has attributed to 'the Pythagoreans' the idea that the centre of the world is the place of punishment, Inchofer (like Thomas Aquinas) blamed this idea on Pythagoras himself. Inchofer agreed that Hell is in the centre of the Earth, and it contains fire. But he said that this would not necessarily cause the Earth to move, because 'if the argument is valid, it proves also that lime kilns, baking ovens and hot fire-grates are animated and in themselves movable.' He also argued that fire naturally goes upward, whereas 'such a motion is conceded neither by Pythagoras nor other more recent followers'. Inchofer agreed that in some sense there is fire in animals that, affected by their soul, causes their bodily motions. But he disagreed that this applies to Earth:

> Thus if Pythagoras attributed soul to Earth because of fire contained within it, he did so in ignorance, and against his own principles, for elsewhere the stars are said to be fiery and animated, and even the Sun is animated by its fiery nature, and hence they would be endowed with motion; then how can the Sun be at rest in the Centre as the Earth moves around it?

Inchofer cited Thomas Aquinas as having proven that Earth is not animated, 'against the arguments of Alexander who followed the errors of the Gentiles, who thus attributed a cult of the Deity to the Earth'. Inchofer said the Earth is not moved by a soul:

> thus clearly the Earth is not animated, nor can its generator be a soul joined to the Earth, as Pythagoras wanted, rather it has a separate mover, which however is not sufficient for Earth in the opinion of the Pythagoreans and other ancients, but [allegedly] requires also a soul conjoined to the Earth as

> a proximate mover; thus, if the Pythagoreans want it to be fire, they have to admit that this fire is produced by a separate form and again that Earth was produced by fire according to this same Philosophy; but everyone sees that all of this is ridiculous and absurd.
>
> Thus clearly, the Pythagoreans have clashed greatly against their own principles, especially in saying with Pythagoras that Earth is made of forty-eight angles, containing six triangles with equal sides, and similarly with the Cube; and as was said above in Chapter 3, giving a completely useless account of [Earth's] motion; nevertheless they endow it with a soul, and affirm that it rotates with an extremely fast motion. This has led their followers into deceptive errors, as not undeservingly ridiculed by the philosopher Hermias in his *Book on the Pagan Philosophers*, and especially given their argument about the Cube and the Earth, they are laughed at.

Inchofer thus cited Hermias against the Pythagoreans. He didn't explain it, but Hermias had ridiculed claims about transmigration, that souls consist of 'number in motion', and many worlds. More importantly – Inchofer denounced the notion that Earth has a soul. Bruno had said this, not Galileo.

Next, the seventh argument, criticized by Inchofer, claimed that the opinion of Earth's motion was so ancient that Pythagoras derived it from Moses, and therefore it was true. Inchofer countered that philosophical opinions are not rejected because they are new or old, but because they are false. He said many old opinions are 'false and disproven', including the claim that 'the Sun is a prison for the souls of sinners.' More to the point, Inchofer said that it is false that Moses originated the Pythagorean theory. He added that it is indifferent whether Pythagoras was really Jewish. He did not specify who said that, but Campanella had stressed it in *Defence of Galileo*.

Inchofer insisted that if good astronomers use any statements that seem to contradict Scriptures, then they should proceed as if such statements are mere hypotheses, not truths. He therefore quoted astronomer Nicholas Mulerius, who in 1611 wrote that Sacred Scriptures should have such a great authority over us 'that we would not dare to fall into the opinion of the Pythagoreans, which is openly contrary to scripture'.[201] Elsewhere Mulerius called the Pythagorean astronomers 'a sect'.[202]

Inchofer construed the theory of Earth's motion as an ancient, Pythagorean idea that had been revived and refined by Copernicus. His knowledge of Christian theology informed him that such ideas had been criticized centuries ago by thinkers such as Aristotle, Hermias, John Chrysostom, Augustine and Thomas Aquinas. Inchofer summarily explained: 'Indeed before Copernicus came to light, the Philosophy of the Pythagoreans was contemplated by many, but it was not introduced into the [monastic] Schools, either because nature opposed it, or divine Faith.'[203]

Inchofer traced the infamous beliefs of Pythagoras to his supposed teacher, quoting Tatian: 'the fictions of Pherecydes were followed by the dogmas inherited by the Pythagoreans, and followed by the imitations of Plato.' Pherecydes of Syros apparently lived around 550 BCE and some writers claimed that at some point he lived on the island of Samos. Several writers claimed that Pherecydes was the first to teach the transmigration of human souls and their eternal existence. Cicero and Augustine, for example, both said that he first taught the 'immortality of the soul'. Hence some writers said that Pherecydes had been the teacher of Pythagoras.[204]

For Catholics not everything in scriptures had to be interpreted literally, but there were clear limits to interpretative liberties. In particular, any Catholic was required to accept the interpretations dogmatized by the Church. Thus it was heretical for Bruno and Galileo to wilfully reinterpret biblical passages as they wished, especially to try to make such passages mean the opposite of what they literally state, or the opposite of what Church Fathers had explained.

Heretics typically said, on the one hand, that they believe everything the Catholic Church teaches, while on the other modifying certain dogmas at will, say, to match pagan beliefs. In opposition, Inchofer complained:

> Christian piety does not conform with the precepts of error. Otherwise never was Philosophy taught better than to defend with clear reasons against these glorious errors: that in each man there are three distinct substantial souls; that the one in all is informed by a rational soul, as in each mortal; also that Angels are corporeal; that other Worlds exist in the Sun

> and the Moon, with other creatures endowed with reason and humanity; and six hundred other monstrosities stirred up, introduced into the [monastic] Schools, and defended with obstinate speculations; when many [philosophers] of such kinds, excluding the Faith, dispute philosophically, and despite a certain verisimilitude, cannot prove even one such discourse.[205]

Inchofer said that Thomas Aquinas had written, 'it is not possible that there exists an Earth other than this one.'[206]

Inchofer's critiques against Pythagorean ideas are remarkable for how much they echo ideas that the Roman Inquisition had denounced in the books of Giordano Bruno. Inchofer rejected the following Pythagorean ideas:

(1) The Earth moves.
(2) The Earth is animated, alive.
(3) The Earth has a soul.
(4) Souls transmigrate from body to body.
(5) There are many worlds, including the Sun and the Moon.
(6) Such worlds are inhabited by rational, humanlike beings.
(7) The Sun too is animated.
(8) The stars too are animated.

The Roman Inquisition had censured *all* of these ideas in the works of Bruno! No historian previously had pointed this out because they usually treat Galileo's trial separately from Bruno's. This is revealing because it shows that many of the main and controversial concerns in Bruno's trial were in fact part of the systematic critique against Galileo three decades later. Inchofer thought that such pagan falsehoods were intimately linked to the propositions Galileo defended. Although Inchofer mentioned the latter two notions only briefly, that 'the stars are said to be fiery and animated, and even the Sun is animated by its fiery nature', he rejected such ideas. In addition to these critiques, Inchofer also rejected other so-called Pythagorean ideas:

(9) The Sun does not move.
(10) The Sun is at the centre of the universe.
(11) The Sun is a prison for sinners.

The propositions about the Sun being immobile and central had been affirmed by Copernicus, Galileo and Kepler, and denounced by the Inquisition as 'Pythagorean' and heretical. The claim that the Sun is a prison stems from Aristotle's early accounts of the beliefs of the Pythagoreans. It is essentially a religious proposition.

The majority of these eleven Pythagorean propositions were not merely about astronomy or nature, they were *about souls* and the afterlife. Galileo, by attributing the Copernican theory to the Pythagoreans, had unwittingly pushed it closer to pagan doctrines that theologians rejected as profoundly perverse.

Inchofer required that the 'Pythagorean philosophy' should not be taught. He ridiculed the Pythagoreans for believing that Earth has a soul and that hellfire causes it to move. As we have seen, the idea that Earth has a soul had been advanced by 'Plutarch', Origen, Philostratus, Plotinus, Abelard (allegedly), Ficino, Bruno, Campanella and Kepler. And the idea that hellfire causes the Earth's motion had been discussed by Campanella.

The Imprimatur for Inchofer's book specifies: 'Three theologians of the Society [of Jesus] . . . reviewed it and approved it for publication.'[207] The Commissary General of the Roman Curia, Lucas Waddingus, promptly approved Inchofer's book for publication, remarking: 'This theologian has given a Christian refutation of these Pythagoreans. And he shows rightly that mathematics and the human sciences should be subordinated to the rule of Sacred Scripture.'[208] This book too was approved for publication by Inchofer's friend Riccardi.

Like Holste's book on Porphyry and Pythagoras, Inchofer's book bears on its first page the official emblem of the Barberini family: three bees. The globe of the Earth is illustrated inside a triangle, and on each corner a bee holds the Earth in place with its front legs. Overhead, a banner proclaims: 'Fixed by these, it rests.' The emblematic bees show that Inchofer's book was approved by the *highest* authorities.

Thus I confirm that Inchofer's *Summary Treatise* is a neglected but revealing window into the motivations of the most critical expert in the infamous trial against Galileo. Historians would be thrilled to have a comparable treatise by, say, Bellarmine, explicitly explaining his personal critiques and concerns in the earlier proceedings against Galileo and Bruno. We are fortunate to have Inchofer's account for the trial of 1633. Furthermore, it turns out that there exists a more extensive account by Inchofer, which historians have entirely neglected.

No Life in Other Worlds, No Living Earth

Galileo was perplexed when he heard about Inchofer's *Summary Treatise*. In July 1634 Galileo complained about it to a close friend:

> Froidmont restrained himself to submerge just below the mouth that the mobility of the Earth is a heresy. But recently a certain Jesuit Father has printed in Rome that such an opinion is so horrible, pernicious and scandalous, that although we do allow in teaching, in [discussion] circles, in public disputes and in publications to convey arguments against the foremost articles of faith, against the immortality of the soul, the Creation, the incarnation, and so on, yet it cannot be allowed that there be dispute nor even argument against the stability of the Earth, such that only this article above all can thus be held as sacred, that in no way can there be anything against it, nor any dispute in any way, but for its corroboration. The title of this book is *Summary Treatise of Melchior Inchofer, of the Society of Jesus*.[209]

Galileo exaggerated, yet he realized that the Earth's motion was far more disturbing to Inchofer than he could comprehend.

Inchofer's *Summary Treatise* consolidates a circumstantial argument about the role of Pythagorean notions in Galileo's trial. Inchofer denied Earth's motion while linking it to pagan beliefs known as 'Pythagorean'.

However, this is what his book does *not* do: it does not specify whether this Jesuit knew that Church Fathers, Saints and theologians had condemned such pagan beliefs as *heretical*. Like Giordano Bruno, Inchofer linked the Earth's motion to the theory that Earth has a soul and that many worlds exist – but did he know those beliefs were heretical? Apparently Bruno himself did not know. So did Inchofer know?

After all, we have traced an intricate web of developments in which numerous Catholic authorities denounced Pythagorean beliefs across the centuries. Were such old pronouncements remembered in the 1630s? Did any of the consultants or Inquisitors in Galileo's trial fret about the ancient judgements of Philaster, Isidore of Seville or Pope Zacharias? Did anyone connect the Copernicans to the heretical Gnostics or to Origen Adamantius? Did anyone worry about the pagan poetry of Orpheus, Ovid or Virgil? Did anyone care about the Spirit that nourishes within?

If any of the consultants or Inquisitors had cared about such things at the time of Galileo's trial we might expect that they would have actually written something about it. If so, such statements would constitute a 'smoking gun', evidence still simmering after having fired at Galileo. Does such evidence exist? Remarkably, *it does* – and I found it five years after beginning my research, right when I was nearly finished.

While Inchofer was composing his *Summary*, he began writing a longer manuscript explaining why 'the Holy Tribunal' condemned the Copernicans. I first learned about this unique, unpublished manuscript in a book by Domenico Berti, published in 1876. I later noticed it mentioned in a footnote in Blackwell's translation of Inchofer's *Summary*. Blackwell later told me that he had never read it.

Its title drew my attention: *Vindication of the Holy Apostolic See, the Sacred Tribunal and Authorities against the Neo-Pythagoreans' Moving Earth, and Stationary Sun.*[210] This nearly 400-year-old manuscript is in an archive at the Biblioteca Casanatense, Rome, which houses the papers of many cardinals and clergymen. As far as I know, there exists no scholarly analysis or translation of this manuscript in any language, and I don't know why. The only account of its contents that I have found is the old, very brief summary by Berti, no more than a single paragraph. This is my translation of what Berti wrote; his text is in Italian with some Latin phrases (which I here render in italics). According to Berti, Inchofer argued:

> Mathematicians tread slippery terrain when they try to walk without theology by adopting the opinion of Pythagoras which was more than once rejected as impious *and for so many ages buried in the grave*, and which will never be reconciled with religion, *with religion it can never cohere*. There does not exist any other but only one world, the Earth, on which was created the first man, from whom all men are descended;

> *and by one man sin was introduced into the world*. Since God ordered the apostles to teach all peoples, it is necessary that all peoples are contained on the Earth: *conclusion, all the human race is in this world*. But if there were more worlds we would not know how to determine which are the men who in those other worlds should be immune from original sin and to which of them extends the well-being of redemption. Therefore the things said by Kepler and Galileo about the similarity of the Earth to the other stars are nothing but dreams that will not take long to dissipate if they have not dissipated already. This refutes the opinion of Gilbert that the Earth is a magnet, and it studies with various arguments to prove that the Copernican doctrine is heretical, impious, philosophically and mathematically erroneous. In ending, the book provides the history of the prohibition of the same, the Decree of 5 March 1616, then that of 22 November 1619, in which the corrections are introduced in the book *On the Revolutions* of Copernicus, *as requested by the Supreme Pontiff*.[211]

The Gospel of Matthew includes a statement whereby Jesus Christ, having resurrected, orders his apostles to take his message of salvation to *everyone*: 'Go therefore and make disciples of all the nations, baptizing them in the name of the Father and the Son and the Holy Spirit, teaching them to observe all that I commanded you' (Matthew 28:19–20). Thus Inchofer concluded that all peoples really must live on Earth alone, because if anyone lived in other worlds, like the Moon or Jupiter with its moons, then such persons could not receive the apostles' crucial message.

Therefore, the Moon and planets were not worlds. Earth was the only world. This shows again that Inchofer knew that the theory of Earth's motion entailed more offensive, ancient notions: multiple worlds exist, they are inhabited, the Creation story in Genesis is challenged, and therefore the Christian doctrine of salvation becomes confused. These beliefs were associated with Bruno.

Strangely, historians have not analysed Inchofer's manuscript. In the recent literature the main instance in which it has been just briefly mentioned was in connection with two much shorter manuscripts. Famously, in 1982 Pietro Redondi discovered an undated three-page manuscript in the *Protocolli* of the Index. That brief manuscript, known as 'G3' (a cryptic tag on its first page), denounces the doctrine of atomism in Galileo's book of 1623, *The Assayer*. The anonymous author of the three-page manuscript argued that Galileo's atomism was incompatible with the Catholic doctrine of transubstantiation. On the basis of G3, Redondi famously developed a reinterpretation of Galileo's trial, advocating a provocative but very speculative conjecture that Galileo was accused of Copernicanism only because the Pope protected him against the greater accusation voiced in G3. Most historians rejected Redondi's conjecture for lack of evidence, in addition to the Pope's hostility towards Galileo. In 1999 Mariano Artigas discovered another document in the Archives of the Index, which were finally opened to scholars in 1998. This new document was just one and a half pages, immediately preceding G3 in the same archival volume.[212] This document, known as EE 291 (archival volume 'EE', plus its folio number), is also about transubstantiation, but it does not specify its author or date of composition.[213] EE 291 is a draft or an internal report in which someone recommends that the Inquisition conduct an investigation against Galileo.

Artigas's collaborator, Rafael Martínez (no relation to me), soon discovered that the author of EE 291 was Melchior Inchofer. The handwriting matches Inchofer's manuscripts: two signed documents in the same archival volume EE plus two others at the Biblioteca Casanatense.[214] Inchofer's neglected long manuscript *Vindication . . . against the Neo-Pythagoreans* has thus been used merely to confirm his authorship of EE 291. Some historians conjectured (with no certainty) that the anonymous G3 was drafted around 1624 and that Inchofer composed EE 291 sometime between 1631 and 1632.[215]

In a survey of censures against Galileo, including EE 291 and G3, Francesco Beretta too looked at Inchofer's long manuscript. Beretta mentioned this in an article, but gave even less indication of the manuscript's contents than Berti had done in 1876. Beretta alludes to its contents in five sketchy sentences, noting: 'Inchofer engages a series of questions pertaining mainly to biblical and historical controversies.'[216]

There has been much discussion about documents G3 and EE 291, documents that constitute less than five pages of text. Ironically, the very long manuscript by Inchofer remains almost entirely neglected: a fully finished and polished treatise explaining why the Inquisition acted against the Copernicans!

To analyse Inchofer's arguments in depth, I travelled to Rome in 2014 and 2015 to study his unique manuscript at the Biblioteca Casanatense. The handwritten but bound text is 210 pages long. It consists of two parts or 'books' bound as one. The title page shows that its original title was *Vindication of the Sacred Tribunal and Authorities against the Neo-Pythagoreans' Moving Earth and Stationary Sun*. In the margin Inchofer subsequently added the phrase, *the Holy Apostolic See*, above the title, as if his work vindicates the views of the Papacy itself. Recall that his *Summary Treatise* had the image of the three Barberini bees.

The title page lacks a date, but Inchofer finished the manuscript in 1635, because in the Preface he referred to his *Summary* as having been written 'two years ago', and because by January 1636 the internal reviewers of the Jesuits finished reviewing it. The *Summary* has 94 pages of printed text, constituting roughly 33,000 words. The *Vindication* is much longer, it consists of roughly 49,000 words. The *Summary* does summarize some material contained in the *Vindication*, but each work includes much material that is not mentioned in the other.

The sophistication of Inchofer's manuscript warrants a full scholarly analysis of its contents. However, I now confine myself only to discussing its primary contents, which remarkably are directly relevant to the historical developments and Pythagorean context that I have reconstructed at length. By the time I finally inspected Inchofer's manuscript, my present book manuscript was virtually complete, so I am delighted that the *Vindication* connects directly to the issues at hand and unites them as a capstone. I have to discuss Inchofer's manuscript succinctly, owing to space limitations, but this will still be at greater length than any works previously discussed.

AS IN HIS *Summary Treatise*, Inchofer reiterated that it was necessary to 'militate' against the Pythagoreans because their vulgar arguments were 'temerarious' and embodied 'manifest errors'. Inchofer complained about Pythagoras and his early followers, and their damned opinions that had been 'buried in the grave for many ages'. But his main critiques were about the recent theorists whom he repeatedly called 'New Pythagoreans'. In almost every manuscript page, and sometimes even five times on a single page, Inchofer repeatedly wrote one damning phrase: *New Pythagoreans*.

In his *Summary Treatise* Inchofer had not overtly named Galileo, Campanella or – God forbid – Bruno. After all, it was unseemly to discuss heretics, but I have pointed out how Inchofer clearly criticized their arguments. However, in his *Vindication* Inchofer did mention Galileo by name several times. At far greater length, Inchofer named other 'New Pythagoreans' who had not been formally declared to be heretics: Copernicus, Kepler, William Gilbert and Philippe van Lansberge, a Dutch minister who had authored a short treatise in favour of the Earth's motion (1630), which was published in a 65-page Latin translation in Middelburg, the capital of the Dutch province of Zeeland.[217] By the time Inchofer wrote, all of these so-called New Pythagoreans were dead, except Galileo. Dr Gilbert had died in 1603, apparently of the bubonic plague in London; Kepler died in 1630; and Lansberge had died most recently, in 1632. All of the recent New Pythagoreans named repeatedly by Inchofer were Protestants, except for Galileo. He did not discuss Bruno, Foscarini, Zúñiga or Campanella. It is impossible to imagine that he was unaware of them – he just chose not to discuss them.

Inchofer cited numerous ancient and authoritative theologians against the New Pythagoreans. For example, in his Preface he noted that St Augustine had criticized Manichaeus (or Mani) because 'his teaching concerning the heavens and stars, and the motions of Sun and Moon, was false.'[218] Augustine said that the arrogance of Manichaeus was sacrilegious, by affirming things of which he was ignorant, and by perverting ideas with egregious vanity and pride, as if he were divine. Inchofer also cited St Epiphanius, Origen's critique against Celsus, St Basil against Origen, and others.

The primary importance that Inchofer attributed to very old ecclesiastical authorities is evident in his first chapter. That chapter discusses how Pope Zacharias had opposed and 'condemned' Virgilius in the year 748. For Inchofer, this was the very 'First Argument' against the New Pythagoreans. He avoided the interpretation that this ancient conflict had merely been a misunderstanding and disagreement about whether there are other people living on the opposite sides of the spherical Earth, the antipodes, specifically because Zacharias had also denounced the plural notion that there are 'suns and moons'.[219] This was directly relevant against the views of Bruno, Galileo, Kepler and others. It is very noteworthy that in his *Dialogue* Galileo had briefly echoed one of Bruno's beliefs in a comment voiced by Salviati, the character that spoke for Galileo himself, saying, 'and as the fixed stars, *which are so many Suns*, in accord with our Sun are perpetually at rest' (emphasis added).[220]

Inchofer said, 'The postulate of many worlds is ridiculous.'[221] He said that the professor of theology at Leuven, Libert Froidmont, had piously and learnedly condemned the heresies of the 'recent Pythagoreans'.[222] Inchofer cited the accounts by Baronius and Aventinus about the dispute between Zacharias and Virgilius. Regarding the account by Aventinus, who had spoken favourably of the theory of Virgilius, Inchofer noted that Froidmont had rightly remarked that the book by Aventinus had been censured by the Congregation of the Index.[223] Inchofer cited Lactantius and Augustine for having criticized the theory of the antipodes. In *City of God*, Augustine denied the antipodes as a 'fable' that had not been proved, because, he said, even if the Earth indeed were spherical, one would have to prove that it has lands throughout, not just bare waters, plus, one would have to prove that there were people there, and descended from Adam.[224] Augustine preferred biblical and historical evidence over scientific conjectures. Similarly, Inchofer argued that the Earth's motion was imaginary and false.[225]

In addition to citing Pope Zacharias as the principal authority against the notion that there exists more than one world, Sun and Moon, I was surprised to see that Inchofer also cited another pope (although not of Rome): Dionysius of Alexandria.[226] (I mentioned him earlier precisely because I found him cited in Inchofer's manuscript.) Around 260 CE Dionysius had criticized Epicurus and Democritus, partly for arguing that 'infinite worlds' are randomly created by the random interactions of atoms.[227] Inchofer's arguments lead me to think that his title *Vindication of the Holy Apostolic See* referred not to Pope Urban VIII but to the popes of antiquity in particular and to the Apostolic See or Chair in general. After all, the Pope was supposedly the designated representative of God on Earth, capable of making infallible pronouncements. The Inquisitors' rulings against Copernicus, Galileo and others had vindicated the Papacy.

Next, Inchofer discussed the 'various opinions about many Worlds'. He discussed the beliefs of ancient philosophers, Talmudic writers and 'recent' theorists. Inchofer affirmed that there exists only one world, as shown by faith in the scriptures and in the Holy Fathers. Inchofer cited the Pseudo-Plutarch's *Placita* to name several ancient philosophers who in one way or another posited more than one world: Anaximander, Anaximenes, Aristarchus, Xenophanes, Leucippus, Democritus, Epicurus and Metrodorus.[228] He noted that the Pythagoreans believed in many or infinite worlds. He also cited 'Cleombrotus, who in a triangular figure constituted three and eighty worlds'.[229] Inchofer complained that 'the recent Pythagoreans' had readily adopted similar opinions.

Against them, Inchofer cited various authorities, including Augustine, Theodoret and Tertullian. In his *Cure of Greek Maladies* (*c.* 430) Bishop Theodoret had ridiculed the opinion of many worlds, which he attributed to Anaximander, Democritus and others, adding that 'Heraclides among others of the Pythagorean sect said that each of the stars are individual worlds, which contain land and air.'[230]

Inchofer mentioned and dismissed Hebraic claims that there were 'eighteen thousand worlds'.[231] For example, according to tradition, in the first century CE, in ancient Israel, Rabbi Simeon bar Yochai allegedly composed the Zohar, the chief work of Kabbalah, and in that work we find the claim: 'When the words come forth from the mouth to the lips to murmur in all the eighteen thousand worlds: until all are connected together in the twelve paths and roads.'[232] Inchofer dismissed such notions as 'Talmudic fables'.[233]

Inchofer cited Tertullian for having once said that change should not be construed merely as a mutation of garments, 'as when Anaximander made many worlds'.[234] Inchofer also cited an ancient letter in which St Jerome criticized Origen for believing in many worlds. Jerome there explained a difference between Origen's conception and that of other philosophers:

> In his second book he asserts that worlds are innumerable, not according to Epicurus where at one time many similar [worlds] exist; instead, after the end of the world, another begins; and before this world of ours, there was another world; and after this one another will be in the future, and after it another; again another after another.[235]

Moving along, Inchofer referred to the things that 'Galileo's observations' with a telescope had allegedly revealed on the Moon: woods, fields, mountains, valleys, rivers and seas. Since Galileo himself had not affirmed so much, Inchofer turned to criticize Kepler's *Discussion with the Starry Messenger* for having posited men who live on the Moon and hide in its shadows.[236] Regarding Galileo and Kepler, Inchofer disdained 'their deviant imagination of many worlds', insisting that theologians rightly teach that there is only one world.[237] He said that the deviant fictions were necessarily 'erroneous and temerarious'.[238]

To consolidate his argument, he said that the plurality of worlds had been declared absurd by St Philaster in his book on heresies, and in conflict with scriptures. And there it is! The explicit evidence that Inchofer knew 'that worlds are innumerable' had been categorized as *heretical*. In 1616, an important date, Bellarmine had similarly noted that he knew well about ancient pagan heresies thanks to the works of Philaster and others. And now Inchofer made an even more explicit statement.

Inchofer added that St Isidore of Seville had also condemned the 'postulate of infinite worlds' as a heresy. He also cited Augustine, Thomas Aquinas and the apostles John and Paul. Echoing Augustine, Inchofer admitted that in a Christian sense it was admissible to say that there exist two worlds. Augustine had borrowed the expression from Plato: that there is a sensible world and an intelligible world. Like other writers, Inchofer justified this notion on the basis of John 18:36, where Jesus said, 'My kingdom is not of this world', as if there exists another. Still, for Inchofer there existed only one physical, sensible and present world: the Earth.

Inchofer argued: 'Therefore, allow that it is a Catholic assertion that the world is one, to deter those who resolutely contend multiple, from achieving heresy.'[239] Inchofer noted that Origen held the opinion of many worlds, although it was not easy to believe.[240] Inchofer further attacked the notion of many worlds and systems, and ridiculed those who 'very stupidly' affirmed it by misinterpreting scriptures. Right then he again dismissed 'the imaginary wisdom of Galileo and Kepler', elicited by the new telescopic phenomena, but which was actually just 'trash and dreams'.[241] Inchofer argued that we can

> appreciate all errors acutely, as they do not understand this nonsense at all, because they [Galileo and Kepler] have trusted their imagination so much, that which others condemn from the sense of reasoning, about the Earth's quiescence and the Sun's motion, from sight and errant imagination daring to affirm many worlds, animals and lunar men.[242]

Inchofer then reiterated that such 'New Pythagorean arguments' had already been denounced as fallacious when Pope Zacharias condemned the notion of the antipodes.

Moving along, Inchofer argued that the errors of the New Pythagoreans resembled the heresies of the Manichaeans and the Valentinians (Gnostic heretics). He mentioned questions about souls, the Earth's magnetism and the generative power of the Sun's heat. He touched upon whether 'the universe has life and sense, the metempsychosis in the opinions of the Pythagoreans and of the Manichaeans, as thought commonly by the innovators, and also by the institutes of Kepler and Galileo that humans and other animate beings inhabit the orb of the Moon'.[243] Thus Inchofer linked Galileo to the heretical theory of many worlds, and lunar beings, and even the transmigration of souls (metempsychosis) – but again, without mentioning Bruno or Campanella.

Next Inchofer discussed an ancient confrontation between Pope Gelasius and the Manichaeans. Recall that the Manichaeans believed that Earth is infused by a soul and that there are many worlds. Many of the Manichaeans also refused to eat meat or drink wine. Hence Melchior Inchofer recounted that around 494 CE Pope Gelasius had persecuted the Manichaeans by decreeing that everyone had to partake of communion by ingesting both the bread and the wine. Thus the many members of this sect who had infiltrated the Christian Roman community would be exposed as heretics. Similarly, Inchofer argued that now the Copernicans or New Pythagoreans had to be exposed.

Furthermore, St Augustine had also criticized the Manichaeans for worshipping the Sun and the Moon as divine.[244] In one of his writings, Augustine replied to a Manichaean who claimed to believe in Mani because 'He taught us why there is day and why there is night. He taught us about the course of the Sun and the

Moon. Since we have not heard this in Paul or in the writings of the other apostles, we believe that Mani is the Paraclete.'[245] Augustine replied that God wanted to make Christians, not astronomers, and that knowledge about the Sun and the Moon did not pertain to Christian doctrine. Similarly, Inchofer warned about the misleading pronouncements of Pythagoras and the New Pythagoreans about the Sun and the Moon.[246]

Inchofer insisted that the Earth's motion was false, just 'Pythagorean dreams, that are eternally false and imaginary, and plainly opposed by the true contents of divine Scriptures'.[247] The 'doctrine of Pythagoras' seemed to be the product of a depraved intellect. Inchofer praised Tycho Brahe for having rightly asserted that there are five planets and that both the Sun and the Moon circle the central Earth.[248]

Furthermore, Inchofer complained that some of the New Pythagoreans had cast doubt on the traditional Catholic notion of how long the Earth had existed. He therefore pointed out that Philaster had listed, as heresy number 84, to be 'uncertain about the number of years since the origin of the world, and to be ignorant of the course of human time, from Adam to the flood, there being ten generations and two thousand two hundred and forty-two years'.[249] Still, Inchofer also cited various time frames specified by other ancient Christian authorities.

Inchofer associated the New Pythagoreans to other heretics or heresiarchs, such as Jan Hus and John Wycliffe.[250] Like Bruno, Jan Hus had been burned alive, whereas Wycliffe was already dead when he was condemned, so the Catholics exhumed his body and burned it. Inchofer also accused heretics who had been condemned of pantheism, and he cited the Fifth Ecumenical Council of 553 for condemning Origen for having believed that the heavens are animated.[251]

Next Inchofer turned his wrath to Philippe Lansberge, the Dutch Calvinist minister who had published a booklet in 1630 endorsing Copernicus. Although Lansberge had recently died, Inchofer criticized him as if he were still a major threat – after all, his book had become very popular. Inchofer associated Lansberge's beliefs with those of Origen, and he wrote that 'the mystical theology of Lansberge is plainly ridiculous' and vain. He praised Libert Froidmont for having derided Lansberge's 'barbaric speculations' and 'inane New Pythagorean arguments'.[252] Inchofer then defended the traditional, literal interpretation of the miracle of Joshua at Gibeon and the valley of Aijalon. He also cited Job 26:7, that God 'suspends the Earth over nothing'.[253]

Inchofer tenaciously belittled Lansberge, Kepler and Gilbert, and he traced the pagan lineage of their Pythagorean heresy:

> Orpheus, then Aglaophamus, who Pythagoras follows, [and then] Plato's teacher Philolaus [following] Pythagoras, and we omit others, but Pythagoras rather badly *departing* from the example and doctrine of his teachers, with Philolaus, [was] the first of anyone who asserted the heresy of the Earth's motion, certainly promoting impiety (just as Cleanthes of Samos later), unless he were wary.[254]

The expression 'and we omit others' leads me to think about who else taught the 'heresy' of the Earth's motion. By now I instantly think of someone in particular: Bruno. Inchofer also discussed 'the sermon of Hermes to Asclepius'. In that ancient work, Hermes Trismegistus claimed that 'the Spirit agitates or governs all species in the world, as distributed from God to each of them according to their natures ... Therefore the world nourishes bodies, spirits, and souls.' Inchofer did not quote these words, yet he rejected Hermes's notion of the soul of the world, and complained that 'other New Pythagoreans wonder and adore this, as with [David] Origanus, in their contempt, perhaps not laughing at their own heresy'.[255]

In case his passing allusions to the soul of the world were not enough, Inchofer further dedicated an entire chapter to attacking this notion. He argued that the notion that the Earth has a noble soul was defeated by scriptures, St Augustine and St Anselm. He said that Origen's notion of an animated heavens had been condemned. Again, Inchofer knew well which 'philosophical' notions were theologically illegal for a Catholic person.

Inchofer ridiculed Anaxagoras for having been cited by Gilbert as an ancient authority who claimed that magnets have portions of the soul of the world. Anaxagoras was also problematic because he, like Orpheus, had reportedly taught that the Moon is similar to the Earth. In his biography of Anaxagoras, Diogenes had reported that: 'Anaxagoras ... was the first who attributed soul to matter . . . He said that the Sun was a mass of burning iron, and greater than Peloponnesus; but some attribute this doctrine to Tantalus; and that the Moon has dwellings, and also hills and canyons.'[256] Hence Gilbert affiliated his own conjectures to those of Anaxagoras; but Inchofer complained that Gilbert's philosophy was ridiculous. At length Inchofer ranted that the alien doctrine of the 'New Pythagoreans' was 'absurd', 'a-Catholic' and 'contrary to the faith'. Inchofer listed and celebrated the foremost critics of the New Pythagoreans: 'the Great Mersenne, Libert Froidmont, Christopher Scheiner, Joannes Costeus and others'. Inchofer complained that the ancients believed in a universal soul, and then he quoted disapprovingly the relevant passages from Gilbert's *On the Magnet*, which was published in 1600, the year the Roman Catholics killed Bruno.

It should be recalled that Gilbert had not mentioned Bruno in that book, but he echoed several of Bruno's beliefs, such as that the Earth moves because it is a living animal with a soul, and that it gives parts of its soul to humans, to vivify them, and that the Sun, Moon, planets and stars all have souls – and even that God is a soul. Gilbert was active in London at the same time that Bruno lived there, and therefore historians argue that Bruno helped to inspire Gilbert's cosmology. Gilbert died in 1603, but he left a manuscript, *On the World*, in which he mentioned Bruno three times. He discussed two kinds of motion that 'the Nolan' had attributed to the Earth 'when he was younger'.[257] In this work too Gilbert wrote about the souls that animate the Earth and the heavenly bodies.[258] None of this was orthodox Catholic doctrine.

In his *Dialogue*, however, Galileo had discussed Gilbert at length.[259] Galileo wrote about whether the Earth is really magnetic. He then praised Gilbert for 'the progress in his manner of philosophizing, with a certain similarity to my own', and he urged that Simplicio (or the reader) should actually read Gilbert's book.[260] Galileo highlighted 'Gilbert's progress upon philosophizing' in the margin of the page as well as in the index of his book.

Instead, in his *Vindication*, Inchofer quoted Gilbert's awful claim: 'we think that the whole world is animated, all the globes, all stars, and also Earth, each have their own distinct souls to govern from the start, to maintain their motions.'[261] He also quoted Gilbert's claim that the Earth is an animal. Inchofer replied: 'We laugh at him more than at the Stoics, as in a contradictory way he affirms and denies that which is a soul, he identifies it with what has vigour or form.'[262] Inchofer remarked that here readers could well see 'how evil is impiety'.[263] Similarly, this was how Schoppe and Mersenne had disparaged Bruno.

Again Inchofer cited more Christian authorities – St Hilary, Augustine, Gregory of Nyssa, St John of Damascus and John Chrysostom – all to argue that misinterpretations of scriptures had given rise to heresies and irrationality. For example, in his book *On the Orthodox Faith*, John of Damascus had discussed philosophers' various opinions about the heavens, to show their confusion and uncertainty, and he there remarked:

> It must not be supposed that the heavens or the luminaries are endowed with life. For they are inanimate and insensible. Thus when the divine Scripture says, *Let the heavens rejoice and the Earth be glad*, it is the angels in heaven and the men on Earth that are invited to rejoice. For the Scripture is familiar with the figure of personification, and is wont to speak of inanimate things as though they were animate: for example . . . again, *the heavens declare the glory of God*, does not mean that they send forth a voice that can be heard by bodily ears, but that from their own greatness they bring before our minds the power of the Creator.[264]

Accordingly, Inchofer quoted multiple passages from scripture, which, to him, were also not meant to be taken literally, for example, 'Listen, you heavens, and I will speak' (Deuteronomy 32:1) and 'while the morning stars sang together' (Job 38:7). Inchofer stressed that there are many other such phrases: 'six hundred others of this kind, of which the explanation is so easy and obvious, and if the New Pythagoreans strongly oppose this, they should be ashamed of their ineptitude, but being guided by Hermes Trismegistus, Orpheus and Pythagoras dims the constitution'.[265] Inchofer argued that if the Sun and the Moon have a soul and a mind, and if the Earth truly moves, then in the Joshua miracle the Earth would need to have hearing organs in order to obey. Inchofer pleaded, 'Turn off the dementia.'

He quoted Origen for falsely asserting that the 'heavens and the Sun and the Moon and the stars and the waters that are above the heavens, have a soul'. This falsehood, Inchofer said, had been echoed by Gilbert, but rightly condemned by Augustine, Anselm and Ambrosius Boethius. Moreover, Inchofer quoted the reputedly official words of the Fifth Ecumenical Council of Constantinople condemning Origen for this very heresy, that 'the heaven and Sun and Moon and stars and the waters that are above the heavens, are animated and have material powers.'[266] Inchofer then pointed out the Copernican affiliation with Origen's heresies, while quoting from the Fifth Ecumenical Council:

> '*Anathema to Origen who is Adamantius, who promoted this with his nefarious and entirely execrable doctrine, and anyone* (note the New Pythagoreans) *who claims or defends that statement, or presumes to protect it in any way or at any time.*' It's all over for the New Pythagoreans, or as settled according to their own opinion, [they are] Heretics, for if through all philosophy, like heaven and the stars so too they think that the Earth is animated, not to say a monstrous animal, having a soul throughout its entire body, clearly to some degree or other they are refuted by positing an animated Heaven and even more so by affirming this about the Earth.[267]

Furthermore, Inchofer complained about the theory that God is the soul of the world, which governs the motions of heavenly bodies. Inchofer said that St Basil, in his *Hexameron*, had rejected the claim that the heavens and the stars are animated, in Basil's words, as 'an ancient imposture, ruinous and rotting'.[268] Inchofer further supported this denunciation by citing John of Damascus, Cyril, Ambrosius and Lactantius.[269]

Inchofer continued to repeatedly attack Gilbert. He said that Gilbert's absurd New Pythagorean philosophy seemed to revive the ancient 'Idolatry or Egyptian paganism', in professing that the Sun is divine. Such beliefs waged war against Catholic philosophy. Inchofer said that Gilbert's claim that magnets and the Earth have souls was clearly and plainly derived from the awful Hermetic philosophy. And right then Inchofer complained that Galileo had greatly supported Gilbert's philosophy.[270]

Inchofer said that it was absurd for a stone, magnetic or not, to have sense. He further complained about the notion that the magnetic Earth is animated by a spirit. He also ridiculed the role of pagan poetry: 'according to the fables of poets, we would think that rivers too are animated; [but] there is nothing very applicable that pertains to the New Pythagorean philosophy, which is so very similar to the fictions of poets.' Inchofer said that since Gilbert had discussed the sounds of the 'Pythagorean harmony', hence 'Gilbert is another Orpheus.'[271]

Once Inchofer had concluded his long chapter against Gilbert's magnetic theory and the soul of the world, he presented a chapter titled 'NeoPythagoreanism is Absurd in Mathematics'. He argued, for example, that it is false to feign that the poles and circles of motion belong to the Earth instead of the heavens. He argued that Lansberge's attempted demonstrations of phenomena about the Earth and the planets were ridiculous, as well as his claims about the prime mobile. Inchofer also argued that nothing confirms that the New Pythagorean explanation of the phenomenon of the precession of the equinoxes is better than the geocentric account.

Whatever might change about astronomy, Tycho Brahe's scheme seemed much more credible than that of Copernicus. Inchofer explained that Brahe had rightly tried to detect any parallax in the orbit of Mars, that is, if indeed the Earth is moving then throughout the year one should observe apparent shifts in the orbit of Mars. But Brahe detected no such shifts, so he concluded that Copernicus was wrong. Similarly, at some point in the fifth century the educator Martianus Capella had proposed an astronomical system in which not all bodies circle the Earth. He had argued that most of them do, except for Mercury and Venus, which circle the Sun. Hence his system was similar to what Brahe proposed a thousand years later. Accordingly, Inchofer quoted Capella for having argued that 'the Earth is not the centre of all the planets', while retaining the Earth's centrality overall.[272]

Again, instead of focusing his critiques on the Catholic Copernicans, Inchofer attacked the Protestants. For example, he wrote: 'This is essentially Lansberge, who with his erratic mind precipitated, I know not what apparently confused [him], this ridiculous man continued to exclaim that the Copernican hypothesis is true, and audaciously affirmed that not the Sun but the Earth moves.'[273] Inchofer also discussed whether the world in Gilbert's account was finite or infinite. Like Bruno and Digges (and unlike Copernicus), Gilbert had claimed that the stars are not all in one sphere, but extend outwards far beyond the planets. In his book of 1600 he did not specify whether the universe is infinite, yet Inchofer inferred that indeed 'the philosophy of Gilbert has accepted, [that] the world is infinite.' But Inchofer countered, 'We say certainly finite, not denying that it can move.'[274]

Inchofer complained that the New Pythagoreans posited 'six hundred other things' against the Ptolemaic System and the Tychonic System, but that all such things were not demonstrated but were merely assumed and postulated with their 'abortive imagination'.[275] He concluded that the notions that the Earth has magnetism, and various motions and inclinations were 'figments' that were just 'not credible'.[276]

Next, he titled the last chapter of his manuscript 'The Censure of the Sacred Tribunal against the Propositions of the New Pythagoreans'. Inchofer wrote that they were 'abusers of the sacred philosophy, who on the authority of Pythagoras try to vindicate the entire sect, contrary to religion'.[277] Inchofer then quoted the relevant passages from the Decree of the Index of 1616 that censured Copernicus, Zúñiga, Foscarini and the 'false Pythagorean doctrine' of the Earth's motion and the stationary Sun. Inchofer commented, 'this was the first Decree after which the New Pythagoreans began to be heard about outside the [monastic] schools.' He said that, however, this censure alone did not suffice to restrain the 'evil wit' of the false imagination that could lead some Catholics to being rash, having 'temerity', and that, consequently, the Church had to issue a second sanction.[278] He then quoted the 'Warning' or 'Reminder' of the Holy Congregation against the Copernicans published on 15 May 1620, which Inchofer dated to 22 November 1619. That document not only required the censorship of specific passages in Copernicus's book, but it reiterated a stern warning to the readers of Copernicus, as Inchofer quoted: 'the principles about the location and motion of the terrestrial globe, which are repugnant to the true and Catholic interpretation of Holy Scripture (which is minimally tolerated in a Christian man)'.[279]

Next, Inchofer summarized the third official action against the New Pythagoreans: the Sacred Congregation's condemnation of Galileo in June 1633, and that the Index had decreed that his *Dialogue* should be prohibited in August 1634. Inchofer then quoted the key points of the proceedings against Galileo:

> And from what has been said, the Judgment and decree of the S. Congregation, built upon the authority of the Supreme Pontiff, we have this about the NeoPythagorean opinion, '*it is false*', in the first place, '*and entirely opposed to the divine Scripture, slithering perniciously into the Catholic truth*'. Then, '*It is repugnant to S. Scripture, and the true Catholic interpretation,* [to be] *minimally tolerated in a Christian man*', and finally, '*totally prohibited*'.[280]

The quoted lines echo the Decree of the Index of 1616, and the subsequent rulings, but the wording is slightly distinct.

Furthermore, Inchofer remarked that the author of the Preface of Copernicus's book, namely Andreas Osiander, was a known heretic. Recall that Bellarmine had listed Osiander as a heresiarch. Inchofer also accused Kepler of 'strenuously cultivating impiety', and a 'hard and frigid heresy'.[281] Like Froidmont, Inchofer quoted the proclamation of Pope Sixtus v, to the effect that any propositions 'which pertain to the dogmas of the Faith' are reserved to being exclusively interpreted by the Holy See.[282] He also echoed Froidmont for saying that the Copernican theory had set a foot on the threshold of heresy, but Inchofer added that this was not just in part, 'but completely'.[283]

Inchofer ended his book by enumerating the various ways in which, in his expert opinion, the questionable theory transgressed Catholic considerations. He titled that last section 'The Author's Censure against the NeoPythagorean Opinion'. He said that, first, it was 'offensive' to say that Earth moves and not the Sun. Second, he said that it was 'scandalous', for not having any justified rationale. Third, it was rash or 'temerarious'. Fourth, he said that it was 'ill sounding'. 'Fifth, we think that it is erroneous. Affirming the false as if true.' Sixth, he said that it was a 'conscious heresy'. And lastly, 'Seventh, we think that without ambiguity it is openly Heretical: because it fights against the explicit truth in the S. Scriptures.' All these terms – *offensive*, *scandalous*, *temerarious*, *ill-sounding*, *erroneous* and *heretical* – were standard Catholic categorizations for distinct kinds of transgressions.[284]

Did Inchofer's superiors agree with his critiques? A relevant piece of evidence appears in a page that Inchofer appended at the start of his manuscript. The Jesuits required that every Jesuit should submit any manuscript to an internal review, before it was submitted to the Roman censors or possible publishers. Thus the Jesuits tried to secure a doctrinal uniformity within their works. In his note addressed to such Jesuit reviewers, Inchofer explained that he had described the theory of the Earth's motion around the Sun as heretical only because he had consulted those to whom the right to make that judgement belonged.[285] Hence, Francesco Beretta remarks, 'This text by Inchofer gives us an important indication concerning the intention of Urban VIII himself.'[286]

Inchofer declared that the New Pythagorean beliefs were a 'Heretical depravity'. It all deviated from the 'unanimous consensus' of the Fathers and the Sacred Tribunal. Inchofer noted that he himself had submitted his opinion to the Holy Apostolic See and the Sacred Tribunal. He complained that some writers tried to accommodate scriptures to their own *mathesi*, that is, their mathematics or science, and he noted that Philaster had denounced this practice as 'Heresy 43'. Inchofer did not quote that heresy, but it denounces Colarbasus, an obscure heretic and follower of Valentinus and Pythagoras, who tried to use letters and numbers to pronounce about the stars and the lives of men. Inchofer then specified that it was very closely related to 'the Valentinian Heresy' and 'the philosophy of the Gnostics'. It should be recalled that St Hippolytus had complained that Valentinus had derived his heresies from Pythagoras and Plato, and therefore was 'a Pythagorean and Platonist, not a Christian'. Inchofer did not explain this at length, but it was said that the Gnostics boldly committed a typical error of heretics: while purporting to be Christians they modified orthodox Christian beliefs by adding notions that were foreign to Christianity, such as the transmigration of souls. Accordingly, Inchofer also cited the Manichaeans once again.

Finally, on the very last page of his long manuscript, Inchofer again emphasized the offensive core of the Pythagorean heresies:

> but even if the pagan Heathen wanderers degenerate, attributing perceptions and comprehension to degenerate the mute nature of things, affirming that the universal world possesses a soul, even the Earth on which we walk, not knowing what kind of strange soul, latching onto the more ridiculous or stupid. Such truly impious arguments, and just

> as formerly the heads of the Church already condemned it in sacred laws, and even in Councils, as we show here, amply repeating it, we do not seem to exaggerate the case, we think not.[287]

Summing up, Inchofer's manuscript *Vindication* is a striking document for clearly showing how Galileo's judges could and did link his seemingly particular indiscretions with greater pagan heresies. It was their obligation to know well the realm of Catholic improprieties, and to study the standard canons and history of heresies.

Inchofer finished his manuscript in 1635. Did he already know in 1633 that these pagan beliefs were offensive? Yes, because he mentioned them in his *Summary Treatise*. He there quoted Augustine's brief work on heresies, which includes the heresy of many worlds. Moreover, in a book published in 1629 he had already criticized the 'deliriums' and 'madness' of philosophy, including 'that stars are animated', and 'that in the immensity of things many seeds exist, which can constitute immense worlds, according to Anaximander'.[288] Inchofer also blamed Anaxagoras for saying that 'animals arose from the seeds falling from the heavens.' He mentioned also Thales, Pythagoras, Anaxagoras, Democritus, Epicurus, Valentinus and the Gnostics.[289] Inchofer thus quoted Irenaeus's book *Against Heresies*, and referred also to Jerome, Theodoret and Epiphanius – all before Galileo's trial.[290]

Galileo's philosophical or scientific concerns, despite his precautions, were easily linked with pagan beliefs contrary to Catholic beliefs. Moreover, Galileo's beliefs seemed to be a subset of the heresies of Giordano Bruno. We must be careful when trying to analyse topics about which individuals chose to remain silent, yet it is necessary to identify tacit knowledge, because it too influenced Galileo's judges. In writing, Inchofer did not mention the heretic from Nola, but Bruno's shadow seems to infuse his manuscript text. Inchofer rebuked many notions that appear prominently in Bruno's writings, trial and condemnation, including the plurality of inhabited worlds, the soul of the world, the Earth's motion and the eternity of the world. Remarkably, Inchofer even quoted phrases that had a deep significance in Bruno's world view.

For one, Inchofer mentioned Ecclesiastes to deny the notion that the world is eternal, 'as asserted by some'. He then specified his reference to Ecclesiastes by adding these words in the margin, '*what is it that was, that which the future is*'. Then in the main text he added Solomon's words 'there is nothing new under the Sun'.[291] Remarkably, these are the very lines from Ecclesiastes 1:9 that Bruno had edited into his personal motto about Pythagoras, which to him signified the immortality of the soul and the eternity of primal substance. These lines previously had also been appropriated by Origen, who construed them as referring to the succession of many worlds. Bruno had praised Origen as the only theologian who had dared to voice the truth of eternal transmutations.

Also, in the midst of his long diatribe against Gilbert's theory of magnetism and souls, Inchofer briefly touched upon the key phrase of pagan poetry that for decades had inspired some eccentric philosophers and astrologers, including Bruno, Kepler and Campanella. Inchofer briskly quoted and disdained the infectious words of Virgil:

> But really, the animating power of the Earth, and as the Poet says, *spirit nourishing within*, to ennoble, and nevertheless it is always fixed in place, clinging like an oyster or a sponge, benumbed, this implies several more things contrary to nature, more than can be easily explained. Since all of this lies open to inspection, we need not pursue it more here.[292]

Thus Inchofer abruptly and strangely quoted the words that seemed to convert ancient poetry into subversive secret knowledge about God or polytheism. The cryptic allusion to an oyster or sponge seems reminiscent of an argument voiced by Campanella in his book of 1620, *Sense of Things and Magic*. Campanella had there argued that the human soul cannot be transmitted into oysters or sponges because they are utterly different from humans.[293] Campanella and Inchofer both critically departed from Bruno, who in one of his early discussions of transmigration had specifically said that the substance of human souls 'is the same in essence specifically and generally with that of flies, marine oysters and plants, and of whatever thing that is animated, or having soul'.[294] Bruno boldly affirmed the equivalence of human souls with those of very different beings. The same soul of the world inhabited all living things.

Inchofer's attacks against the pagan religious notions show that indeed he was not merely concerned about philosophy, astronomy or physics. Instead, he was most upset about a clear and present danger: an apparent heretical challenge to Catholicism. It was what Inchofer called the 'NeoPythagorean theology'.[295]

I suspect that other consultants and Inquisitors in Galileo's trial were similarly aware of how Galileo's views seemed akin to pagan claims that saintly Catholic authorities had repeatedly denounced in the Pythagoreans, Democritus, Porphyry, Origen, Virgilius, Abelard and Bruno. We have already seen that Cardinal Barberini was very annoyed by what Galileo's work implied: *that the Earth is a star* – a Pythagorean belief denied by Aristotle but which Bruno and Foscarini had defended, and which the cardinals had censored in Copernicus. To find out whether other participants in Galileo's trial criticized Pythagorean notions, it would be necessary to inspect all of their extant writings. For example, soon after Galileo's trial, another one of the critical consultants, Zaccaria Pasqualigo, did reject the 'vain' opinion of Pythagoras about the transmigration of souls, and he echoed Aristotle's argument that 'there cannot exist another world.'[296]

Furthermore, I wonder which censors, consultants and Inquisitors associated the Galileo affair with the transgressions of Giordano Bruno. At the time the documents from Bruno's trials were still available to the Inquisitors. I should point out one intriguing trace of evidence. In the 1920s Enrico Carusi found new documents from Bruno's trial in Rome. Between two sheets of 'The Decree of the Congregation of the Holy Office' regarding Giordano Bruno, dated 20 January 1600, Carusi found a slip of paper that reads 'From the sheets of F. Mo . . . of this Congregation there remains only to add the 2nd decree of friar Giordano Bruno, all the others have been posted.'[297] Luigi Firpo conjectured that this referred to a compilation of documents prepared for the 'Father Monster', Niccolò Riccardi. This sliver of evidence showed that Riccardi had consulted the proceedings against Bruno. Unfortunately, as Luigi Firpo notes, the slip of paper found by Carusi is now missing.[298]

Campanella's Exile and Death

Meanwhile, Campanella's hyperbolic explanations of Pope Urban's poems failed to win the approval of the Curia for publication. Protestant reformers had argued that Urban's Roman ideas were proof of his perversion, so Campanella's emphasis on classical sources would hardly help the Papacy, as historian Peter Rietbergen remarks: 'The Pope, afraid he himself might fall under suspicion of harbouring Copernican, and worse, Pythagorean and neo-Platonic views, resignedly accepted the fact that the Campanella-edition of his poems would not be after all.'[299]

Soon after Galileo was sentenced to live under house arrest, Campanella too was discredited. Members of the Spanish government complained about his presence in the Vatican, because they suspected his involvement in acts of insurgency in Naples.[300] In addition he was subjected to persecution by the Master of the Sacred Palace, Riccardi. In 1634 Campanella fled to France, at least with the apparent support of letters from the Pope. Campanella then stayed at a convent in Paris with fellow Dominicans. Fortunately he was also well received by some French savants and by Cardinal Richelieu, and even received a few kind words from King Louis XIII.

Campanella then tried to retaliate against Riccardi by attacking his writings, unsuccessfully. He promptly published new books and second editions of his works in Paris and Lyon. These were initially sanctioned by the Catholic faculty of the Sorbonne. However, one of the Pope's emissaries in France and Cardinal Francesco Barberini both warned in letters that Campanella's theology was dangerous, requesting that he be discredited before Cardinal Richelieu and the French court, and that the publication of his works should be opposed.[301] Francesco Barberini stressed that it was important to monitor Campanella with vigilance because he was 'temerarious, and presumes to know much', he was prone to 'writing extravagances', and worse – capable of 'being a Heresiarch'.[302] In another letter he warned that Campanella was 'volatile, inconstant and cannot be trusted', because he might wilfully make himself the leader 'of some pernicious doctrine', and therefore that it might even be necessary to remove him from Paris.[303] Hence the initial, warm welcome that Campanella had received in France, from Richelieu, the liberal Sorbonne and even the King, quickly crumbled. Early in May 1636 the faculty of the Sorbonne retracted all of the approbations it had inscribed in Campanella's recent books.[304] His request to publish *Theologica*, his new manuscript, was also denied.[305]

For years Campanella had heard complaints about his book *On the Sense of Things and Magic*. He had planned to publish a second edition, which would include an addendum defending his claims. Although the Sorbonne had retracted its authorization to publish his books, Campanella had previously secured permissions to publish his *Defence of On the Sense of Things* in Rome, in 1631, and also in January and April of 1636 in Paris. Therefore his *Defence* was finally printed in 1637, with such permissions, together with a full copy of *On the Sense of Things*. Campanella opened the book with a dedication to Cardinal Richelieu, exhorting him actually to build the City of the Sun that he had written about many years ago.[306]

In his *Defence* Campanella tried to prove several controversial propositions, as if many great authorities clearly supported them. His main two theses were that things have sense and that the world has a soul! Campanella waged an attack on many fronts, which he called 'phalanges'. He gave dozens of arguments to insist that things really do have sense, quoting profusely from the works of many authorities, including many saints, such as John Chrysostom, Ambrose, Basil, Gregory of Nyssa, Clement, Irenaeus, Augustine, Albertus Magnus and Thomas Aquinas. He also quoted several other Christian authors, including Lactantius, Clement of Alexandria, Eusebius of Caesarea and even Origen. Contrary to Melchior Inchofer, Campanella construed any author's metaphor about things or parts of the world somehow perceiving as literally true. Hence the world was alive.

The 'corporeal world', Campanella wrote, is an image of God or, more exactly, 'a living statue of God'. He said that in articulating this view he was following Orpheus, Timaeus and Origen.[307] He noted that Orpheus believed that the world is a rational animal, with sense and reason, as reported by Eusebius.[308] Then Campanella added a very eccentric and wilful 'conclusion', as he wrote: 'Therefore it is an error, and at least a material heresy, to deny the soulful life of material things. But a formal heresy against all the Scriptures to deny life, and sense in every way.'[309] He then claimed to be following the rules of Augustine, Thomas Aquinas and the Sacred Canon, yet none of those sources attributed soulful life to material things.

Years earlier, in his original edition of the *Sense of Things*, Campanella had said that Copernicus ascribed both motion and sense to the Earth. Now that the Copernican or Pythagorean doctrine was clearly heretical, however, Campanella did not repeat or elaborate his earlier claim. In that first edition, Campanella's comments on the soul of the world appeared in only six pages, in subtle and understated ways. But in his *Defence* he added fifteen more pages discussing this thesis openly and vigorously, and articulating it in detail. To Varro, Anaxagoras and the Stoics he attributed the opinion that God actually *is* the soul of the world.[310] He contrasted their opinion to the notion that, instead, God created an intelligent soul of the world, and Campanella credited this notion to 'Timaeus the Pythagorean', and to Plato, Porphyry, Plutarch, Plotinus, Proclus, Iamblichus, Virgil, Calcidius, Numenius, Ficino, Christopher Landinus, Pico della Mirandola, St Basil and Gregory of Nyssa. He could have added Giordano Bruno to the list. Furthermore, Campanella quoted the ringing lines from the sixth book of Virgil's *Aeneid*, which he wrote as:

> In the beginning, the Sky and Lands, and regions of water,
> Shining globe of the Moon, and Titanic stars
> the Spirit nourishes within; the totality infused through
>   the limbs,
> a Mind agitates the Mass, and intermixes itself with
>   the great body.[311]

Campanella remarked that Virgil rightly said 'that a mind is infused in the limbs of the world, totally agitating the mass, which forms all bodies, and animates, and produces for its own ends'.[312] He also quoted the alleged words of Pythagoras, as reported by Clement of Alexandria, that God is not outside the world, but is within everything, and is the mind and mover of all. Campanella immediately explained the alleged words of Pythagoras, saying: 'Therefore God did appoint the world intrinsically to give; and in his light, and contemplation, and it produces the vivification of souls.'[313]

To justify his belief that the Earth somehow vivifies all organisms, Campanella also quoted passages from the Bible. For example, he quoted that God commanded: 'waters produce living reptile souls, and volatile' (Genesis 1:20), and that God commanded: 'Earth produce living souls' (Genesis 1:24).[314] Thus Campanella construed many biblical statements very literally, wherever it suited his purpose, and contrary to orthodox Catholic interpretations. Campanella insisted that souls do not come from nothing nor from a substance, but from the soul of the world. He argued that heat animates all, that there is a spirit in the bodily humours, a kind of animal heat: *in humore spiritus (calor animalis)*. He quoted Aristotle on how animals are generated from solar heat. He did not mention that others such as Kepler and Julius Caesar Scaliger had echoed the same argument. Campanella clarified that although the universal mind animates the world, animals gain their senses from the etherial heat. Again, to substantiate his argument that the soul is conceived in the air and diffused in the body, he cited passages from ancient and religious authorities, including Moses, Augustine, Lactantius, Aristotle, Pliny, Averroes and Galen.[315]

By stressing the sense and soul of all things, Campanella claimed to oppose atheism. Meanwhile, still in 1637, a Greek priest living in Paris, Athanasios Rhetor, drafted a book attacking Campanella's 'errors'. Rhetor promptly sent his manuscript to Rome for approval and publication. He also extracted an abridged version, which the Sorbonne approved for publication in 1638, titled *Anti-Campanella, A Summary Written against the Book On the Sense of Things and Magic*. Rhetor explained that he had written his tract to defend the Catholic truth and to warn Christians against a nefarious doctrine. He argued that the thesis that 'things have sense' was a heresy. His main complaint was against the notion of the soul of the world. He complained that this was a *dogma ethnicum*, that is, a pagan doctrine.

Campanella had said that the conscious and intelligent world was 'the most noble, the best, and the most beautiful of beings', and that it was 'the son of the sovereign good', and even the first instrument of God's wisdom.[316] Therefore Father Rhetor argued that this doctrine was heretical, because it conflated the world with Christ, the only true son of God.[317] It was also heretical to claim that the world was a being with a soul superior to those of humans, he said, because humans were the only beings who had been created in the image of God. The Greek priest further argued that God, in order to create and act, did not need any mediator or instrument such as the soul of the world.[318] Bellarmine had made that argument. Rhetor also criticized Campanella's *City of the Sun* as indecent and impious, and noted the disrespectful way in which the author had named Jesus Christ and the Prophets merely in a list containing impious and atheist legislators. Rhetor said that Campanella actually taught atheism while pretending to do the opposite.[319]

In his dedicatory preface to the Chancellor of the Sorbonne, Father Rhetor asked the Chancellor to rage against Campanella's pestilent doctrine potentially capable of dragging many souls to perdition. Rhetor demanded that Campanella be punished, and that the reading, printing and sale of his books be rigorously prohibited.[320]

Meanwhile, despite all misfortunes, old Campanella remained optimistic. One of his Parisian friends, Nicolas Chorier, wrote down an anecdote that conveys Campanella's positive outlook and his notion of the soul of the world as the warm, divine breath:

> He [Campanella] was a joyful and festive man: when he walked in the farmland or in the garden, all he did and said was that he was glad to have remissions from serious labours, which relax the mind. He chased birds, trying to run, with the heavy mass of his old and obese body, tossing his hat into the air, he turned to them and said, 'Inhale, inhale, life from the life of the world. Air is the life of the world', he said, 'which is the soul of nature.'[321]

In May 1639 Campanella suffered colic pains for three weeks. He then died, long before Rhetor's critique was finally printed.

FOUR

WORLDS ON THE MOON AND THE STARS

Like Inchofer and Campanella, other writers associated Galileo with Pythagorean beliefs. In 1638, while Galileo was living under house arrest, an Anglican clergyman and Fellow of the Royal Society, John Wilkins, published a treatise titled *The Discovery of a World in the Moon*. Wilkins sought to prove that the Moon was a habitable world. He explained that many ancient and modern mathematicians had defended this thesis. He noted that some attributed this belief to the very ancient Greek poet Orpheus, followed by Xenophanes, Anaxagoras, Democritus and Heraclitus. Wilkins then added 'unto these agreed Pythagoras, who thought that our Earth was but one of the Planets which move round about the Sun, (as Aristotle relates it of him) and the Pythagoreans in general did affirm, that the Moon also was terrestrial, that she was inhabited as this lower world.'[1]

Then Wilkins echoed 'Plutarch's' account in the *Placita*, about the creatures and plants living there. Wilkins wrote that Plato too had agreed with 'this opinion of Pythagoras', and that Aristotle had opposed it but that it had never been refuted with any solid reason. Wilkins said that Nicholas of Cusa and Giordano Bruno had asserted that every star is a world, but that the narrower proposition, about the Moon being an inhabitable world, had been more directly substantiated by Maestlin, Kepler *and Galileo*.

Wilkins argued that although some readers dismissed Kepler's and Galileo's remarks about the Moon being another world, by construing them as jests, such comments were really meant seriously, because otherwise Campanella would not have had to apologize for Galileo, and 'Galileo would never have suffered so much for it as afterwards he did.' Wilkins celebrated the invention of the telescope, he praised Galileo as 'the new Ambassador of the Gods furnished with this perspective to unfold the nature of the Stars, and awaken the ghosts of the ancient philosophers'.

Furthermore, Wilkins tried to prove the proposition that 'a plurality of worlds does not contradict any principle or reason of faith.' To do so, he attacked Aristotle's inferences and he listed alleged reasons why 'some say' that such a proposition is contrary to scriptures: that Moses spoke only about one world, that St John had said that God made the world, in the singular, as Thomas Aquinas had argued, and that 'The opinion of more worlds has in ancient times been accounted a heresy, and Baronius affirms that for this very reason, Virgilius was cast out of his Bishopric, and excommunicated from the Church', and hence he was 'condemned'.[2] But Wilkins thoroughly disagreed, giving multiple arguments, even clarifying that Pope Zacharias had actually been concerned about there being a habitable world *within* the Earth.

How Heretical, Really?

In the end, the Pythagorean claims of the Copernicans are noteworthy in light of the pagan heresies that early Christian authorities had damned. An intriguing question remains: to what extent were the so-called Pythagorean beliefs considered heretical in the 1610s and thereafter? This broad question is difficult to answer mainly because there were disagreements about what constitutes a heresy. There was no single inquisitorial list that comprehensively itemized all propositions that were deemed heretical by all Catholics. Multiple compendiums had different lists, and individuals disagreed about which departures from orthodoxy were grave enough to be heretical. Yet we can get a sense of the answer by first discussing a narrower question: to what extent was the Earth's motion around the Sun viewed as heretical? This question is a subset of the first, since the Inquisition and the Index had called it a 'Pythagorean doctrine'.

It was not merely a notion from astronomy. Instead, it was also a part of an eccentric religious conception that the Earth has a soul, and therefore it is a living animal that moves like other *animated* beings, as argued by Ovid, Origen, Ficino, Bruno, Gilbert, Kepler, Campanella and others.

Here are some of the disagreements, the wide spectrum of opinions, on whether this Pythagorean doctrine was heretical. In 1584 Diego de Zúñiga said that 'the statement of the Pythagoreans' could be used to explain Holy Scriptures. Yet in 1597 the Catholic Consultors censured the proposition that Earth moves in the works of Giordano Bruno. In response, Bruno insisted that Earth spins and moves but the firmament does not, and he denied that this undermines divine scriptures. However, any Catholic who refused to abandon an opinion that contradicted the Bible was considered a heretic.

Catholic writers referred to claims about Earth's motion and the Sun's immobility in various ways. The Jesuit Serarius said that it was heretical. The astronomer Nicholas Mulerius wrote that 'we would not dare to fall into the opinion of the Pythagoreans, which is openly contrary to Scriptures'. Fathers Zuñiga and Foscarini argued instead that the 'Pythagorean opinions' actually do not disagree with scriptures. But then Cardinal Bellarmine warned Foscarini that actually it 'is a very dangerous thing, likely not only to irritate all scholastic philosophers and theologians, but also to harm the Holy Faith by rendering Holy Scripture false'.[3] In private Bellarmine seems to have held an even stronger opinion, as Prince Cesi wrote to Galileo: 'Bellarmine told me that he considers it heretical, and that the motion of the Earth, without any doubt, is contrary to Scripture.'[4] Galileo disagreed. In a deposition, Father Caccini said that it was 'nearly heretical'. But in private he too was apparently more critical: Giannozzo Attavanti reported that Caccini had told him that the Sun's immobility was indeed 'a heretical proposition'.[5] Father Ferdinando Ximenes said that it was 'false and heretical'.[6] The Tuscan ambassador Piero Guicciardini wrote that Bellarmine and Pope Paul v had called it 'erroneous and heretical'.[7] Alessandro Tassoni said 'heretical'.[8] The official Notice from Rome 'to be published everywhere' said that the Congregation of the Index ruled that 'the Pythagorean opinion by which the Earth moves and the Sun stays fixed is contrary to Sacred Scripture.' Father Antonio Querengo said that the opinion 'manifestly dissents from the infallible dogmas of the Church'.[9] In 1631 Froidmont and Jean Morin hesitated to condemn it as heretical, pending Pope Urban's explicit judgement, but they called it 'temerarious', and entering 'the threshold of heresy'.[10] By 1635 Melchior Inchofer judged that this Pythagorean opinion was offensive, scandalous, temerarious and heretical.

From this sample of more than fifteen Catholic individuals we see that there were various disagreements, describing the Earth's motion and Sun's immobility as consonant with scriptures, or temerarious, erroneous, false, contrary to scriptures, nearly heretical or heretical. Likewise, other so-called Pythagorean doctrines were variously viewed.

Consider the thesis that multiple worlds exist. In 1631 Froidmont cited Pope Zacharias at length to finally say that the notions that there is another Sun, another Moon or another inhabited Earth are 'heretical, or are nearly so'.[11] Inchofer referred to the idea that there are worlds in the Moon or the Sun as monstrous 'errors'. In his long manuscript he condemned it as 'heretical', following Philaster, Zacharias and others. Previously, in his *Defence of Galileo*, Campanella too had noted that St John Chrysostom had said that it is 'heretical and contrary to scriptures' to assert that there are 'many heavens and orbs'. Campanella was not the only one; for example, the Jesuit Nicolas Caussin also cited Chrysostom to note that heretics believe in 'other starry heavens and worlds'.[12] But Campanella and Wilkins disagreed with them and celebrated what they admired as a great discovery by Galileo.

Domenico Gravina voiced a critical and authoritative opinion. Gravina was a consultor of the Inquisition, and in 1630 he published the *Catholic Prescriptions against Heretics of our Times*. Gravina listed and denounced many heresies, and among those he argued that 'it is very clearly against the orthodox faith to dogmatize' that 'matter is coeternal with God', and to 'defend the transmigration of souls', and 'to dream of many worlds preceding the formation of our first parents'.[13] Bruno had been accused of these very claims, but as usual he was not mentioned in this book. Gravina denounced such claims right alongside heresies about adultery, being re-baptized, having sex with demons, and denying that souls were created by God.[14] As a consultor of the Inquisition, Gravina knew the Inquisitors in Galileo's trial, Inquisitors whom he named and praised in his writings.[15]

Likewise, in a book of 1629, an authoritative Spanish jurist rejected the plurality of worlds by referring to St Augustine's work on heresies, along with the writings of Aristotle, Tertullian, St Isidore and Justus Lipsius.[16] Similarly, editions of St Jerome's works also denounced the notion as heretical.[17] Another book of 1631 cited Augustine to argue that it is false that the world is eternal and that 'innumerable worlds' exist.[18] New editions of Augustine's authoritative and selective list of heresies continued to include the '77th heresy': that innumerable worlds exist.[19] Furthermore, the 'opinion of innumerable worlds' was also listed as a heresy in the authoritative compendium of *Canon Law* issued by Pope Gregory XIII, which was printed in 1582, 1591, 1605 and 1622 – which was still the official code of law in 1633.[20]

Furthermore, the Jesuit theologian Antonio Rubio had decried the belief in many worlds in a book first published in 1617 and reprinted in 1620, 1625 and 1626. Rubio identified it as a heresy and cited critics of this belief, including Augustine.[21] As noted before, Rubio did not attribute this belief to Pythagoras. But multiple writers attributed the theory of more than one world to Pythagoras or the Pythagoreans, including 'Plutarch', Hippolytus, Lucian, Iamblichus, Theodoret, Bruno, Piccolomini, Kepler, de Nancel, Lagalla, Galileo, Jean Tarde and Campanella.

Moreover, when the Calvinist minister Johann Heinrich Alsted authored a 'Chronology of Heresies, Sects and Schismatics', in the 1620s, he listed ancient and recent heresies, including 'inventing innumerable worlds'.[22] Did Alsted know about Bruno? Yes, and he had studied Bruno's art of memory.[23] Some other books, though, were not as explicit. One theologian in 1624, for example, discussed the notions of pagan philosophers such as Thales, Pythagoras, Plato and Anaximander, and noted that 'even in the Church the errors of such philosophers have induced heresies.' Among the various questionable philosophical views, the author included the claim that 'innumerable worlds' exist.[24] Here the claim of heresy was ascribed to philosophers in general, not overtly to the one thesis in question in particular. Regardless, other more authoritative works did categorize it as a heresy, as we have seen.

Next, consider the thesis that Earth has a soul. Bruno affirmed it as compatible with scriptures, calling it a 'Pythagorean doctrine' and he advocated this 'Pythagorean way' of reading the Bible in order to understand what Christians mean by 'the Holy Spirit'. Bruno's Inquisitors disagreed with his departure from the Council of Trent: only the Holy Church had the right to interpret or explain scriptures, not the Pythagoreans and not Bruno. Apparently Bruno did not know that it was a heresy to say that the Holy Spirit is the soul of the world.[25]

By 1611 Campanella cited Origen to support 'the Pythagorean dogmas' that Earth is alive and soulful, like the stars. Yet any Inquisitor who knew Origen's infamous heresies knew that, 'If anyone shall say that the Sun, the Moon and the stars are also rational beings', that person is a heretic.[26] Bishop Tempier too had censured the belief that the heavenly bodies have souls. Hence, in 1613 one of Galileo's critics, the clergyman Francesco Ingoli, declared (in a dinner hosted by Prince Cesi) that the opinion 'that the heavens are animated' had been 'condemned as erroneous by the Sorbonne of Paris'.[27] Yet Galileo secretly thought that a vivifying spirit emanates from the Sun.

Accordingly, in 1627 one theologian warned that Christians should approach Plato's doctrines with caution, rejecting the belief that the world is a soulful animal, 'that which is denied by our law', because 'if it were allowed, it would give the opportunity to Heretics who hold (wrongly) that the heavens and the elements are animated, an opinion that is false & heretical and is legally condemned & reproved.'[28] That same year the Index of Forbidden Books prohibited a treatise by Robert Fludd (who had been physician to King James I of England), which argued that the world has a soul that emanates from the Sun, by the Trinity, which is how God becomes ubiquitous.[29] Fludd argued that this world soul animates and vivifies beings and he attributed this belief to the Platonists, the Pythagoreans and others, and he quoted Virgil's line 'Spirit nourishes within'.[30] Hence, in 1631 Froidmont dismissed Kepler's claim that the Earth has a soul as a delusion. Father Rhetor said that it was heretical. Inchofer categorized it as an 'error' that leads to other 'deceptive errors'. In addition, in his manuscript *Vindication* he duly quoted the Fifth Ecumenical Council and declared that it was a heresy to believe that Earth or the heavenly bodies have souls.

Next, the doctrine of the transmigration of souls was not merely heretical: it was anti-Christian. Irenaeus had denounced the Gnostics for this belief. Theodoret said that Mani and the Gnostics had copied it from Plato and Pythagoras. It was also often attributed to Empedocles and Simon Magus, to Basilides, as an alleged punishment for sinners, and to Carpocrates and his followers. Hippolytus, Tertullian and Epiphanius denounced this notion in Pythagoras, Marcion, Valentinus, Elchasai, Colarbasus, the Gnostics and the Manichaeans. For centuries it continued to be criticized as a common 'error' of the heathen or even the ancient Hebrews.[31] In 1599 it was condemned again in the Synod of Diamper. Any Catholics who asserted this error, especially against fair warning, were regarded as heretics.[32] Accordingly, in 1620 Campanella referred to Pythagorean transmigration as a heresy.[33]

Next, some notions that were often attributed to the Pythagoreans were also classified as heretical even though they were too widespread among various religious and philosophical groups to be identified simply as Pythagorean. In particular the notion that souls are immortal had been frequently attributed to Pythagoras since antiquity. But it was also attributed to other philosophers, such as Plato and Empedocles, both of whom were said to be followers of Pythagoras. Christian writers denounced this notion because it claimed that souls pre-exist human bodies or are themselves corporeal. Instead of naming it in relation to any particular philosopher, this heresy became associated with prominent Christian heretics, such as Origen and Tertullian. Thus the 'Origenists' and 'Tertullianists' appeared in St Isidore's list of heresies, and consequently also in the *Corpus of Canon Law* published by Pope Gregory XIII, which states, for example, that the Tertullianists are those who proclaim that 'the soul is immortal, but preaching that it is corporeal, and they think that the souls of men who are sinners are converted into daemons after death'.[34] Like other entries, this heresy does not specify the Pythagoreans by name but such beliefs had often been attributed to them.

Consider finally the French scholar Gabriel Naudé. He had been a prominent librarian in Paris, until in 1629 he became the librarian for Cardinal Giovanni Francesco Guidi di Bagno in Rome. In 1640 Naudé wrote a letter to the astronomer Ismaël Boulliau, in which he argued:

> I'm afraid that the old theological heresies are nothing by comparison to the new ones, which the Astronomers want to introduce with their worlds, or rather the lunar and celestial Earths. Because the consequence of these will be much more perilous than the previous ones, and will introduce some very strange revolutions. God help, above all, those who would say about Lucian that what he gave us as extravagant fables, and which he professed to not exist and to not be true, that

> it would be verified to be the truth itself, what do you think that he would say? This reminds me of the Antipodes, which nobody could believe for some two or three hundred years without being declared a Heretic.[35]

It should be recalled that Bruno himself had said that Lucian had made the mistake of not realizing that the Moon and planets really are inhabited. Like Inchofer and others, Naudé thought that the belief in other worlds was a heresy. In a previous book, Naudé had complained about recent 'innovators' who followed Telesio, Bruno and Campanella, 'who truly have no other intention but to strike this Philosophy with an elbow, and to ruin this great building that Aristotle and more than twelve thousand others who have interpreted him have struggled to build over a long span of years'.[36] When his employer Cardinal Guidi di Bagno died, in 1641, Naudé became a librarian for Cardinal Antonio Barberini, brother of Francesco, head of the Inquisition. As we will see, other members of the Barberinis' entourage shared the same concerns.

Summing up, from the 1590s until the 1640s there were various opinions about the acceptability, error or gravity of several so-called Pythagorean beliefs. Since we presently have more anecdotal evidence that some individual clergymen explicitly described the Earth's immobility and the Sun's motion as heretical, it might seem that such views were stronger than the opinions against other 'Pythagorean' doctrines. But no, on the contrary, the opinion that many or innumerable worlds exist was more objectionable precisely because it had been explicitly listed as a heresy for centuries, whereas the former opinions had not. Galileo, in particular, was not accused of such a belief because unlike Bruno and Kepler he had specifically abstained from elaborating such notions. Immediately after the proceedings of 1616 Galileo wrote to Cardinal Muti's nephew, to clarify statements Galileo had made in front of the Cardinal, insisting now that it was 'absolutely false and impossible' that there are plants, animals and intelligent human beings living on the Moon. Galileo was covering his tracks, intentionally moving away from the kinds of claims that could get him in trouble. It was to Galileo's advantage that he did not assert that human souls are reborn in other worlds, or that the Earth has a soul, or worse, that its soul is the Holy Spirit.

According to standard practice, if an Inquisitor informed a Catholic person that their opinion was an error, by its opposition to scriptures, that person had to abandon that error or else be treated as a heretic: one who wilfully chooses false beliefs. The belief of a moving Earth was dangerously connected to heretical notions. For Inchofer the *cause* was false: that Earth has a soul, and therefore its alleged *effect* was also false: that therefore it moves.

Thomas Aquinas had defined heresy as a kind of infidelity in men who profess the faith of Christ, yet corrupt its dogmas. In 1616 the Congregation of the Index had described the Pythagorean doctrine of the Sun's immobility as being 'formally heretical'. Such a heresy was 'formal' in that any Catholic who asserted it was wilfully departing from scriptural truth. Obstinate adhesion to false beliefs constituted formal heresy. Or, if instead someone asserted a heretical proposition out of mere ignorance of the true creed, that is, if they could reasonably be excused of not knowing the truth, then such a heresy would be described merely as 'material' or 'objective'.

Since the Earth's *immobility* was not expressly defined as an article of faith, some Catholics described the opposite view, the Earth's mobility, as being a 'nearly heretical proposition'. The Sun's mobility was more directly connected to an article of faith, inasmuch as it seemed to be necessarily implied by the Joshua miracle. How could it stop moving if it were always immobile?

Galileo's attempt to interpret the Joshua miracle as meaning the opposite of what it literally said was just one among several ways to interpret the Bible, in so-called Pythagorean ways. Another such interpretation was Bruno's effort to construe the Holy Spirit as the soul of the world. A further Pythagorean interpretation was to construe the Resurrection of Jesus as a kind of transmigration of his soul. Furthermore Galileo's eccentric advocate, Campanella, insisted that Catholics should use a better philosophy (instead of Aristotle) for interpreting scriptures, arguing that the Bible should be interpreted 'according to the philosophy of Pythagoras'. All such Pythagorean interpretations violated the Council of Trent. Inquisitors viewed heresy as one of the worst possible crimes and their obligation was to purify the faith. Therefore both Bruno and Galileo were condemned as heretics. Whereas Cardinal Bellarmine was converted into a saint.

Bellarmine's Innumerable Suns

Cardinal Bellarmine did not believe in Bruno's innumerable worlds, but he did believe in the world of the afterlife, Paradise, complete with 'innumerable angels'. Echoing the Bible, he described it as a kingdom, a city and 'a house full of mansions'. The year when he confronted Galileo, 1616, Bellarmine published a book about that heavenly kingdom. In 1584 Bruno had written that 'innumerable suns' exist, and now Bellarmine also said that they certainly do, but in heaven itself. He mentioned the biblical miracle in which Jesus Christ shone with the brightness of the Sun. Bellarmine then said that in God's kingdom all persons who are blessed and just will also shine like the Sun. And he said that all the saints, the Virgin Mary and Christ will also shine gloriously: 'What then will it be, to behold innumerable Suns,' he asked – and he replied that one's eyes would be blessed, 'so that they may observe without injury not one Sun, but innumerably many'.[37] At the time, the expression *innumerabiles Soles* was extremely rare: I have found only one earlier instance of it, in Bruno's book of 1584. Therefore I suspect that Bellarmine appropriated the phrase from none other than the heretic from Nola. According to both Bruno and Bellarmine, the innumerable suns 'glorified' God.[38]

Bellarmine also mentioned the superior pleasure of eventually and ultimately, in death, learning directly from God about nature and astronomy, instead of from one's feeble observations and thoughts during life. He wrote:

> But not only is the sight of God promised to holy men in heaven, but also of all things which God has made. Here on Earth we observe by our eyesight the Sun, & Moon, & stars, & lands, & seas, & rivers, & animals, & trees, & metals. But our mind observes nothing, that is, it does not perfectly know any created substance, no essential differences, no properties, no powers; man does not even see his own soul, but like the blind it [the mind] palpates the effects, and by discoursing by reason it acquires some knowledge. Therefore, what a joy will it be, when to our intelligence will be revealed, manifestly, the nature of all things, differences,

> properties, powers? And with what great exultation, amazed, will we see the army of innumerable angels, none of whom resembles each other, and shall clearly see the differences and singularities of all! For such a theatre will be, how desirable, how lovable, when holy men from the beginning of the world until the end will all be congregated with Angels, and we will be allowed to see each of their merits, palms [of victory] and crowns! Not without delight and pleasure will we see the crimes and torments of the damned; in which the sanctity of the pious, and God's justice, marvellously shines; then His hands will be washed in the blood of the sinners, as the Prophet predicted so long ago.[39]

In Bellarmine's vision of heaven, he would finally enjoy seeing the bloody torments of heretics such as Bruno.

The founder of the Jesuits, Ignatius of Loyola, had encouraged them to meditate on the great fires of Hell, to hear the wailings, smell the filth, taste the sad tears of the damned, and feel how the fires burn the soul.[40]

During the last years of his life Bellarmine was head of both the Inquisition and the Congregation of the Index. His last days were described by an English Jesuit who witnessed his death.[41] Bellarmine became very ill in late August 1621, when he was almost 79 years old. After days of illness, he prepared the last confession of his life, but as that English Jesuit reported, 'such was the innocency of the man, that albeit he were in his perfect sense, yet could he hardly find what to confess.' Bellarmine found a few minor defects, for which a Father absolved him. The next day the Pope visited him and Bellarmine reportedly spoke: 'nothing troubles my conscience; for God (His goodness still be thanked therefore) hath so preserved me hitherto, as that I do not remember in the whole course of my life to have committed any scandalous action.' Was his memory clear? He said that he felt no weakness of mind yet.

Bellarmine was notified that he would soon die, so he requested the sacraments. He then ate 'the body of Christ', the wafer, and tasted His blood, the wine, and was anointed. Many men visited him – cardinals, bishops and clergymen – 'not only to see him, but to kiss his hands, his head, or some other thing about him; and when therein they had satisfied their devotion, they would touch his body with their books, their beads, handkerchiefs, crosses, medals and other similar things, and that very reverently on their knees'. Many cardinals asked for his blessing, 'and they seized his hand, and blessed themselves with it'. While he was unconscious or unresponsive, the cardinals, bishops and prelates sent many little caps of silk, which they would wear, and such caps were placed one by one on Bellarmine's head, 'and with them they sent also little crosses of gold and silver, reliquaries, prayer books, and other things to touch him, and that in such multitude, as there were more than one hundred and fifty red, white and other caps put on and taken from his head during this time', and more afterward. Medical attendants applied leeches to reduce his inflammation, and 'they used clean white handkerchiefs whereon the creatures might disgorge, and carried them away, stained with sacred blood, for distribution among their friends.'

The Pope then gifted a plenary indulgence to frail, old Bellarmine to cancel any and all temporary punishment due to sins, to thus send

The embalmed corpse of Cardinal Bellarmine, at the Church of St Ignatius, the chapel of the Collegio Romano. His body wears the red robes, hat and shoes of a cardinal, displayed behind glass and under bright lightbulbs.

The golden mask of Bellarmine's corpse. Beneath the mask, one can see the back of his skull.

'the hammer of the heretics' straight to heaven. Finally they laid 'a great crucifix' upon his lips, and let it rest upon his shoulders.

On the morning of 17 September 1621 Bellarmine was dead. Allegedly some people in the city miraculously heard his voice that morning, saying, *Addio, adesso me ne vado in paradiso*, 'Farewell, I am now going into paradise.' Many people kneeled before his body and kissed his fingers. The Pope's physician took the body to embalm it, and he distributed towels, handkerchiefs and sponges stained with Bellarmine's blood. The physician carved out and kept a piece of bone from the back of the skull, esteeming it 'a peerless jewel and inestimable treasure'. The embalmed corpse was then displayed in a church for veneration, guarded by soldiers. Yet the clergymen and the adoring mob stole nearly all of its clothing. Marvellous tales spread through the city about the miracles done by his relics.

In 1627 Pope Urban VIII nominated Bellarmine to be canonized. Three centuries later, in 1930, he was finally canonized as St Robert Bellarmine by Pope Pius XI. Can he now tell the innumerable angels how the noble Romans treated saints? Can he finally delight in seeing the innumerable suns? But not everyone agreed about his saintliness. In 1853 one Protestant commentator criticized him for sins that Bellarmine did not confess on his deathbed:

> 'Such was the innocency of the man!' Aye, such was his self-satisfaction. No misgiving as to the tendency of his teaching troubled him. No doubt as to the lawfulness of the rebellions and civil wars that he had promoted. Two of

> his disciples had assassinated two Kings of France; but he did not hear the voice of their blood crying from the ground. Victim after victim had he seen bound, weeping, racked, burning; but no image of anguish or death came before his eyes. Prayers from Syrians of India, remonstrances from invaded churches, groans from the pits of Minerva, deprecations of the dying, curses of the living, troubled him not while searching his memory for sin, just for something to be pardoned. Neither cruel deaths nor treasons were sins to his apprehension, if only the victims were heretics. He said that he had no sin. He was a liar, therefore, and the truth was not with him.[42]

Critiques after Galileo's Death

In 1641 a professor at Pisa, Paganino Gaudenzi, managed to publish a book that censors had first sequestered until it could be corrected. They worried that it seemed to endorse pagan ideas. Its title was *On the Pythagorean Transmigration of Souls*. In his manuscript Gaudenzi took the opportunity to mention Galileo, whom he praised as *clarissimus*: most enlightened. One of Galileo's friends at Pisa, Vincenzo Renieri, then informed old Galileo in a letter:

> Finally I end by telling you a beautiful fact. Paganino, in the book he published *On the Pythagorean Transmigration of Souls*, named Your Excellency at a certain point; he had written *most enlightened Galileo*, but the Father Inquisitor did not allow that *most enlightened*, and with difficulty he [Gaudenzi] managed to obtain at least *well-known Galileo*.[43]

The following year Galileo died, in January 1642. Gaudenzi promptly composed sonnets about him.[44] In a later book Gaudenzi returned to the topic of transmigration, to defend Tertullian's claim that the Pythagorean theory that souls take new bodies is false.[45]

Galileo's body was deposited not in his ancestors' tomb, but in a modest chamber under the bell tower of the Church of Santa Croce, Florence. The Grand Duke of Tuscany, Ferdinando II de' Medici, wanted to transfer the body to an impressive monument that would be a counterpoint to the monument of Michelangelo in the great

Galileo's body was initially entombed in this small chamber, located at the back wall of the chapel of the novices at the Church of Santa Croce in Florence.

nave of Santa Croce.[46] But when Pope Urban VIII heard about the plan he summoned Niccolini and voiced objections:

> he [the Pope] wanted to tell me that it would not be a good example to the world for Your Highness to do that [monument], because he [Galileo] had been here before the Holy Office for an opinion so very false and so very erroneous,

> with which he had also impressed many others there, and had given such a universal scandal to Christianity with a doctrine that had been damned.[47]

The same day Cardinal Francesco Barberini gave a message to the Florentine Inquisitor to be delivered directly to Ferdinando II: 'It is not good to build mausoleums for the cadaver of someone who has been punished by the Tribunal of the Holy Inquisition, because it might scandalize good people, with prejudice toward the piety of His Highness.'[48]

WEEKS AFTER Galileo died, a commentary on the Book of Joshua was published. Cornelius Lapide was a Flemish Jesuit who had lived in Rome since 1616 and had written long commentaries on almost all the books of the Bible. In the 1630s he was working on his commentary on Joshua, but he died in 1637, almost sixty years old. When it was finally published in 1642 it was with the approval of Riccardi. The published version has an interesting passage about the biblical miracle by Joshua:

> [Joshua 10:12] '*Sun, do not move over Gibeon, and Moon over the valley of Aijalon.*' In Hebrew is: *Sun settle down*, by which Solomon understood the Pythagorean harmony of heaven. That is, 'Sun I cannot hear the suns, orders are given to be quiet and listen to the voices; mine, and pause a degree.' But truly this is ridiculous and futile: *Sun settle down*, thus being the same as *Sun be quiet*, do not move.[49]

Father Lapide argued that this Pythagorean interpretation is ridiculous, and that instead the true, literal meaning is that by God's omnipotence the Sun really did stop moving.

Was Lapide aware of Galileo's Pythagorean interpretation of the Joshua miracle? Yes, Lapide did not mention Galileo, but he succinctly rejected Galileo's main argument. In his *Dialogue* Galileo had focused on one argument that he obstinately thought was the best evidence in favour of Copernicus: Galileo argued that the Earth's motion causes the tides. Kepler and others had explained that this is false, that instead the tides are caused by the Moon. Accordingly,

Lapide explained that when the Sun and the Moon miraculously stopped moving, the tides also stopped, because, he said, the tides are really caused by the Moon's motion. In one brief passage, without even mentioning Copernicus, even this biblical scholar thought that he could easily refute Galileo. One way or another, the Catholic clergymen rejected Pythagorean interpretations of Sacred Scriptures, whether pertaining to a moving Earth, a stationary central Sun, the immortality of souls, the harmony of the spheres, or many worlds.

Meanwhile, Galileo's last assistant, Vincenzo Viviani, sought a suitable memorial for his master. When his requests too were denied, he constructed the monument in his own house. It seems that Viviani himself cultivated questionable beliefs. In November 1646 Viviani met with a French scientist, who summarized their conversation in his diary: 'I went for a walk with Mr Viviani who had been three years with Mr Galilei. He told me his opinion about the ✹ [sun] that he believed is a fixed star, the necessity of all things, the nullity of evil, the participation of the universal soul, the conservation of all things.'[50]

The margin of the published version specifies that these were the 'Opinions of Mr Viviani', yet historian David Wootton conjectures that such beliefs came from Viviani's mentor, given that in 1615 Galileo had privately written about the spirit that spreads throughout the universe, accumulates in the Sun and vivifies all beings. Wootton argues that since Galileo 'never' wrote about Christ, Galileo did not accept Christ, except perhaps in 1639.[51] Wootton conjectures that Galileo really was not a Christian, but secretly believed in the soul of the world.

Viviani voiced three Pythagorean beliefs that Bruno had defended: that the Sun is a star, that there is a universal soul, and the conservation of all things, or in Bruno's words, that nothing is really new because primal matter is eternal. Furthermore, Viviani embellished his biography of Galileo with a Pythagorean fiction: he misreported the date of Galileo's birth as being not 15 February 1564, but four days later. Thus, it seemed to immediately follow the death of Michelangelo: 18 February 1564. Readers could surmise that the soul of the great artist had transmigrated into Galileo.

When later writers referred to Galileo's death, they often said (some still do) that Newton was born in the year Galileo died. Consider one account:

> It is singular that Galileo died in the year (1642) in which Newton was born. It was almost as if the soul of the one had transmigrated into the other, as if Galileo's spirit, spurning the leaden laws of Italy's faith, and leaving, with a sigh, its golden climate, had sprung unsaddled to the more congenial land.[52]

But the dates are confused. The mistake stems from using the Gregorian calendar to date Galileo's death while using the Julian calendar for Newton's birth.[53]

AFTER GALILEO's death some Catholics continued to denounce Pythagorean heresies. In November 1642 the Jesuit Pierre La Cazre wrote to Pierre Gassendi warning him to reject Galileo's theories. Gassendi had discussed the Earth's motion hypothetically. La Cazre complained that such a belief entailed that the stars are inhabited, which was dangerous to Christianity:

> Now consider not what you might think, but what others will conjecture, who on your authority or reasons will be persuaded that the Earthly globe moves among the Planets. First they will conclude that without doubt Earth is one of the Planets, which since it has inhabitants, then it will be easy to also believe that in the other Planets too, and even in the fixed Stars, inhabitants are not lacking, and how much preferable, if the other Stars exceed the Earth in magnitude and perfection. Here Genesis would become suspect, for saying that Earth was made before the other Stars, established only on the fourth day [of Creation], to illuminate the Earth, to measure times and years. Thus the entire Economy of the Incarnate Word and the truth of the Gospel would become suspect, indeed and the whole Christian faith, which presupposes and teaches that all Stars are not dwellings [*habitationem*] for men, or other creatures, but only to illuminate the Earth with their light, fertilizing, to be produced by the creator God. You see therefore how dangerous it is to divulge this in public, and especially by men whose authority seems to generate faith; and how not undeservedly since the time of Copernicus the Church has

Marble statue of Pope Urban VIII, by Gian Lorenzo Bernini and his assistants, at the Musei Capitolini, Rome, *c.* 1635–40.

always opposed itself to this error, even very recently, not only some Cardinals (as you say) but the supreme head of the Church, condemned Galileo in a Pontifical decree, and that teaching it be solemnly prohibited in future words or writing.[54]

This Jesuit thus explained that the Pope and cardinals had condemned the theory of the Earth's motion because it led to dangerous anti-Christian 'absurdities', such as that the stars are inhabited.

In a manuscript published in 1658, posthumously, Gassendi discussed whether many worlds exist. He credited such views to many authors, from the Pythagoreans to 'Jord. Brun.', and he quoted critiques by Hermias and Augustine.[55] Gassendi also discussed the notion of the soul of the world. He said that Pythagoras and the Stoics believed that if anything is warm it absolutely has a soul, a mind that penetrates all.[56] And Gassendi quoted Virgil:

> Spirit nourishes within, totally infused in the limbs
> a Mind moves the mass, and intermixes itself with the
> great body.[57]

Gassendi argued that it was permissible to think of God as the 'soul of the world' only inasmuch as God is the powerful presence that penetrates all things, governs and thus animates them, but not as a material component. Gassendi tolerated the expression 'soul of the world' also to describe the heat that propagates throughout the universe. But Gassendi noted that these two meanings were just 'an improper sense, as an analogy; but to be willing to properly say, or admit that in the whole World there is a soul, without abusing the word soul, namely which is vegetative, or sensitive, or reasoning, it does not seem easy, wherein lawful'.[58] Two centuries later, Christian readers debated whether Gassendi believed in the soul of the world, 'the monstrous doctrine' that embodied 'the fatalist and immoral pantheism that weighed upon antiquity from Pythagoras to Plotinus', or whether this was a 'deplorable calumny' against Gassendi, that 'in the 17th century, a proper savant, a Christian, a priest, Gassendi . . . will remain plunged into one of the most shameful errors of paganism.'[59]

Meanwhile in 1633 the Jesuit astronomer Christopher Scheiner finished a manuscript of a book criticizing Galileo's errors, but it lingered unpublished for many years until 1651.[60] Similarly, Inchofer's *Vindication . . . against the Neo-Pythagoreans* was never published. In January 1636 the internal reviewers within the Society of Jesus at the Collegio Romano did not approve it for publication.[61] Most importantly, they claimed that the subject seemed to require a mathematical approach, but that Inchofer had instead provided a theological and historical argument, which might not meet the potential readers' expectations.[62] The Jesuit reviewers also complained that the proposed book title suggested that Inchofer officially spoke for the institutions that he defended, whereas it was just his own work. Finally, the Jesuit reviewers complained that Inchofer had not written with the 'dignity', 'seriousness' and 'solidity' used by other writers in their Society. Inchofer promptly responded. He offered to shorten the title to *Vindication of the Holy Authorities*.[63] Inchofer acknowledged that his arguments were theological and historical, not mathematical. He said that his new book would vindicate the claims made in his *Summary Treatise*, which for the most part was not mathematical either. He explained that he followed 'the Holy Fathers, Augustine, Jerome, Justin and others', as well as historical writers such as Cardinal Baronio, and writers against heretics, such as Bellarmine. Inchofer denied that there was any lack of solidity in his historical and theological accounts. He expected that 'matters of faith' should trump alleged mathematical demonstrations.[64]

Despite his command of theology and Church history, Inchofer was out of his depth in the area of mathematical astronomy. In addition, his tone was too bitter and combative. For whatever reason, Inchofer did not edit and resubmit his long manuscript. Regarding the difference between Inchofer and some of his fellow Jesuits, historian Michael Gorman remarks that 'Inchofer's style of argumentation from scripture and the Church Fathers' deviated from astronomy and logical demonstrations and thus threatened 'to place the mathematicians of the Jesuit order outside a global republic of astronomical practitioners that they had helped to create'.[65]

Still, Inchofer became increasingly close with Cardinal Francesco Barberini, who consulted him on several occasions.[66] Between 1637 and 1647 Inchofer served as 'one of the Cardinal's intimate advisors on two matters: Spanish intrigues in Sicily and the Jansenist movement in Belgium'.[67]

Meanwhile Inchofer became increasingly disgruntled with some fellow Jesuits. In 1645 an anonymous volume criticizing the Jesuits was published in Venice. For one, it accused them of indulging in excessive freedom in matters of philosophy. In 1648 the Jesuits put Inchofer on trial for having co-authored or contributed to that offensive book.[68] He confessed to some of the charges and was incarcerated. He died a few months later. Blackwell writes: 'This story could hardly be more ironic. Galileo's foremost critic at his trial, who was the sharpest and most damaging in his denunciation of the *Dialogue*, was fated himself fifteen years later to undergo a trial, condemnation, and sentence which had eerie similarities to what happened to Galileo.'[69]

CATHOLIC ORTHODOXY extended also to the Spanish portions of the New World. Perhaps nothing is quite as compelling in showing that a particular belief was truly heretical as to find the record that someone was executed for such a belief. For decades writers have mistakenly downplayed the importance of Bruno's belief in many worlds in leading to his execution. Thus it is essential not only to know that this belief was in fact heretical, but that apparently another clergyman in Rome had been previously executed for believing in 'many suns', as we saw. But furthermore, subsequently, the theory of the plurality of worlds was involved in the execution of yet another heretic. It happened in 1659, in a faraway land: Mexico. Just as the Catholics had denounced the transmigration of souls in distant India, they also denounced it in Mexico City.

The case in question involved Sebastian Alvarez (or Albarez), born in Bayona de Galicia, Spain, but residing in Mexico. The record of 1659 stated that he was then 'more than sixty-three years old', so he had been born around 1595. He had been incarcerated allegedly 'for being a sectarian of Luther, the Sacramentarians, and other heretics, and for inventing many and new heresies'. In his third deposition, Alvarez said that thirty years earlier (roughly around 1629) he had written some texts that had many errors; but that once he finally read the Bible he realized that he himself was actually Jesus Christ. The Inquisitors inferred that sometimes he was possessed by the Devil. On the night before his auto-de-fé, the Inquisition's public ritual of penance, Alvarez insisted that he truly was Christ and would resurrect in three and a half days to judge both the living and the dead. And then immediately, as the Inquisition's report notes:

> He said that there exist thousands of worlds, and that in each of them Jesus Christ would die twice; and that having once died on the cross, it was now his turn to die a death by fire,

> and he added: 'Look, fathers, if after three days you do not see me resurrect, then don't believe me', and that he rejoiced to die in order to be resurrected. He insisted on the heresy of the transmigration of souls from some bodies to others, and for the many heretical blasphemies that he said, the clergymen asked that he wear a gag even while in prison.[70]

Alvarez claimed that God had given him the ability to interpret scriptures, and that he actually had within him the soul of Solomon.

Like Bruno, Alvarez was turned over to the government so that he would be burned alive without using a garrotte to first kill him. This was the most severe kind of execution, for the gravest heresies. However, when Alvarez was being taken to the funeral pyre, he was moved by seeing a priest who cried for him. The priest then said that Alvarez was 'walking toward Hell'. Alvarez asked him: 'Father, why are you crying?' The priest replied that it was for the loss of his soul. Alvarez then said: 'Well, what do you want me to do, Father?' The priest begged him to recant his errors. So finally Alvarez agreed, and retracted and confessed everything that they asked him. Afterward, the executioner killed Alvarez with a noose, and then threw him into the bonfire.[71]

Records show that from 1540 until 1700 the Spanish Inquisition executed a total of seventeen individuals in Mexico, out of the 950 who were put on trial for heresies. In the same time period, the Secretariat of Aragon (which included Mexico), plus the Secretariat of Castille, both constituting all of Spain and its distant lands, executed a total of 826 persons, out of more than 44,000 who were put on trial.[72] The case of Alvarez holds remarkable similarities to that of Bruno. Both men allegedly voiced blasphemies about Jesus; both were accused of being possessed by the Devil or by demons, and of being Lutherans. They both spoke about many worlds and transmigration, both reportedly accepted their sentence fearlessly, and both were gagged and sentenced to execution by fire. Whether in Rome or in distant lands, the Catholic Inquisition suppressed impenitent heretics.

In the 1680s Daniel Morhof was a German professor of history and the chief librarian at the University of Kiel. He wrote an authoritative historical encyclopaedia in which he included a chapter on Bruno and other 'recent innovators'. Morhof's first paragraph

Statue of Giordano Bruno, in Mexico City, Mexico. Sculpted by José Ortiz and Josafat Chavez and unveiled 1 February 1991.

plainly stated why the Romans killed Bruno, and he linked Bruno with Galileo:

> Giordano Bruno, asserted a multitude of innumerable Worlds and Suns, and he paid for this opinion by being burned alive. A plurality of Worlds, and Moons, other Planets, homogeneous to the Earth, was defended before him by Nic. of Cusa and Nic. Copernicus, and after him by Galileo Galilei (forced to voice a Recantation), Johannes Kepler, Athan. Kircher, Tomm. Campanella and other mathematicians.[73]

Also, while discussing the followers of Epicurus, Morhof noted again and again that Bruno was burned for his belief in other worlds, 'although there are not lacking some who think that he was executed for some other reason'.[74] After summarizing Bruno's opinion that many worlds exist, Morhof immediately remarked:

> Everywhere Galileo subtly argued and presented this opinion, as other Mathematicians too who, although they do not voice all the words of the speaker, tacitly conspire with him. But this miserable man [Bruno] paid for his audacious opinion by burning alive; which greatly injured that belief; for it made his writings become suspected of the crime of Atheism.[75]

Morhof noted that Campanella had written an *Apology* to defend Galileo, in which he tried to use scriptures to prove the existence of men on the Moon.[76]

In a book of 1705 a German librarian and theologian, Johann Albert Fabricius, also discussed the claim that the Moon and the stars are inhabited. He said that Orpheus was the first to assert this belief, followed by the Pythagoreans and others. Fabricius noted that Philaster, Augustine and Praedestinatus declared that this belief was a heresy, and he added that, 'in the [Corpus of] Canon Law, Cause 24, question 3, column 39, it is ascribed as a heresy, the *Ophic* heresy which some write as *Orphic*'. Fabricius added that Praedestinatus 'called this the heresy of the *Ametritarum*, namely that which to an infinite power attributes infinite effects, introducing infinite worlds. The same was taught by the Manichaeans, about which see Archelaeus with Epiphanius, vol. 1, p. 645.'[77]

Immediately Fabricius listed the 'recent writers who in some sense said that the stars are inhabited', namely, 'Nicolas of Cusa, the miserable Giordano Bruno, Tycho Brahe, Tommaso Campanella, William Gilbert, René Descartes, and those who follow him, and Johannes Kepler, Galileo Galilei, and David Fabricius, those who with their own eyes have dared to see the inhabitants of the Moon'. About Bruno, he added a footnote: 'Burned in Rome on 9 February Year 1600. His teachings were first expounded in his eight books on the innumerable, the immense and shapeless or the universe and the worlds.'[78]

Finally, in 1829 and 1830 an English antiquarian, Algernon Herbert, published historical conjectures that included alleged links between the heresies of the Pythagoreans and the Copernicans. Herbert, who was Dean of Merton College at Oxford University, argued that Pythagoras was 'a charlatan' who travelled widely to learn mystic secrets, including the idea that Earth moves around a central fire; he was a 'corrupted mystic' who pretended to be the god Apollo.[79]

Herbert wrote that 'Melchior Inchofer was privy to all the hidden motives which had animated Galileo and his friends, in republishing (at the time they did) the doctrines of Pythagoras; and from thence conceived such a hatred of those doctrines as he evinced in his *Treatise de terrae solisque motu*.'[80] Furthermore, Herbert denounced Campanella for being 'addicted to atheism and magic', he called him 'the Anti-Christ', inspired by an 'infernal muse'.[81] Herbert criticized Campanella's dream of a Philosophical Republic:

> it superabundantly shows how justly the schemes and doctrines of the Solipses are reprobated by Father Inchofer; and how closely those impious and immoral notions were at the time interwoven with the Galilean astronomy. It avows that they intended to worship the Sun, inculcate the law of nature, and renew the dæmonolatrous atheism of the city and tower of Babel. And moreover that a certain man was to reign over all nations, calling himself, and pretending to be, the Sun, as Pythagoras pretended to be Apollo; not, however, to be the Sun circumvolving but the Sun central, which was also the secret of Pythagoras.[82]

Herbert used the name 'Solipses' to refer to those mystics accused of being Mithraics, that is, worshippers of the Pythagorean Sun, atheists and alchemists like Pythagoras, who believed in the transmigration of souls.[83]

Thus ends my account of how the Copernicans used Pythagorean ideas, not entirely to their own advantage. By the 1590s such ideas seemed very offensive to Catholic censors who struggled against the tide of Protestant beliefs. In 1616 the Inquisition denounced the Pythagorean doctrines about the Earth and the Sun. Later, some prominent writers, including Rusca, Campanella, Holste, Froidmont and Inchofer, knew well the questionable or heretical beliefs of the Pythagoreans, including the beliefs that Bruno had tried to defend. But Galileo seems mostly oblivious to the questionable ideas that were blamed on Pythagoras and his followers, pagan ideas that Christians had damned for more than a thousand years.

Conclusion

Eventually the Catholic Church changed its views on Galileo.[84] All censures against books discussing the Earth's motion were removed. Yet the Church did not forgive Giordano Bruno. Even nowadays most astronomers hardly appreciate Bruno's importance in the history of cosmology. Likewise, most historians of science don't mention Bruno when discussing Galileo's troubles with the Inquisition. Likewise, astronomers and historians discussing the Copernicans sometimes mention Pythagoras merely as an ancient mathematician or astronomer who pioneered the theory that the Earth moves. That story is fiction, like many other claims about Pythagoras. But historical fictions exert powerful effects on people.

By pulling together accounts by Aristotle, Diogenes, Porphyry, Hippolytus, Macrobius and other ancient writers, we can now sketch an ancient Pythagorean cosmology and theology:

> The universe is eternal. It was not created by God. Many gods exist – the Sun, Moon and the stars are all gods. The stars are worlds in infinite space. The Earth has a soul. It is alive. It moves around a central fire: the guardhouse or prison of Jupiter. The Sun is the Creator and a Great Geometrician, and the Moon and planets are worlds like the Earth. Human souls descend from the Moon and are reborn successively, even in animal bodies. Souls spend time on each planet, and those who philosophize rise to a kindred star. Finally, souls go to the Sun. The infernal regions start at the Milky Way, where souls have withdrawn from the heavens, but can slip into Earthly bodies. Pythagoras was related to the Sun god Apollo. Pythagoras was the first philosopher. His soul was reborn multiple times, and he remembered his past lives: he came from Jupiter and also lived on the Moon, on Earth and in Hell. He performed miracles and spoke to gods and the dead. He practised silence to teach that philosophy's treasures should be kept secret. Measures and numbers can be used to predict future events and to explain religion. Knowledge by reason is superior to knowledge by faith. Apollo exposed the corruption of the Christians, a confused and vicious sect

> that suffers for its emphasis on faith and unreason. Jesus of Nazareth was not God incarnate.

It is important to note that this synthesis is not found in any one source, and that these beliefs evolved over time. Nevertheless, this aggregate of Pythagorean views conveys a sum of ancient accounts that, for the most part, were available and widely read in the Renaissance. Hence parts of it came together in the minds of readers back then.

Had it not been for the Protestant Reformation, we might imagine that such pagan notions could have remained as fanciful, harmless legends, much as they sound to our ears today. But Rome in the 1590s was a very different world. Similarly, the times of the early Church Fathers, around 200 to 400 CE, were tumultuous. Pagan notions threatened Christianity so they were vigorously suppressed.

Those early confrontations between Christians and the Pythagoreans were not forgotten in the Renaissance. Catholic clergymen adhered to many writings by the ancient Church Fathers, including Irenaeus, Tertullian, Athanasius, John Chrysostom, Jerome, Augustine and Isidore. The Church Fathers had vehemently denied that the soul of Pythagoras had been reborn or had lived in Hell. They accused Pythagoras and his disciples of faking their deaths and resurrections. They denied the 'Pythagorean doctrine' that living men are formed from dead ones, and that souls are imprisoned in bodies. Such lies 'were sent forth by Satan to bring dishonour upon the Church'. They complained that the belief in transmigration was shameful, indecent, and had caused 'profane corruptions of Christianity' contrary to the Christian promise of Resurrection. Such 'perverse falsehoods' threatened people's faith in Jesus Christ.

Tertullian blamed philosophers as 'the patriarchs of heretics', saying that every heresy originates from them. The Church Fathers complained that pagan philosophy was full of contradictions and was the source of all mischief. Allegedly, philosophers disparaged faith by seeking knowledge from too many sources, rather than having complete faith in the words of God and Christ. The Church Fathers denied the powers of numbers and monads, along with the supposed eternity of the world, and the belief in unconditional immortality.

The alleged feats of the Pythagoreans were false imitations of the miracles of Christ.

The Christians denied the claim that souls travel between the planets, the Earth or the stars. They vilified the claim that there are many worlds, some with suns and moons, some inhabited. Epiphanius scorned Pythagoras for saying that the Sun, the Moon, stars and planets are parts of God's body. John Chrysostom reviled Pythagoras as extremely evil, saying that Pythagoras 'practised ten thousand kinds of sorcery', to deceive fools with false doctrines about the soul: 'this is the snare of the Devil.'

Still, Porphyry defended Pythagoras as divine, saying that he was the most eminent man ever. Porphyry believed that many gods and demons exist, and that the universe was eternal, not created. He said that human souls spend time on the Moon and the planets before finally going to the Sun. He said that Pythagoras taught that the people of dreams are souls collected in the Milky Way. By comparison to his combative *Fifteen Books against the Christians*, it might seem that Porphyry's *Life of Pythagoras* was less poisonous to the early Christians. However, one historian made the opposite point. By the 1830s the English Reverend George Waddington explained that by contrast to *Against the Christians*,

> that which being more insidious, may have been more pernicious was his [Porphyry's] 'Life of Pythagoras'. Early in the third century, one Philostratus, a rhetorician at Rome, had composed a fabulous account of Apollonius of Tyana, a celebrated philosopher and magician; and so wrought out the supposed extraordinary incidents of his life, as to establish a close resemblance between them and the miracles of Christ. Porphyry imitated this example; and he represented the peaceful Pythagoras as having worked by his own power many stupendous prodigies – and having, moreover, imparted the same power to his principal disciples, Empedocles, Epimenides and others. Such is the weapon, which as it proceeds from the imagination, and eludes the grasp of reason, has proved at all times the most dangerous to Christianity.[85]

To the early Christians, the fictions of Ovid, Virgil and Porphyry were not as harmless as they now seem. St Augustine criticized

Pythagoras for being a necromancer who tried to divine the future by communicating with the dead or with demons pretending to be gods. Augustine repudiated Porphyry for communicating with demons, for believing in transmigration and for being the bitterest enemy of the Christians.

Such were the extensive religious accusations against the Pythagoreans. It was against this background that Catholic clergymen in Renaissance Italy assessed claims about the confirmation of Pythagorean doctrines. To say that Pythagoras was right about the Earth's motion or the existence of other worlds was intimately connected to heresies that had been denounced by the Church Fathers.

Copernicus said that the Pythagoreans knew that the Earth moves, he wrote as if planets are worlds, and for decades abstained from publishing, arguing that philosophy's secrets should be shared only with friends, following Pythagoras. Copernicus was impressed by the ancient letter of Lysis the Pythagorean, which portrayed Pythagoras as godlike. The views of Giordano Bruno were even closer to those of the Pythagoreans. He praised 'saintly' Pythagoras as the first philosopher, and argued that knowledge is superior to faith, and that a new science should be created based on Pythagorean principles. Bruno argued that infinitely many stars exist, and that stars and planets are worlds. He said that the universe is spatially infinite and eternal; that it has a soul, and that it was not created. He wrote that stars have souls, the Earth is a star, and it moves because it has a soul. Apparently Bruno believed that human souls are repeatedly reborn, even in animal bodies. Thus Bruno defended ancient heresies, and he died for his Pythagorean convictions.

In 2008 I drafted an account titled 'Galileo's Pythagorean Heresy', which became a chapter in a book published in 2011. Back then I did not know the many interconnections between Pythagorean topics and the Copernican Revolution. I knew some relevant points about Kepler and Galileo, but I had not analysed Bruno's case at all. Later, I was stunned to discover the extent to which so-called Pythagorean notions featured in Bruno's works and trial. It led me to three important findings. First, that the ten propositions the Inquisition censured in Bruno's works during his trial were not a disorganized assortment of eccentric beliefs; they were actually a coherent body of claims tightly interconnected by so-called Pythagorean notions. Second, that among dozens of accusations, the ones that Bruno refused to recant were known as Pythagorean beliefs. Third, that such beliefs had been officially listed as heresies for centuries.

For a long time Bruno's trial has intrigued writers in regard to the Copernican theory. Historians know that Bruno's partial support of Copernicus was not the principal point that irritated the Inquisition in the 1590s. Still, the doctrine that Earth moves was one of ten censured propositions in Bruno's works. Surprisingly, nearly all of those censured propositions were widely viewed as Pythagorean: that nothing is created (all transforms), that primal substance is not vanity but is eternal like God, that the world has an eternal soul (it was not created), that the universe is infinite like God, that human souls are derived from a general principle, souls are not form (but are captives in bodily prisons), even the Earth has a soul and is alive, so it moves, and that there are many worlds. I know of no evidence that such beliefs were originally held by Pythagoras – so in that sense they were not really Pythagorean at all. However, by the 1590s each of these beliefs had been prominently and repeatedly characterized as Pythagorean, by writers, philosophers and clergymen, even by Bruno himself. Because so little about Pythagoras was known, the evolving misrepresentations became far more influential than the real man.

Some of the final accusations against Bruno are unknown, five or six of eight heresies selected by Cardinal Bellarmine. But we do know which beliefs Bruno repeatedly refused to recant: the Pythagorean beliefs about souls and the universe. Since he again refused to repudiate the next and final list of heresies, it seems that it included some of those Pythagorean beliefs. Indeed, after Bruno's execution Schoppe reported the heretical doctrines of which Bruno had been accused, starting again with five Pythagorean beliefs: worlds are innumerable, souls transmigrate to such worlds, one soul can take two bodies, the Holy Spirit is the soul of the world, and the world exists eternally. Therefore Bruno's crime was that he obstinately advocated pagan heresies that earlier Christians had denounced.

Other writers also entwined the Copernican and Pythagorean beliefs. Kepler conjectured that the Pythagoreans truly knew the structure of the heavens. He poetically wondered whether the soul of Pythagoras had transmigrated into Copernicus or into himself.

Kepler too imitated the Pythagoreans by keeping his opinions private. He too believed that the Moon and the planets are worlds like the Earth, and that beings live in such worlds. Writing to Galileo, he praised Pythagoras and Plato as 'our true masters'. Kepler tried to decipher the Pythagorean harmony of the planets. He believed that mathematics contains hidden divine knowledge, and he said that Geometry is God, similarly as Hippolytus had said that the Pythagoreans saw God as a Great Geometrician.

Campanella agreed that Earth and the planets orbit the Sun. He conspired to establish a similar heavenly order in society, in the Italian province of Calabria, which included the ancient Pythagorean city of Croton. Campanella envisioned a city in which citizens would worship the Sun, without fearing death because souls are immortal, and some are reborn. Such 'Solarians' would have only one book: Wisdom, to read it like the Pythagoreans.

Galileo's stunning discoveries gave credibility to the Pythagoreans. His telescope provided evidence of other worlds. Galileo noted that mountains and valleys on the Moon seemed to revive the 'Pythagorean' belief that the Moon is another world. Kepler speculated about its inhabitants. Galileo wrote that 'with absolute necessity' the Pythagoreans were right that the planets orbit the Sun. While Copernicus had credited some Pythagoreans, Galileo took it a step further: he gifted credit to Pythagoras. Galileo also saw that there are countless many more stars than previously seen, which led some thinkers to theorize that innumerable stars exist, as the Pythagoreans and Bruno had claimed. Galileo also found that Jupiter might be another world, because it has moons. Philostratus had described Pythagoras as the man who came from Jupiter. Bruno had asserted that Jupiter is another world, and now Kepler agreed.

Like Porphyry, Bruno, Foscarini and Campanella, now Galileo too argued that the Church is not infallible in philosophical matters. Galileo argued that the Bible is not all literally true, like ancient pagans and recent Protestant reformers, along with other advocates of Pythagorean doctrines: Copernicus, Zúñiga, Bruno, Kepler, Foscarini and Campanella. They all transgressed Catholic authority, committing the same 'vanity' that the Church Fathers had condemned in the Greek philosophers: they believed that philosophy can independently yield true knowledge.

Some believers in the Earth's motion also believed that the Earth has a soul that is nourished by a universal spirit that vivifies all beings. Versions of this belief were advocated by Gemma, Bruno, Gilbert, Kepler and Campanella. Even Galileo secretly held some such notion. Pagans and heretics had in fact or allegedly advocated this belief, including Pythagoras, Porphyry, Origen, Novatian and Peter Abelard. Many drew inspiration from Virgil's *Aeneid*. Hence the Copernicans faced a staggering opposition, including critiques by Jerome, Augustine, Anselm, William of St Thierry, the Council of Sens, Pope Innocent II, Alfonso de Castro, Thomas Aquinas, Bishop Tempier, Petreto, Piccolomini, Padiglia, Bellarmine and Inchofer.

Against Catholic authorities, the Copernicans thought that they could uphold the Bible alongside the Pythagorean doctrines. They thought that the ancient wisdom of Pythagoras could explain the true meaning of God's Word. This sinful lack of piety resembled the heresies of the early Gnostics, who dared to set a portrait of Pythagoras alongside the portrait of Jesus. But the Catholic Church required that the Bible should be interpreted in light of the writings of the ancient Church Fathers. The Copernicans thought that they could choose whichever authorities best suited their interests. Catholic theologians and censors had denounced this very practice in the Protestants.[86]

Bruno and Galileo committed the kind of brazen acts that the Bishop of Paris had condemned in 1277 as 'manifest and execrable errors, or rather vanities and false insanities'.[87] The Bishop warned that anyone trying to speak wisely should exercise great care to not 'disrupt the unity of his listeners', especially when supporting apparently convincing pagan beliefs. He complained that philosophers disguised their claims:

> For they say that these things are true according to philosophy, but not according to the Catholic faith, as if these were two contrary truths, and as if against the truth of Sacred Scripture were the truths voiced by the condemned pagans, of whom it is written, *I will destroy the wisdom of the wise* [1 Corinthians 1:19], because true wisdom destroys false wisdom.[88]

Bruno and Galileo both claimed that philosophers and mathematicians had a special dispensation from theologians' authority, that as experts on nature they could ascertain the true meaning of scriptures. Thus Bruno interpreted the Holy Spirit as the soul of the world, in accord with what he called 'the Pythagorean doctrine'. Such temerity offended the Inquisitors who required that all Christians should submit to Catholic authority on any subject. Bruno wilfully transgressed the limits of philosophy to make pronouncements about God's creation, scriptures and souls. Galileo wilfully transgressed the limits of mathematics and astronomy to subvert the literal meaning of scriptures, in favour of Pythagorean interpretations. By refusing to abandon their beliefs after such notions had been pinpointed as serious 'errors' by Catholic authorities, both men became heretics.

REGARDING THE Galileo affair, a traditional story is that: 'In ancient times, Pythagoras argued that the Earth and the planets orbit the Sun; his theory was later adopted and refined by Copernicus, and it led Galileo to clash with the Church.' This story is wrong because half of it is fictitious. Instead we can say that Galileo attributed the heliocentric theory to Pythagoras – but this association entailed pagan connotations that increased its offensiveness among Catholics. For centuries theologians construed the Pythagorean outlook as overtly anti-Christian. In 1616 the Index 'completely prohibited and condemned' Foscarini's scriptural defence of the 'New Pythagorean System of the World', a rejection more forceful and damning than the injunctions against Copernicus, Zúñiga and Galileo. Thus it is understandable that some Jesuits decried Galileo's *Dialogue* as more vile and harmful than the writings of Luther and Calvin. At least the Protestants did not defend a heretical, pagan cosmology. The ancient Pythagoreans offered a polytheistic or pantheistic world view.

Pope Urban VIII bitterly complained that Galileo dared entering into 'the most severe and dangerous matters' involving 'great harm to religion', 'the most perverse subject-matter that one could ever handle'. The Pope specified that he did not refer to mathematical theories about astronomy, so his words make sense if he worried about the implications of supporting a pagan world view at the time of the Reformation, the splintering of Christianity. Compared to Catholicism, the ancient Pythagorean cosmology was far more radical than any heresies by Luther or Calvin.

If this is the case, then some writers should have actually noted, explicitly, that the Pythagorean heresies, say those by Bruno, were worse than those of the Protestants. Some did, such as Schoppe, Emmius, Mersenne and Naudé. While describing Bruno's transgressions, Schoppe explained that the Inquisitors viewed Bruno's heresies as far worse than those of the Lutherans and the Calvinists, who were actually treated well by Catholics in Rome. Bruno's unusually 'horrendous' beliefs included that worlds are innumerable, souls go from body to body, souls migrate to other worlds, the world has a soul, the world is eternal, and magic can produce miracles such as those of Jesus. These beliefs had been voiced by the Pythagoreans and criticized by Church Fathers. Schoppe was a Catholic convert who ruthlessly criticized the Protestants, yet he explained that most people did not recognize that the impenitent Bruno was not 'a Lutheran', but something far worse, a 'Monster'.[89]

Galileo seems unaware that, by attributing the heliocentric theory to Pythagoras, he connected it to heresies. It is amazing that the one person who most strongly linked Galileo's transgressions with the heresies of the Pythagoreans was not a distant commentator, gossiper or a courtesan – instead it was the Jesuit theologian and consultant for the Inquisition who had rendered the most damning expert testimony against Galileo: Melchior Inchofer. The Pythagorean thread goes right through the eye of this most prickly needle.

To what extent were other clergymen in the proceedings against Galileo concerned with such connotations? That remains a direction for future research, to seek concerns such as were voiced by Inchofer. But at least it is clear that some individuals, such as Inchofer, Froidmont, Waddingus, Dormeuil and Riccardi, sensed that the new Pythagoreans were not just philosophers or mathematical astronomers – they were a dangerously heretical 'sect'.

Evidence has turned up far more abundantly than I ever imagined. I should note that in my search for sources I found Holste's book on Porphyry and Pythagoras only late in my research, so it was startling to find not only that a work published in the Vatican, in 1630, discussed the old critiques of Pythagorean heresies by Tertullian, Eusebius, Augustine, Lactantius and others, but that it had been authored under the sponsorship of Cardinal Barberini, of all people, and that it had been approved for publication by

Riccardi, the same censor who one year later delayed the publication of Galileo's *Dialogue*. Likewise, I afterward found Campanella's relevant second edition of *On the Sense of Things*, which remarkably highlights and connects many of the key themes. I was finally able to examine Inchofer's manuscript *Against the Neo-Pythagoreans* even later. My book ends not because I have run out of sources, by no means, but because it is already too big, editors told me to cut it, and I have told my long story.

The evidence reviewed does not show that the Pythagorean religious notions were the major factor in Galileo's trial of 1633, but it reveals that such connotations and concerns persisted strongly at the time. These associations were of crucial importance in the long proceedings against Bruno, and they constitute a long-suspected missing link between the proceedings against Bruno and Galileo. The connection was not merely because Inquisitors censured both of them for affirming that the Earth moves, but because Galileo's advocacy of Pythagorean notions seemed to support Bruno's heretical views.

As I have shown, there is a neglected continuity between Bruno's trial and Galileo's admonition of 1616, given the subject-matter and the important participation of influential Inquisitors in both: Bellarmine, Sfondrato, Borghese and Taverna. Bellarmine was a relative latecomer in the protracted proceedings against Bruno, as his participation seems to have begun in March 1597 (not all the names of the previous consultors were specified). He first served as Consultor and then as cardinal Inquisitor, enacting in both capacities important roles in Bruno's trial: composing a list of principal heresies and approving his final sentence. Sfondrato, a longstanding cardinal Inquisitor, participated in the trial in Rome from 1594 until Bruno's sentencing in early 1600. Taverna was trained as a lawyer and served as Consultor in the proceedings from late 1593 to early 1595. He became governor of Rome in 1599 and therefore presided over Bruno's final imprisonment and brutal execution.

By 1616 Sfondrato had become head of the Inquisition, and it was he who first received the earliest complaints against Galileo. It was Sfondrato who initially sensed their gravity and who convened Inquisitors to investigate. The proceedings promptly involved Pope Paul v, who likewise had served for four years (1596–1600) during Bruno's trial. Galileo then met with Bellarmine, Taverna and other

Inquisitors in 1616. Bellarmine gave him the official warning and order to stop believing in Earth's motion and the Sun's immobility, under threat of imprisonment.

Thus at least four of the most important Inquisitors in the early proceedings against Galileo knew well the heresies that previously Bruno had refused to recant. They knew that Bruno had obstinately defended what he called the Pythagorean doctrine that the universe is infused by a spirit, such that the Earth itself is a soulful animal and therefore it moves. They knew that Bruno claimed to follow Pythagoras in the belief that the Earth is a wandering star, like the planets, and therefore such stars are worlds. Bruno defended the idea that there are countless stars, suns or worlds, to his death.

Those four Inquisitors learned that Galileo entertained 'the old opinion of Pythagoras that the Moon is another Earth'. They saw that he too rashly claimed that the Earth moves – a theory that he attributed to 'Pythagoras and all his sect'. And Galileo proclaimed that he had discovered uncountably many stars with his telescope. Furthermore, Galileo declared that we must conclude that the planets orbit the immobile Sun 'in agreement with the theories of the Pythagoreans'. Thus Galileo himself canonized the Sun's immobility as a Pythagorean doctrine.

Galileo and his readers did not distinguish clearly between recent conjectures and the ancient claim that Earth orbits a central fire. Clergymen such as Campanella likewise claimed that it was Pythagoras who had first taught that the Earth circles the Sun. Hence in 1616 the Congregation of the Index prohibited the 'false Pythagorean doctrine' that Earth moves and the Sun does not. The consultors of the Inquisition concluded that the 'foolish' idea of Earth's motion was 'at least erroneous in faith' and that the Sun's immobility was 'formally heretical' for literally contradicting the Bible.

A problem with some old interpretations of Bruno's trial is that writers tried to make sense of it in terms of the subsequent trial of Galileo. However, we should proceed in the opposite direction, following chronology: analyse the proceedings against Galileo in terms of the previous case against Bruno. Doing so, I find that the two proceedings were more closely connected than I have anywhere read. If we disregard the many sundry accusations against Bruno that he denied, along with the many spurious accusations against Galileo that the Inquisitors promptly dismissed, it turns out that the doctrine of Earth's motion as censured in 1616 was essentially a subset of Bruno's heretical beliefs.

There is an interesting phrase in the Decree of the Index of 1616: the books of Copernicus and Zúñiga would be suspended and Foscarini's work would be banned 'in order that such an opinion [the "false Pythagorean doctrine"] does not slither further to harm Catholic truths'.[90] This phrase makes sense if the cardinal Inquisitors feared that the potential harm of allowing one or two Pythagorean claims might entail the proliferation of others, such as those advocated by Bruno. By 1635 Inchofer quoted, in his *Vindication*, this complaint against Galileo: that the Pythagorean opinion was '*slithering perniciously into the Catholic truth*'. Thus I suggest that the strong critiques against Galileo's claims in 1616 and 1633 were partly motivated by the knowledge that such claims had arisen before. They were part of the 'philosophical' and heretical world view of Bruno. It makes perfect sense that the Inquisitors in 1616, some of whom had served in Bruno's trial, judged that it was necessary to censure the Pythagorean doctrine that the Earth moves, so that such errors would not creep or 'slither' towards other related, graver Pythagorean heresies, such as that the Earth is a rational, soulful animal, that its soul is the Holy Spirit, that the stars too have rational souls, that the universe was not created, and human souls are reborn in other worlds (such as the Moon or Jupiter), and that worlds are innumerable. For years, when Bellarmine systematically attacked the 'horrendous, terrible effects' of the plague of heresy, he had stressed 'the great speed with which this venom diffuses into the heart', and 'how rapidly it slithers; what invades one home today, soon fills an entire city with corpses'.[91]

Bellarmine instructed Galileo to not hold or defend the Pythagorean ideas. Galileo promised that he would acquiesce. Nonetheless, he patiently waited and schemed until he finally betrayed the agreement, subverting his promise. But as St Thomas Aquinas had argued, 'It is a more serious sin not to perform what one has promised.'

In 1633 Galileo lied to the Inquisition. He said that he had written his *Dialogue* to demonstrate that the Earth does not move, and that he had never believed in Earth's motion – whereas his own writings and letters admitted that he did. His judges were unconvinced.

His behaviour was visibly short of what they expected from a pious Catholic. To show this, we can try to characterize Galileo's transgressions in terms of the sins of which clergymen were mindful. The Book of Proverbs 6:16–19 specifies seven things or sins that are 'an abomination' unto God. Of these, we might say that Galileo had committed five: pride, lying, having 'a heart that devises wicked plots', being 'a deceitful witness that utters lies', and sowing discord among others. The Epistle to the Galatians 5:19–21 specifies more sins, including four more that Galileo committed: variance, emulation, sedition and heresy. The apostle Paul there warned, allegedly, that persons who enact such sins 'shall not inherit the Kingdom of God'. Yet Galileo pretended to surpass or emulate the Church Fathers in the ability to understand scriptures. He wilfully departed from the intellectual unity of the Church. Galileo excelled at boasting and hubris; his self-esteem led him to ridicule those whose intellects he deemed inferior. His friends admired Galileo as brilliant, bold and sincere, but to his enemies he seemed arrogant and full of vainglory. Among the 'seven deadly sins', pride was often regarded as the worst one, the root of the others. In his *Defence of Galileo*, Campanella said that one critique against Galileo was that those who taught unusual ideas were viewed as guilty of 'pridefully placing themselves above others', whereas Proverbs 22:28 cautioned: 'Do not go beyond the limits set by your fathers.'[92]

The Inquisitors recognized these underlying motives: intellectual pride and excessive reliance on individual insight, the exaggerated pretence of religious zeal. Thus in 1616 Galileo grandiosely bragged: 'my own behaviour in this affair has been such that a saint would not have handled it either with greater reverence or with greater zeal toward the Holy Church.'[93] Likewise in 1635 Galileo wrote:

> I have two sources of perpetual comfort: first, that in my writings there cannot be found the faintest shadow of anything to diminish piety and reverence for the Holy Church; and second, my own conscience, which only I completely know on Earth, as God in Heaven, for He well knows that in this cause for which I suffer, many may have been learned, but none, not even the Saintly Fathers, have spoken or proceeded with more piety or with greater holy intention for the Holy Church than me.[94]

Most of his indiscretions were minor compared to Galileo's greatest sin, for which he was judged: the sin of heresy. According to most authorities, any Christian who wilfully departed from Catholic dogma, especially after having been instructed not to do so, was a heretic. As St Thomas Aquinas emphasized, 'the sin of heresy is the greatest sin in the whole range of human perversity.' It was viewed as the act of an evil will, one who knowingly chose to oppose the will of God.

In Catholicism, true believers were those who accepted all the dogma formulated by the Church, in its entirety – especially during the Reformation. Any Catholic who instead accepted only the parts of Catholic dogma that he or she personally selected or modified was a heretic. And that was what Bruno and Galileo did. They each selected at will the particular parts of Catholic dogma that they disliked: those that they thought were insufficient, ambiguous or misinterpreted. They philosophized as if they could rightly ascertain knowledge without the guidance of the Catholic Church. As St John Chrysostom had complained, philosophers disparaged faith because they obstinately tried to figure things out by themselves and therefore found nothing.[95]

In the Renaissance it was a common practice to justify established or innovative beliefs on the basis of ancient authorities. Thus Galileo ascribed the theory of Copernicus to Pythagoras, a mystic philosopher that could hardly help his case. Galileo's rhetorical devices failed to convince the Catholic clergy; the questionable pagan sage did not give enough prestige to the idea of Earth's motion.

Eventually, however, Galileo's rhetoric had an unforeseen effect: it served to canonize Pythagoras as a great scientist and mathematician. In his *Dialogue* Galileo not only claimed that Pythagoras was a great astronomer, but he included an imaginative claim that Pythagoras proved the so-called Pythagorean theorem:

> There is no doubt that a long time before Pythagoras had discovered the proof, for which he did the hecatomb, he was sure that the square on the side opposite the right angle of the right-angled triangle was equal to the squares on the other two sides; and the certainty of the conclusion helps not a little in the discovery of the proof.[96]

In the previous two thousand years, writers mainly noted the alleged finding. It was not common to claim that Pythagoras *proved* the hypotenuse theorem. Galileo was not the first to make such a claim, but through his famous work this assertion spread and became a pervasive myth in scientific and mathematical education for the following centuries.

Galileo's *Dialogue* of 1632 stands at the middle point in the historical, mythical transformation of Pythagoras: the ancient philosopher's fame as an eccentric mystic and pagan cult leader gradually diminished, while at the same time he became increasingly portrayed as a pioneering scientist. While Galileo inadvertently succeeded in canonizing Pythagoras as a mathematician and astronomer, we have now seen how the esoteric image of Pythagoras contributed to the converse: Galileo, mathematician and scientist, was converted into a heretic.

In hindsight, the Pythagorean improprieties of Bruno and Galileo were nothing compared to the crimes of the Church. The God-fearing urge to create a Christian empire of souls led the Popes, bishops and cardinals to implement a repressive state, as if it would be a utopia where good Catholics would all live by piety and faith. We might well say that the Roman Inquisition was totalitarian inasmuch as it not only prohibited Catholics from expressing certain thoughts, or from thinking them, but it dictated what individuals should think. It sought to control the thoughts of its subjects by controlling their actions. Philosophy and science could not grow freely in that environment. Instead, writing became constrained by the kind of hypocrisies and measured silences that we see in Galileo's writings and depositions.

Totalitarian regimes treat writers as dangerous. The Catholic authorities seized, censored and destroyed certain books. Librarians such as Agostino Steuco, Cardinal Baronio, Gabriel Naudé, Lucas Holste and even Francesco Barberini noticed ancient heresies that were not known by everyone. Militant fanatics such as Bellarmine and Inchofer responded sternly to such beliefs. Ancient philosophical ideas became scorned as dangerous heretical novelties. The Inquisitors criminalized and persecuted such opinions, as if paranoid that such fictions might possibly overpower their own. The miracles of Pythagoras were denied as devilish lies, while the miracles of Jesus Christ were hailed as ennobling truths. Hence the

Inquisition imprisoned and murdered Catholics who openly dared to disagree with its prescriptions for what to believe. By banning books and imprisoning heretics, they tried to suppress freedom of thought. Thus the Catholic clergymen won, in Rome, for at least a while. But many years later they finally lost in the court of public opinion, the open society, by failing to obliterate every trace of the arguments of Bruno, a martyr for freedom of thought.

Bruno and Galileo earned our admiration because they dared to disobey the rules of the Church. No matter how deeply the clergymen were offended by relatively imaginative ideas – that the Earth is alive, that it has a soul, that it moves in the heavens, that the stars are worlds – the courage of heretics became an inspiration for those who seek to find the truth about anything, rather than obey the instructions of institutions.

REFERENCES

Introduction

1 Ronald L. Numbers, ed., *Galileo Goes to Jail and Other Myths about Science and Religion* (Cambridge, MA, 2009).
2 Maurice A. Finocchiaro, 'Philosophy versus Religion and Science versus Religion: The Trials of Bruno and Galileo', in *Giordano Bruno: Philosopher of the Renaissance*, ed. Hilary Gatti (Burlington, VT, 2002), pp. 51–85.
3 Peter Godman, *The Saint as Censor: Robert Bellarmine between Inquisition and Index* (Leiden, 2000), pp. 214–15. See also William Shea and Mariano Artigas, *Galileo Observed: Science and the Politics of Belief* (Sagamore Beach, MA, 2006).

1 The Crimes of Giordano Bruno

1 'Decretum Sacrae Congregationis', 5 March 1616, Rome.
2 Flaminius Adrianus, 'Decreto della Congregazione del Sant'Uffizio', 4 February 1599, in Luigi Firpo, *Il processo di Giordano Bruno* (Rome, 1993), p. 314.
3 Galileo Galileo [Galilei], *Sidereus Nuncius* (Venice, 1610), p. 9. See also Galilei, *Sidereus Nuncius, or The Sidereal Messenger*, trans. Albert Van Helden (Chicago, IL, 1989), p. 43.
4 Urban VIII, quoted by Francesco Niccolini to Andrea Cioli, 5 September 1632, in Antonio Favaro, ed., *Le opere di Galileo Galilei: Edizione Nazionale*, 20 vols (Florence, 1890–1909, repr. 1929–38), XIV, pp. 383–4. See also, for example, Maurice Finocchiaro, *The Galileo Affair: A Documentary History* (Berkeley, CA, 1989), pp. 229–30.
5 Quoted by Galilei to Elia Diodati, 15 January 1633, in Finocchiaro, *The Galileo Affair*, p. 225.
6 Jole Shackelford, 'Myth VII: That Giordano Bruno was the First Martyr of Modern Science', in *Galileo Goes to Jail and Other Myths about Science and Religion*, ed. Ronald L. Numbers (Cambridge, MA, 2009), p. 67.
7 Few historians have discussed Pythagorean notions in Bruno's works. See Ramon G. Mendoza, 'Metempsychosis and Monism in Bruno's *nova filosofia*', in *Giordano Bruno: Philosopher of the Renaissance*, ed.

Hilary Gatti (Burlington, VT, 2002), pp. 273–98; Hilary Gatti, '"The Pythagorean School and Our Own": Bruno and the Philosopher from Samos', in *Giordano Bruno and Renaissance Science* (Ithaca, NY, 1999), pp. 13–28; Dario Tessicini, *I dintorni dell'infinito: Giordano Bruno e l'astronomia del Cinquecento* (Pisa, 2007). My present book provides original contributions: tracing controversial Pythagorean beliefs throughout centuries, the discovery that such beliefs were denounced by Catholic authorities, and evidence that they were centrally important in Bruno's trial.

8 Steven Dick, *Plurality of Worlds: The Origins of the Extraterrestrial Life Debate from Democritus to Kant* (Cambridge, 1982); Michael Crowe, *The Extraterrestrial Life Debate, 1750–1900* (Cambridge, 1986); and Michael Crowe, ed., *The Extraterrestrial Life Debate, Antiquity to 1915: A Source Book* (Notre Dame, IN, 2008).

9 Walter Burkert, *Lore and Science in Ancient Pythagoreanism,* trans. Edwin Minar (Cambridge, 1972); Alberto A. Martínez, *The Cult of Pythagoras: Math and Myths* (Pittsburgh, PA, 2012).

10 Iacobi Peletarii, *In Euclidis elementa geometrica demonstrationum Libri Sex* (Lyon, 1557), I, pp. 47–8.

11 Aristotle, *De caelo* (*c.* 340 BCE), II, sec. 13.

12 In Apollonius of Tyana's biography of Pythagoras, now lost, he said that Pythagoras was the son of Apollo; quoted in Porphyry, *Life of Pythagoras* (*c.* 300 CE), in *The Pythagorean Sourcebook and Library,* trans. Kenneth Sylvan Guthrie (New York, 1919), revd edn, ed. David R. Fideler (Grand Rapids, MI, 1987), pp. 123–36.

13 Heraclides, quoted in Diogenes Laertius, *Βίοι και Γνώμαι των εν Φιλοσοφίαι Ευδόκιμηεαντων και των Εκασθηι Αιρέσει Αρεσκόντων* (*c.* 225 CE), in Diogenes, *Lives of Eminent Philosophers*, ed. Tiziano Dorandi (Cambridge, 2013), VIII:4, p. 603.

14 Ovid, *Metamorphoses* (*c.* 8 CE), XV.

15 *De placita philosophorum*, I, chap. 3; IV, chaps 4 and 7. This work, known in Greek as *Peri ton areskonton philosophois, physikon dogmaton* and falsely attributed to Plutarch, was actually based on a work by Aetius (*c.* 50 BCE), as noted by Theodoret of Cyrus (*c.* 393–*c.* 458/466 CE). It was also falsely attributed to Qusta ibn Luqa by Ibn al-Nadim; see Hans Daiber, ed., *Aetius Arabus: Die Vorsokratiker in arabischer Überlieferung* (Wiesbaden, 1980).

16 Diogenes, *Lives of Eminent Philosophers*, VIII:4, p. 602, and VIII:21, p. 612.

17 Jerome to Ctesiphon (415 CE), in Philip Schaff and Henry Wace, eds, *St Jerome: Letters and Select Works*, Select Library of Nicene and Post-Nicene Fathers, 2nd ser., vol. VI (New York, 1893), Letter 133, sec. 1, pp. 272–3. Ioannes Chrystostomos, Homily II (*c.* 395 CE), in *The Homilies of S. John Chrystostom, Archbishop of Constantinople, on the Gospel of St John*, I: *Hom. I–XLI*, trans. G. Stupart (Oxford, 1848), pp. 11–14. See also Chrysostom's Homily I, on Philippians 1:1–2, in *The Homilies of S. John Chrysostom, Archbishop of Constantinople*, trans. James Tweed (Oxford, 1843), pp. 8–9; Tertullian, 'XIII. Adversus Hermogenem', ed. E. Kroymann, in *Tertulliani Opera*, I, ed. E. Dekkers, Corpus Christianorum, Series Latina (Turnhout, 1954), p. 404, and 'XVII.

De Anima' (*c.* 215 CE), ed. J. H. Waszink, in *Tertulliani Opera*, II, p. 785.
18 'Jesuit Rules on Theology and Philosophy', Decree 41: 3 November 1593–18 January 1594, in *Behind the Scenes at Galileo's Trial: Including the First English Translation of Melchior Inchofer's Tractatus Syllepticus*, ed. Richard Blackwell (Notre Dame, IN, 2006), pp. 207–9.
19 Thomæ Aquinatis, *Summa totius Theologiæ* (*c.* 1274), 8 vols (Cologne, 1640), I, p. 29.
20 Thomas Aquinas, *Summa Theologica*, part I, Quest. 47, article 3, in *Opera Omnia*, I (Paris, 1871), p. 315.
21 Thomæ Aquinatis, *Tertia Pars: Summæ Theologicæ* (*c.* 1274) (Venice, 1585), pp. 241–2; Aristotle, *De Anima* (*c.* 350 BCE), I.
22 Dante, *L'amoroso convivio* (*c.* 1305) (Venice, 1529), p. 49.
23 Niceforo Callistos, *Ecclesiasticae Historiae III* (*c.* 1320), ed. J. P. Migne, *Patrologia Graeca 145* (Paris, 1904), cols 891–976. Gianfrancesco Pico della Mirandola, *De rerum praenotione* (1507), in Joannis Francisci Pici Mirandulae, *Opera Omnia* (Basel, 1519), pp. 664–74. Porphyry, 'On the Philosophy Derived from Oracles' (*c.* 270 CE), quoted in Eusebius, *Praeparatio Evangelica* (*c.* 314 CE); trans. Amos Berry Hulen, *Porphyry's Work Against the Christians* (Scottdale, PA, 1933), p. 16.
24 Nicolai Copernici, *De revolutionibus orbium cœlestium, Libri VI* (Nuremberg, 1543), Prefatio, fol. iv reverso. Lactantius, *Divine Institutes*, III, chap. 24.
25 Copernici, *De revolutionibus*, fol. 3 reverso. Copernicus listed Philolaus, Hicetas of Syracuse, Heraclides and Ecphantus (who actually was not a Pythagorean). *De placita philosophorum*, III, chap. 7. Some of the works consulted by Copernicus, such as the Pseudo-Plutarch *De placita* and Lysis' Letter, were not genuine. See Bronisław Biliński, *Il Pitagorismo di Niccolò Copernico* (Wrocław, 1977), p. 111.
26 Copernici, *De revolutionibus,* fol. 9 reverso.
27 The writer 'Lysis' referred to the cult of goddesses Demeter and Persephone at Eleusis in Greece.
28 Copernicus, quoted in Owen Gingerich, 'Did Copernicus Owe a Debt to Aristarchus?', *Journal for the History of Astronomy*, XVI/1 (1985), p. 38.
29 Nicolas de Cusa, *De docta ignorantia* (1440), in Nicolai Cusae, *Haec accurata recognitio trium voluminum / Operum clariss.* (n.p., 1514), II, folio 22 verso and reverso.
30 Biliński, *Il Pitagorismo*, p. 111. Edward Rosen, 'Was Copernicus a Pythagorean?', *Isis*, LIII (1962), pp. 504–8. Biliński and Rosen didn't mention heretical dimensions of Pythagorean beliefs.
31 Giovanni Maria Tolosani, 'De coelo supremo immobili et Terra infima stabili' (1546): appendix in *De Purissima Veritate Sacrae Scripturae* (MS., Florence, Biblioteca Nazionale Centrale, 1546), pp. 339*v*, 340*v*. Also in Eugenio Garin, 'Alle origini della polemica Copernicana', *Studia Copernicana*, VI (Wrocław, 1973), pp. 31–42; Garin, *Rinascite e Rivoluzioni: Movimenti culturali dal XIV al XVIII Secolo* (Bari, 1975), pp. 283–95.
32 Hilary Gatti, 'Why Giordano Bruno's "Tranquil Universal Philosophy" Finished in a Fire', in *Ideas under Fire: Historical Studies of Philosophy and Science in Adversity*, ed. Jonathan Lavery, Louis Groarke and William Sweet (Madison, NJ, 2012), p. 109.

33 Anna Foa, *Giordano Bruno* (Bologna, 1998), p. 37; Ingrid Rowland, *Giordano Bruno: Philosopher/Heretic* (New York, 2008), pp. 71–86.
34 Desiderius Erasmus, *The Apophthegms of the Ancients*, II (London, 1753), bk 8, sec. 20, p. 268; bk 7, sec. 201, p. 255.
35 Erasmus, 'Adages': IIIXI/LBII 935A, in John Grant, ed., *Collected Works of Erasmus: Adages, III iv 1 to IV ii 100*, trans. Denis Drysdall (Toronto, 2005), p. 356.
36 Erasmus, *The Praise of Folly and Other Writings*, trans. Robert Adams (New York, 1989), pp. 34–5.
37 Erasmus, *Apophthegms*, bk 8, sec. 60, p. 283.
38 Erasmus, *Adages*, p. 408.
39 Anonymous authors, late 1500s, Oxford, Bodleian Library, MS. Marshall 15 (5266), fol. 66*v*. Christofo de Cattan, *La Géomance du Seigneur Christofe de Cattan, . . . Avec la roüe de Pythagoras*, trans. Gabriel du Preau (Paris, 1558); *The Geomancie of Maister Christopher Cattan,* trans. Francis Sparry (London, 1591).
40 Ioanne Wiero, *De praestigiis daemonum, et incantationibus, ac veneficiis, Libri V* (Basel, 1563), II, pp. 139, 153; see also pp. 98, 113, 123, 463. Johan Weyer's treatise was also published in German (1566 and 1586), and in Latin (1564, 1577 and 1583).
41 Thomas Digges, 'A Perfit Description of the Caelestiall Orbes according to the most aunciente Doctrine of the Pythagoreans, latelye reuiued by Copernicus', in Leonard Digges, *A Prognostication Everlastinge of Righte Good Effecte*, ed. Thomas Digges (London, 1576), pp. 48, 49, 51.
42 Ibid., pp. 43, 49.
43 Bruno, *Sigillus Sigillorum* (1583), sec. 36; Hilary Gatti, *The Renaissance Drama of Knowledge: Giordano Bruno in England* (New York, 1989), p. 80.
44 Miguel Granada, 'Bruno, Digges, Palingenio: Omogeneità ed Eterogeneità nella Concezione dell'Universo Infinito', *Rivista di Storia della Filosofia*, XXXXVII/1 (1992), pp. 47–73.
45 Giordano Bruno, *La cena delle Ceneri* (London, 1584), part 3, p. 165; see also Paul Henri Michel, *The Cosmology of Giordano Bruno,* trans. R. Madison (Ithaca, NY, 1973), pp. 214–15.
46 Lucretius, *De rerum natura* (*c*. 50 BCE), in Ronald Latham, trans., *On the Nature of the Universe* (Baltimore, MD, 1951), p. 91.
47 Giordano Bruno, *De l'infinito universo et mondi* ([Venice], 1584), Dialogue 3, pp. 74–5.
48 Giordano Bruno, *Candelaio* (1582), ed. A. Guzzo (Milan, 2004), p. 118; Bruno, *Sigillus Sigillorum* (1583), in Jordani Bruni, *Opera latine conscripta*, ed. F. Fiorentino, vol. I/I (Naples, 1879), p. 181.
49 Didaci a Stunica [Diego de Zúñiga], *Erimitæ Augustiniani in Iob commentaria* (Toledo, 1584), p. 206.
50 [Giordano Bruno] Nolano, *Cabala del cavallo Pegaseo. Con l'aggiunta dell'Asino Cillenico* (Paris, 1585), p. 87 (*L'Asino*) and end of Dialogue 2, part 2. Most scholars agree that Bruno believed in transmigration (metempsychosis), based on his writings, though he denied it to the Venetian Inquisition. However, Roger Mendoza argues that the soul/ body dualism implied by Pythagorean transmigration is incompatible with Bruno's monism, which involved corporeal matter and incorporeal

matter, such as the soul of the world. Such distinctions don't affect my argument, which requires only that readers recognized support for some kind of transmigration in Bruno's writings, and he repeatedly credited Pythagoras. Also, there did not exist a single version of Pythagorean metempsychosis as Mendoza implies, which results from not reviewing, in his article, the various Pythagorean sources over the centuries; see Mendoza, 'Metempsychosis and Monism', pp. 272–97.

51 Giordano Bruno, *De gli heroici furori* (Paris, 1585), n.p. (end of the Fourth Dialogue).

52 Giordano Bruno, *Articuli centum et sexaginta adversus huius temporis mathematicos atque philosophos* (1588), trans. William Boulting, in *Giordano Bruno: His Life, Thought, and Martyrdom* (London, 1914), p. 212.

53 Iordani Bruni, *De monade, numero et figura liber consequens quinque de minimo magno & mensura: item de innumerabilibus, immenso, & infigurabili, seu, De universo & mundis libri octo* (Frankfurt, 1591).

54 Ibid., in Bruno, *Opere Latine*, ed. C. Monti (Turin, 1980), p. 305.

55 Jordani Bruni, *De triplici minimo et mensura* (Frankfurt, 1591), chap. 3, p. 13.

56 'Denuncia di Giovanni Mocenigo all'Inquisitore di Venezia Giovan Gabriele da Saluzzo', 23 May 1592, in Firpo, *Il processo di Giordano Bruno*, pp. 143–5. See also Giulio Monterenzi [apparent author], 'Sommario' (2 July 1597), in Luigi Firpo, 'Il Processo di Giordano Bruno', *Rivista Storica Italiana*, LX (Naples, 1948), pp. 542–97, and LXI, pp. 5–59.

57 Christopher Black, *The Italian Inquisition* (New Haven, CT, 2010), p. 57; and regarding denunciations, see p. 54.

58 Thomas Mayer, *The Roman Inquisition on the Stage of Italy, c. 1590–1640* (Philadelphia, PA, 2014), p. 117.

59 Bruno's 3rd Deposition to the Venetian Inquisition, 2 June 1592, in Firpo, *Il processo di Giordano Bruno*, pp. 167–8.

60 Bruno, 3rd Deposition, 2 June 1592, in Firpo, *Il processo di Giordano Bruno*, p. 254.

61 Virgil, *Aeneid* (*c.* 20 BCE), VI, 724–7.

62 Firpo, *Il processo di Giordano Bruno*, p. 169.

63 Virgil, *Aeneid*, VI, 685–755.

64 Giordano Bruno, *De la causa, principio et uno* (London, 1584), Dialogue 5, p. 124; reissued in Giovanni Gentile, ed., *Opere Italiane di Giordano Bruno*, I (Bari, 1908), p. 253.

65 Dilwyn Knox, 'Bruno: Immanence and Transcendence in *De la Causa, Principio et Uno*, Dialogue II', *Bruniana & Campanelliana*, XIX/2 (2013), pp. 466, 473. See also Émile Namer, *Les aspects de Dieu dans la philosophie de Giordano Bruno* (Paris, 1926), pp. 35–7, 52–5, 63–4, 70, 87–91, 98.

66 Bruno, *De magia mathematica*, in Bruno, *Opere Magiche*, ed. M. Ciliberto et al. (Milan, 2000), p. 22. All the texts recently republished as *Opere Magiche* remained in manuscript, presumably among the papers of Bruno's pupil and secretary Jerome Besler (1566–1632), until their publication from 1891 and therefore were not available to Bruno's critics in the 1590s. See Iordani Bruni, *Opera latine conscripta*, ed. F. Tocco and H. Vitelli, III (Florence, 1891).

67 Bruno, *Camoeracenis Acrotismus*, in Jordani Bruni, *Opera latine conscripta*, ed. F. Fiorentino, vol. I/1 (Naples, 1879), Article LXV, p. 177.
68 Bruno, *De gli heroici furori*, 'Argomento' and Dialogues 2 and 4 [unnumbered pages]. See also Bruno, *Gli eroici furore* (Milan, 1864), pp. 19, 44, 90.
69 Bruno, 4th Deposition, Venice, 2 June 1592, in Firpo, *Il processo di Giordano Bruno*, pp. 176–7.
70 Giulio Capaccio, *Del trattato dell'impresse* (Naples, 1592), p. 5.
71 Gasparo Contarini and Donato Giannotti, *Della Republica et Magistrati di Venetia: Libri V* (Venice, 1591), p. 243. Giulio Valentino, *Trattato del Conseglio et de' Conseglio, et de Conseglieri de Prencipi* (Venice, 1599), p. 188.
72 Ferdinandi de Mendoza, *De confirmando Concilio Illiberritano ad Clementem IIX Sanctae Romanae et Catholicae Ecclesiae Pont. Opt. Max. Libri III* (Madrid, 1594), p. 42; see also pp. 34, 37, 39.
73 Giovanni Andrea dell'Anguillara, *Le Metamorfosi di Ovidio*, rev. Gioseppe Horologgi (Venice, 1584), p. 255; see also p. 271.
74 Bruno, 4th Deposition, Venice, 2 June 1592, in Firpo, *Il processo di Giordano Bruno*, p. 190.
75 'Verbale di Seduta dell'Eccellentissimo Collegio di Venezia', 28 September 1592, ibid., p. 202.
76 'Verbale di Seduta', 22 December 1592, ibid., p. 207; and 'Verbale di Seduta', 7 January 1593, ibid., pp. 210–11.
77 Mayer, *The Roman Inquisition on the Stage of Italy*, pp. 66, 119. Firpo notes that Cardinal Francesco degli Albizzi, who published an Inquisition manual, *Risposta alla Historia della Sacra Inquisitione* (Rome, 1678), found only eight extraditions from Venice to Rome. Firpo, *Il processo di Giordano Bruno*, pp. 38–9.
78 Mayer, *The Roman Inquisition on the Stage of Italy*, pp. 118, 120.
79 'Denuncia di Fra Celestino da Verona', in Firpo, *Il processo di Giordano Bruno*, pp. 47–8.
80 Bruno, in Angelo Mercati, ed., *Il Sommario del Processo di Giordano Bruno*, Studi e Testi, Biblioteca Apostolica Vaticana CI (Vatican City, 1942), p. 85. Firpo, *Il Processo di Giordano Bruno*, p. 255.
81 In September 1596 the list was submitted to three reviewers: Master [Garcia?] Guerra, Pedro Juan Zaragoza (a Consultor of the Index since 1592), both Dominicans, and a Jesuit presbyter named Gallo. Firpo, *Il processo di Giordano Bruno*, p. 235; Mayer, *The Roman Inquisition on the Stage of Italy*, pp. 121, 289. Apparently nothing else is known about Guerra or Gallo.
82 'Visita dei Carcerati nel Sant'Uffizio Romano', 16 December 1596, in Firpo, *Il processo di Giordano Bruno*, p. 241.
83 Ibid., pp. 304, 80–85.
84 Maurice A. Finocchiaro, 'Philosophy versus Religion and Science versus Religion: The Trials of Bruno and Galileo', in *Giordano Bruno: Philosopher of the Renaissance*, ed. Hilary Gatti (Burlington, VT, 2002), p. 61.
85 Bruno, 4th Deposition, in Firpo, *Il processo di Giordano Bruno*, p. 284.
86 Frances Yates, *Giordano Bruno and the Hermetic Tradition* (London, 1964, repr. Abingdon, 2002), p. 388.

87 ‘Summarium quarundam responsionum Fratris Iordani ad censuras factas super Propositionibus quibusdam ex libris elicitis’, in Mercati, *Sommario*, pp. 113–19.
88 Bruno, *De triplici minimo et mensura.*
89 Bruno, *De l’infinito universo et mondi.*
90 Bruno, *De la causa, principio et uno*, *De l’infinito universo et mondi,* and *Cabala del cavallo Pegaseo.*
91 Bruno, *De la causa, principio et uno.*
92 Bruno, *La cena delle Ceneri*, and *De l’infinito universo et mondi.*
93 Bruno, *La cena delle Ceneri, De l’infinito universo et mondi*, and *De magia mathematica.*
94 Bruno, *La cena delle Ceneri.*
95 Bruno, *De la causa, principio et uno.*
96 Ibid. and *De gli heroici furori.*
97 Bruno, *La cena delle Ceneri, De l’infinito universo et mondi*, *De la causa, principio et uno*, *Spaccio de la Bestia Trionfante* (Paris, 1584), *De gli heroici furori*, *Cabala del cavallo Pegaseo*, *De immenso*, and so on.
98 For accounts of how notions in mathematics, astronomy, religion and alchemy were misattributed to Pythagoras over time, see Martínez, *Cult of Pythagoras*, pp. 1–28, 125–7, 201–24; Alberto A. Martínez, *Science Secrets: The Truth about Darwin’s Finches, Einstein’s Wife, and Other Myths* (Pittsburgh, PA, 2011), pp. 13–42, 55–7, 70–76, 229–30, 261–80.
99 *De placita philosophorum*, IV, chap. 7, and II, chap. 14.
100 Porphyry, *On the Abstinence of Animal Food*, II, *Select Works of Porphyry*, ed. and trans. T. Taylor (London, 1823), sec. 37, p. 74.
101 Hippolytus, ‘*Ο κατὰ πασῶν αιρέσεων ἔλεγχος* (*c.* 225 CE)’, in Hippolytus, *Refutatio omnium haeresium*, ed. M. Marcovich (Berlin, 1986), IV, sec. 14; VI, sec. 23.
102 Étienne Tempier, *Tredecim errores a Stephano episcopo Parisiensi condemnati, 1270*, Prop. 5, and *Sequuntur errores annotati in rotulo, 1277*, Props. 87, 98; in H. Deinfle, *Chartularium Universitatis Parisiensis*, I (Paris, 1889), pp. 487, 548.
103 Tempier (1277), Prop. 109, in Deinfle, *Chartularium*, p. 549.
104 Faculty of Divinity of Paris, Censures of 1 August 1553, quoted in L. Dupin, *A New Ecclesiastical History of the Sixteenth Century*, II (London, 1706), p. 441.
105 Sebastiano Medice, *Summa omnium haeresum* (Florence, 1581), p. 647. Sebastiano Medicis, *Summa omnium hæresum* (Venice, 1587), pt 1, p. 37 rev.
106 Justin Martyr [apocryphal?], ‘Justin’s Hortatory Address to the Greeks’, chap. XIX, in *Justin Martyr and Athenagoras*, II of *Ante-Nicene Christian Library*, ed. A. Roberts and J. Donaldson (Edinburgh, 1867), p. 305.
107 Philostratus, *The Life of Apollonius of Tyana* (Cambridge, 1948), III, chap. 34, p. 308.
108 Augustine, *De Civitate Dei*, VII, chaps 6, 22.
109 Jerome, *Commentariorum Hieronymi in Matthaeum Evangelistam, Liber Primus*, in *Operum divi Hieronymi Eusebii Stridonensis*, IX (Paris, 1534), chap. 8, p. 12 rev.
110 Peter Abelard, *Epitome theologiæ Christianæ*, in *Sæculum XII Petri*

Abælardi Abbatis Rugensis opera omnia, ed. J. P. Migne, *Patrologia Latina 178* (Paris, 1855), chap. 18, cols 1720–21.

111 Peter Abelard, *Introductio ad Theologiam*, I, in ibid., col. 1019. Original in Salvianus, *De gubernatione Dei, octo libri dati ad S. Salonium Episcopum* (*c*. 440 CE), I, in Salviani Massiliensis, *Opera Omnia*, ed. J. P. Migne, *Patrologia Latina 53* (Paris, 1859), col. 29.

112 Petrus Abaelardus, 'Scholarium', Part I, in *Opera theologica*, III*: Theologia 'summi boni'/Theologia 'scholarium'*, ed. E. Buytaert and C. J. Mews, Corpus Christianorum: Continuatio Mediaevalis 13 (Turnhout, 1987), cols 389–93.

113 G. de Saint-Thierry to the Bishop of Chartres and Abbot of Clairvaux, late 1139, in *Receuil d'études sur Saint Bernard et ses écrits*, ed. J. LeClercq, IV (Rome, 1987), p. 352.

114 J. Morrison, *The Life and Times of Saint Bernard, Abbot of Clairvaux, A.D. 1091–1153* (London, 1894), pp. 301–11.

115 Pope Innocent II, 1141, *Conciliorum Generalium Ecclesiae Catholicae*, IV (Rome, 1612), p. 23.

116 A. de Castro, *Adversus omnes hæreses. Libri XIIII* (Paris, 1543), V, p. 80. See also T. Deciani, *Tractatus criminalis omnium hæresum*, I (Venice, 1590), pp. 236–7.

117 Medicis, *Summa omnium hæresum* (1581), p. 647, and (1587), pt I, p. 62 verso.

118 Bruno, *Camoeracensis Acrotismus*, in *Opera latine conscripta*, Article LXV, p. 177.

119 Firpo, *Il processo di Giordano Bruno*, pp. 169, 254; Wisdom 1:7. See also Bruno, *De la causa, principio et uno*, Dialogue 5, p. 124; reissued in G. Gentile, ed., *Opere Italiane di Giordano Bruno*, I (Bari, 1908), p. 253. See also Knox, 'Bruno: Immanence and Transcendence', pp. 466, 473; Namer, *Les aspects de Dieu dans la philosophie de Giordano Bruno*, pp. 35–98. Michele Ciliberto, *Lessico di Giordano Bruno*, II (Rome, 1979), p. 929. For a longer discussion, see Alberto A. Martínez, 'The Copernicans and the Soul of the World', manuscript. On the Stoics, see T. Gregory, *Anima Mundi* (Florence, 1955), pp. 123–7.

120 Bruno, *De l'infinito universo et mondi*, p. xxix.

121 Bruno, *De immenso et innumerabilibus, seu de universo et mundis* (1591), in Bruni, *Opera latine conscripta*, I/2, pp. 171, 291. See also Granada, 'Bruno, Digges', p. 67.

122 *De placita philosophorum*, II, chap. 13.

123 For discussion, see Granada, 'Bruno, Digges', p. 66.

124 Aristotle, *De Anima*, I.

125 Justin, 'Dialogue with Trypho', chap. 2, II of *Ante-Nicene Christian Library*, pp. 87–8.

126 *De placita philosophorum*, IV, chap. 7.

127 Macrobius, *Commentary on the Dream of Scipio* (*c*. 430 CE), trans. W. Stahl (New York, 1952), p. 134.

128 Tertullian, *De Anima*, in *Tertulliani Opera*, II, chaps 28–32, 54; Lactantius, *The Divine Institutes*, III, chap. 18, in *The Works of Lactantius*, XXI/I of *Ante-Nicene Christian Library* (Edinburgh, 1871), pp. 182–3. Epiphanius, *Panarion* (*c*. 374–7 CE), in *The Panarion of Epiphanius of Salamis: Book I*, ed. Frank Williams (Leiden, 2009), sec. 7, p. 24.

129 Jerome, *Ad Pammachium et Marcellinum apologia Hieronymi adversum Ruffinum*, III, in Hieronymi, *Opera Omnia*, ed. M. Victorii (Paris, 1624), p. 537; see also pp. 20–21, 30, 380, 411–12, 444, 551.
130 Hippolytus, *Refutatio*, I, sec. 14, p. 74, and I, sec. 22, p. 84.
131 Bruno, 2 June 1592, in Firpo, *Il processo di Giordano Bruno*, p. 190.
132 Bruno, *c.* 1587, in Firpo, *Il processo di Giordano Bruno*, plate 5 after p. 86. See also H. von Warnsdorf Family Album, 18 September [1587], in ibid., pp. 169, 301, 304, plate 4, after p. 86; Eugenio Canone, ed., *Giordano Bruno: gli anni napoletani e la 'peregrinato' europea* (Cassino, 1992), pp. 121–5; F. Tocco, 'Un nuovo autografo di G. Bruno', *La Bibliofilia*, IX (1906), pp. 342–5.
133 *De placita philosophorum*, I, chap. 9, p. 123.
134 Ovid, *Metamorphoses*, trans. F. Miller, II (London, 1916), pp. 375–7.
135 Porphyry, *Life of Pythagoras*, sec. 19.
136 Jerome, *Ad Pammachium et Marcellinum apologia Hieronymi adversum Ruffinum*, III, sec. 40.
137 Aristotle, *De caelo*, II, sec. 13.
138 Hippolytus, *Refutatio*, I, sec. 15, p. 74.
139 Porphyry, *On the Abstinence of Animal Food*, II, sec. 37, p. 74.
140 Bruno, *La cena delle Ceneri*, p. 51.
141 Digges, 'Perfit Description', fol. 43; Bruno, *De immenso*, III, p. 377; Bruno, *De magia mathematica*, sec. 1; F. Tocco, 'Le opere inedite di Giordano Bruno', *Atti della Reale Accademia di Scienze Morali e Politiche*, XXV (1892), p. 146; A. von Nettesheim, *De occulta philosophia*, ed. V. Compagni (Leiden, 1992), p. 85. Bruno also linked stars and daemons in W. Lutoslawski, 'Jordani Bruni Nolani opera inedita, manu propria scripta', in *Archiv für Geschichte der Philosophie*, ed. L. Stein, II (Berlin, 1889), p. 541. See also *De magia mathematica*, sec. 17. Regarding Bruno's sources for *De magia mathematica*, see S. Ricci, *Giordano Bruno nell'Europa del Cinquecento* (Rome, 2000), p. 426.
142 Porphyry, *On the Abstinence of Animal Food*, II, sec. 37, p. 74.
143 N. Callisti, *Ecclesiasticæ historiæ libri decem & octo* (Basel, 1553), p. 893; Cesare Baronio, *Annales Ecclesiastici*, VII (Rome, 1596), p. 289.
144 Plato, *Timaeus* (*c.* 360 BCE), in *The Dialogues of Plato*, trans. B. Jowett, II (Oxford, 1871), paragraphs 30–31, pp. 525–6.
145 Hippolytus, *Refutatio*, VI, sec. 21, p. 229.
146 *De placita philosophorum*, II, chap. 9, pp. 136–7.
147 Plotinus, *The Six Enneads* (*c.* 255–70 CE), trans. S. MacKenna (London, 1917–30), 4th Ennead, Treatise 4, chap. 27.
148 Philostratus, *Life of Apollonius*, III, chap. 34, p. 308.
149 Iamblichus, *De mysteriis Aegyptiorum, Chaldæorum, Assyriorum*, ed. M. Ficinus (London, 1549), pp. 108–9, 114.
150 Gregory, *Anima Mundi*, pp. 123–7.
151 Augustine, *De Civitate Dei*, VII, chap. 23.
152 Tempier (1277), in G. Klima et al., *Medieval Philosophy: Essential Readings* (Malden, MA, 2007), p. 184.
153 Ovid, *Metamorphoses*, II, p. 389.
154 Dell'Anguillara, *Metamorfosi*, pp. 255 rev., 271 verso.
155 R. Bellarmini, 'Auctores addendi ad tertiam classem Indicis Pii IV,

1592', in Peter Godman, *The Saint as Censor: Robert Bellarmine between Inquisition and Index* (Leiden, 2000), p. 273.
156 Firpo, *Il processo di Giordano Bruno*, p. 284.
157 Bruno, *Spaccio de la Bestia Trionfante*, p. 20; *Cabala del cavallo Pegaseo*, p. 52; *De gli heroici furori*, 3rd Dialogue, p. 80.
158 Bruno, *Spaccio de la Bestia Trionfante*, p. 20.
159 Bruno, *De triplici minimo et mensura*, I, chap. 3. See also Hilary Gatti, *Essays on Giordano Bruno* (Princeton, NJ, 2011), pp. 76–7.
160 Aristotle, *De anima*, I, p. 8 reverso; Tertullian, *De anima*, chaps 34–5, pp. 835–7. Epiphanius, *Panarion*, sec. 7, p. 24; Augustine, *De Civitate Dei*, X, chaps 24, 29–30; XIII, chap. 19; XXII, chaps 12, 26–8; Thomae Aquinatis, *Tertia Pars: Summæ Theologicæ* (*c.* 1274) (Venice, 1585), p. 241.
161 Firpo, *Il processo di Giordano Bruno*, p. 285.
162 For discussion, see Eugenio Canone, *Il dorso e il grembo dell'eterno: percorsi della filosofia di Giordano Bruno* (Pisa, 2003), pp. 7, 54, 65, 242–3.
163 'Summarium': 8th Censured Proposition with Bruno's reply, and 5th Deposition, in Firpo, *Il processo di Giordano Bruno*, pp. 304, 285. Aristotle, *De anima*, II, pt. I.
164 For example Philostratus, *Life of Apollonius*, II, bk 7, chap. 26, p. 222.
165 Lactantius, *Divine Institutes*, VI, chap. 8, p. 447. Jerome, *Adversus Ruffinum*, in Hieronymi, *Opera Omnia*, ed. Erasmus (Paris, 1534), p. 88.
166 Irenaei, *Adversus haereses libri quinque* [*c.* 180 CE], ed. Ubaldo Mannucci (Rome, 1907), I, chap. 25, sec. 4, pp. 264–5.
167 Philostratus, *Life of Apollonius*, II, bk 7, chap. 26, p. 222.
168 Basilii, *Oratio VIII: De temperantia & incontinentia*, in *Omnia quae in hunc diem latino sermone donata sunt opera* (Antwerp, 1570), p. 819.
169 Erasmus, *Apophthegms*, VIII, sec. 60, p. 283. Erasmus, *Apophthegmatum*, VIII, in Erasmi, [*Opera Omnia*] *Quartos tomus quae ad morum institutionem pertinent complectens* (Basel, 1560), p. 340.
170 Callisti, *Ecclesiasticæ historiæ*, p. 893.
171 Tempier (1277), in *Chartularium*, I (1889), p. 544 (Prop. 7), and p. 554 (Prop. 193).
172 Censures of 1 August 1553, in Dupin, *New Ecclesiastical History*, II, p. 441.
173 Bruno, *De la causa, principio et uno*, 5th Dialogue, p. 90.
174 Bruno, *De gli heroici furori*, 2nd Dialogue.
175 L. Spruit, 'Una rilettura del processo di Giordano Bruno', in *Giordano Bruno: Oltre il mito e le opposte Passioni*, ed. P. Giustiniani et al. (Naples, 2002), p. 225.
176 Alberto Martínez, 'Giordano Bruno and the Heresy of Many Worlds', *Annals of Science*, LXXIII/4 (2016), pp. 345–74.
177 Aristotle, *De caelo*, II, sec. 13.
178 Bruno, 2 June 1592, in Firpo, *Il processo di Giordano Bruno*, pp. 167, 269.
179 Hippolytus, *Refutatio*, I, secs 13–14, pp. 72–4.
180 Dionysius of Alexandria, 'Against the Epicureans', *The Writings of Gregory Thaumaturgus, Dionysius of Alexandria, and Archelaus*, XX of *Ante-Nicene Christian Library* (Edinburgh, 1871), p. 171.
181 Philastrius, *De haeresibus liber* (*c.* 384 CE), in Franciscus Oehler, ed., *Corporis Haereseologici*, II/2 (Berlin, 1860), p. 121.
182 Jerome, *Ad Pammachium et Marcellinum apologia Hieronymi adversum*

Ruffinum, II, p. 511; Jerome, *Confessio Hieronymiana, ex omnibus Germanis B. Hieronymi operibus optima fide collecta*, ed. C. Steinuvichii (Paris, 1585), pp. 41, 45, 236; Augustine, *De haeresibus liber* (*c.* 430 CE), in Oehler, ed., *Corporis Haereseologici*, II/1, pp. 195, 218; Stephen Barney et al., eds, *The Etymologies of Isidore of Seville* (Cambridge, 2006), vol. VIII, sec. v.

183 Pope Zacharias to Boniface, 1 May 748, in Laurentio Surio, *De vitis Sanctorum ab Aloysio Lipomano, Episcopo Veronæ, viro doctissimo olim conscriptis*, III (Venice, 1581), p. 160.

184 Alphonsum Alvarez Guerrero, *Thesaurus Christianae Religionis et Speculum Sacrorum Summorum* (Venice, 1559), p. 261. Lamberti Danei, *Aurelii Augustini Hipponensis episcopi liber De haeresibus, ad Quodvultdeum* (Geneva, 1578), p. [M.v] reverse; Jerome, *Confessio Hieronymiana*, pp. 41, 45, 236.

185 Gregorii de Valentia, *Commentariorum theologicorum tomi quatuor: in quibus omnes materiae quae continentur in Summa Theologica Divi Thomae Aquinatis ordine explicatur*, I (Ingolstadt, 1591), p. 991.

186 A. Guerrero, *Thesaurus Christianae religionis* (Venice, 1559), p. 261; L. Danei, *Aurelii Augustini Hipponensis episcopi liber De haeresibus* (Geneva, 1578), p. [M.v] rev.

187 R. Diaconi, *Contra Acephalos* (*c.* 560), in *Antidotum contra diversas omnium fere seculorum haereses* (Basel, 1528), p. 248; rev. L. Ricchieri, *Haereseologia* (Basel, 1556), p. 715.

188 *Decretum Gratiani emendatum et notationibus illustratum, unà cum glossis, Gregorii XIII. Pont. Max. iussu editum* (Rome, 1582), pt. 2, Causa 24, Question 3, col. 1895; repr. *Corpus iuris canonici emendatum et notis illustratum: Gregorii XIII. Pont. Max. iussu editum*, additions by P. Lanceloti (Lyon, 1591), col. 877. See also N. Eymerici, *Directorium Inquisitorium* (1378), with commentaries by F. Pegñae (Venice, 1595), p. 246.

189 Firpo, *Il processo di Giordano Bruno*, pp. 341–2. These were standard expressions in court documents. Prior to 1582 they referred to various canons, including those that were later included in the *Corpus*. See J. Wickersham, *Rituals of Prosecution: The Roman Inquisition and the Prosecution of Philo-Protestants in Sixteenth-century Italy* (Toronto, 2012), pp. 56–8. Inquisitors commonly had degrees in both canon and civil law.

190 Origen, *Peri Archon*, III, trans. Rufinus, in Origenis, *Origenis Adamantii operum*, III–IV (Lyon, 1536), p. 205.

191 Ruperti [Rupert of Deutz], *In Librum Ecclesiastes*, I (*c.* 1120), in D. Ruperti, *Opera* (Mainz, 1631), p. 1200.

192 Bruno, *De gli heroici furori*, p. 20. See also Gatti, *Essays on Giordano Bruno*, p. 272.

193 Bruno, *De gli heroici furori*, n.p., end of the Fourth Dialogue. For a discussion of Bruno's departures from Aristotle, see Ricci, *Giordano Bruno nell'Europa*, e.g. pp. 257–61, 281–98, 358–60.

194 Roberti Bellarmini, 'Præfatio', *Disputationes de controversis Christianae Fidei, Adversus hujus temporis hæreticos*, I (1575) (Venice, 1721), p. x.

195 Godman, *The Saint as Censor*, p. 456.

196 'Visita dei Carcerati nel Sant'Uffizio Romano', 24 March 1597, in Firpo, *Il processo di Giordano Bruno*, p. 244.

197 Io. Baptistae Crispi, *De ethnicis philosophis caute legendis disputationum* (Rome, 1594), p. [xi].
198 Iacobi Menochij, *De arbitrarii iudicum quaestionibus & causis Libri Duo* (Frankfurt am Main, 1576), Casus 374; Ioannis Diaz de Luco, *Practica criminalis canonica* (Rome, 1581), p. 225; Quintilliani Mandosij, ed., *Repertorium Inquisitorum pravitatis haereticae* (Venice, 1588), p. 676; Caesare Carena, *Tractatus de Officio Sanctissimae Inquisitionis et modo procedendi in causis fidei* (Cremona, 1641), p. 253.
199 'Decreto della Congregazione del Sant'Ufficio', 14 January 1599, in Firpo, *Il processo di Giordano Bruno*, p. 312.
200 Felice Tocco, *Giordano Bruno: Conferenza Tenuta nel Circolo Filológico di Firenze* (Florence, 1886), p. 86; Finocchiaro, 'Philosophy versus Religion and Science versus Religion', p. 81.
201 Ludovico Madrutius et al., 'Copia Parziale della Sentenza', 8 February 1600, Firpo, *Il processo di Giordano Bruno*, p. 340. Rowland, *Giordano Bruno*, pp. 259–60.
202 Bruno, 'Sommario', in Firpo, *Il processo di Giordano Bruno*, p. 265; he affirmed this belief in his 2nd, 4th and 5th Depositions.
203 Bruno, quoted in 'Copia parziale della sentenza, destinata al governatore di Roma', 8 February 1600, in ibid., pp. 340–41.
204 Flaminius Adrianus, 'Decreto della Congregazione del Sant'Uffizio', 14 January 1599, in ibid., p. 312.
205 Adrianus, 'Decreto della Congregazione', 4 February 1599, in ibid., pp. 314–15.
206 Andrew Graham-Dixon, *Caravaggio: A Life Sacred and Profane* (New York, 2011), pp. 253–4.
207 Alexio de Menezes, Archbishop of Goa, 'The Acts and Decrees of the Synod of Diamper', 20 June 1599: Decree 4, in *The History of the Church of Malabar; wih the Synods of Diamper*, trans. Michael Geddes (London, 1694), p. 137.
208 Adrianus, notary, 'Decreto della Congregazione del Sant'Uffizio', 24 August 1599, in Firpo, *Il processo di Giordano Bruno*, p. 324. This same statement is reiterated in another document, the 'Bella Copia Sommaria', 24 August 1599, in ibid., p. 325.
209 Bruno, 'Epistola esplicatoria scritta al molto illustre, et eccelente caualliero Signor Philippo Sidneo', in *Spaccio de la Bestia Trionfante*, p. [xi]: 'é à punto comme il nocchiero à la nave'. Bruno, *De la causa, principio et uno*, p. 43: 'Questo ne accorda l'istesso Aristotele il qual quantumque neghi l'anima haver quella raggione verso il corpo, che há il nocchiero alla nave.' Bruno, *Lampas triginta statuarum* (*c.* 1590–91), first published in *Opera latine conscripta*, ed. Tocco and Vitelli, III, p. 253: 'Animae cum Diis, a quibus non alia differre videntur ratione, quam ut nauta in navi a nauta extra navem.'
210 Barney et al., eds, *Etymologies of Isidore*, VIII, sec. iii, p. 176.
211 Firpo, *Il processo di Giordano Bruno*, pp. 95, 280–81.
212 Lucia Boschetti, 'Sul processo di Giordano Bruno: Indagini attorno all'eresia Novaziana', *Rinascimento: Rivista dell'Istituto Nazionale di Studi sul Rinascimento*, 2nd ser., XXXXVI (2006), pp. 93–130; see p. 126.
213 Bruno, *De magia mathematica*, p. 22.

214 Novatiani, *Liber de Trinitate* (*c*. 252 CE), in Cæcilii Cypriani, *Opera Omnia*, II (Würzburg, 1782), chap. 2, p. 460.
215 Ibid., chap. 8, p. 472.
216 Boschetti, 'Sul processo di Giordano Bruno', pp. 120–29.
217 Jacob Pamelius, note 33 on Novatian's *On the Trinity*, chap. 7, in Q. Septimii Florentis Tertulliani, *Opera Omnia*, ed. Iacobi Pamelii, I (1597; repr. Rouen, 1662), p. 1056. In 1613 Bellarmine mentions having read Novatian's *On the Trinity*, and he says that it had been wrongly attributed to Tertullian. Bellarmino, *De scriptoribus ecclesiasticis liber unus* (Lyon, 1613), p. 41.
218 'Decreto della Congregazione', 9 September 1599, in Firpo, *Il processo di Giordano Bruno*, p. 329.
219 'Visita dei Carcerati nel Sant'Uffizio', 21 December 1599, in ibid., p. 133.
220 For a discussion of differences between heresies and errors, see Spruit, 'Una rilettura del processo di Giordano Bruno', pp. 217–34.
221 'Decreto della Congregazione', 20 January 1600, in Firpo, *Il processo di Giordano Bruno*, p. 338.
222 Gaspar Schoppe to Konrad Rittershausen, 17 February 1600, in Gaspari Scioppii, 'Epistola, in qua haereticos jure infelicibus lignis cremari concludit' [or 'Epistola, in qua sententiam de Lutheranis tanquam haereticis atram Romae fieri asserit & probat'], in Péter Pázmány, *Machiavellizatio* (Zaragoza, 1621), pp. 30–35; see also Firpo, *Il processo di Giordano Bruno*, pp. 348–51.
223 Bruno, 'Dialogo Quinto', *De la causa, principio et uno*, in *Opere di Giordano Bruno Nolano*, ed. Adolfo Wagner, I (Leipzig, 1830), p. 283.
224 Regarding the prisons in Rome, see Carlo Cirillo Fornili, *Delinquenti e carcerati a Roma a la metà del '600, opera dei Papi nella riforma carceraria* (Rome, 1991). See also Ingrid Rowland, 'Guilt 1600: Case Study Caravaggio/Bruno', available at www.recht-als-kultur.de, accessed 5 August 2017.
225 Black, *The Italian Inquisition*, p. 90.
226 Schoppe, 'Epistola de variis fidei catholicae dogmatibus', quoted in Mario D'Addio, *Il pensiero politico di Gaspare Scioppio e il machiavellismo del Seicento* (Milan, 1962), p. 26.
227 See John Tedeschi, *The Prosecution of Heresy* (Binghamton, NY, 1991), p. 104. See also Domenico Orano, *Liberi pensatori bruciati in Roma dal XVI al XVIII secolo* (Rome, 1904). Orano conjectured that an additional 65 executions should be added to the list, some of which were later subtracted in Luigi Firpo, 'Esecuzioni capitali in Roma, 1567–1671', in *Eresia e riforma nell'Italia del Cinquecento*, ed. Albano Biondi (De Kalb, IL, 1974), pp. 309–42.
228 'Avviso di Roma', 12 February 1600, in Firpo, *Il processo di Giordano Bruno*, p. 338.
229 Archivo Storico Capitolino, Roma, *Archivo Orsini. I. Corrispondenza*, b. 380, n. 385 [Rome, 12 February 1600], excerpt published in Federica Favino, '"Et sta per brugiarsi un relasso ostinato": Una testimonianza inedita intorno a la condanna di Giordano Bruno', *Galilæana: Journal of Galilean Studies*, VII (2010), pp. 85–6.
230 Ibid., pp. 91–3.

231 Black, *The Italian Inquisition*, p. 91.
232 Bruno, *La cena delle Ceneri*, p. 127.
233 Bruno, 'Epistola esplicatoria', in *Spaccio de la Bestia Trionfante*, p. [v].
234 'Giornale' dell'Arciconfraternita di S. Giovanni Decollato in Roma', 16–17 February 1600, in Firpo, *Il processo di Giordano Bruno*, p. 348. In the 1500s and 1600s it was common to refer to the marketplace square as the 'Campo di Fiore' or 'Campo di Fiori'. Now it's known as 'Campo de' Fiori'.
235 'Giornale', in Firpo, *Il processo di Giordano Bruno*, p. 348.
236 Ordine Nuccio, 'E il notaio "fotografò" Giordano Bruno sul rogo. In un disegno la prima imagine del filosofo', *Corriere della Sera*, Section 'Archivio Storico' (17 April 2011), p. 39. Also at http://archiviostorico.corriere.it, 17 April 2011.
237 Gaspar Schoppe to Johann Wackher, 19 February 1600, in *Giordano Bruno, Documents 1: Le Procès*, ed. L. Firpo and A. P. Segonds (Paris, 2000), p. 520.
238 Hippolytus, *Refutatio*, IV, sec. 51, p. 137, and VI, secs 20–21, pp. 228–9. Irenaei, *Adversus haereses*, I, chap. 23, pp. 246–8; Epiphanius, *Panarion*, sec. 2, part 21: 'Against the Simonians', 2:6.
239 'Avviso di Roma', 19 February 1600, in Firpo, *Il processo di Giordano Bruno*, p. 356.
240 'Sommario del Processo', sec. 1, Ioannes Mocenigo, Francisco Gratianus, in ibid., pp. 247–8, 250–51, 252–3.
241 Celestino de Verona, Franciscus Vaia, Gratianus and Matthew de Silvestris, in ibid., pp. 281–2.
242 Mocenigo, Celestino, Iulius de Salodio, Vaia, Gratianus and Silvestris, in ibid., pp. 83, 143, 282–3.
243 Mocenigo, Ioannes Baptista Ciotus, Iacobus Brietanus, Celestino and Silvestris, in ibid., pp. 247–52.
244 Celestino, Gratianus and Silvestris, in ibid., pp. 278–9.
245 Mocenigo and Gratianus, in ibid., p. 279.
246 Mocenigo, Gratianus and Silvestris, in ibid., pp. 274–5, 286–7; Schoppe to Rittershausen, 17 February 1600, in *Macchiavellizatio*, p. 32.
247 Mocenigo, Gratianus and Silvestris, in Firpo, *Il processo di Giordano Bruno*, p. 288.
248 For example, Iulius, in ibid., pp. 280–81.
249 Mocenigo, in ibid., pp. 279, 292–3.
250 Ciotus, Celestino and Bruno (2nd and 5th Depositions), in ibid., pp. 289–91.
251 Bruno (4th Deposition), in ibid., p. 292.
252 Bruno (5th Deposition), in ibid., p. 291.
253 Celestino, Iulius, Gratianus and Silvestris, in ibid., pp. 293–4.
254 Celestino and Gratianus, in ibid., pp. 263, 276.
255 Celestino and Gratianus (12th Deposition), in ibid., pp. 273–4.
256 Mocenigo, Celestino and Bruno (3rd, 4th, 5th and 8th Depositions), in ibid., pp. 253–8.
257 Gratianus, in ibid., p. 261.
258 Mocenigo and Silvestris, in ibid., pp. 287–8.
259 Celestino, Iulius, Vaia, Gratianus and Silvestris, in ibid., pp. 266–7.

260 Celestino, Gratianus and Silvestris, in ibid, pp. 266–7; Gaspar Schoppe to Konrad Rittershausen, 17 February 1600, in *Macchiavellizatio*, p. 32.
261 Ibid.
262 Ibid.
263 Celestino, Gratianus and Bruno (10th Deposition), in Firpo, *Il processo di Giordano Bruno*, pp. 274–5; Schoppe to Rittershausen, 17 February 1600, in *Macchiavellizatio*, p. 32.
264 Gratianus and Celestino, in Firpo, *Il processo di Giordano Bruno*, pp. 260, 275–6.
265 Mocenigo seems to say this: that Christ 'era un mago, et cosí gl'appostoli', in ibid., pp. 143, 261, 276; Schoppe to Rittershausen, 17 February 1600, in *Macchiavellizatio*, p. 32.
266 Celestino, Gratianus and Silvestris, in Firpo, *Il processo di Giordano Bruno*, pp. 277–8.
267 For example, Gratianus and Silvestris, in ibid., pp. 251–3.
268 Mocenigo, in ibid., p. 289.
269 Mocenigo and Iulius, in ibid., pp. 276–7.
270 Mocenigo, in ibid., pp. 279–80.
271 Gratianus, in ibid., pp. 267, 271.
272 Mocenigo and Silvestris, in ibid., pp. 258–9, 262.
273 Mocenigo, Celestino, Vaia and Gratianus, in ibid., pp. 259–62.
274 Schoppe to Rittershausen, 17 February 1600, in *Macchiavellizatio*, p. 32; Mocenigo and Silvestris, in Firpo, *Il processo di Giordano Bruno*, pp. 259–62.
275 Celestino, Iulius, Gratianus and Silvestris, in ibid., pp. 263–4.
276 Mocenigo, Gratianus (quoting Vaia) and Silvestris, in ibid., pp. 264–5.
277 Mocenigo, Celestino, Iulius, Vaia, Gratianus and Silvestris, in ibid., pp. 282–3.
278 Mocenigo, Celestino, Iulius, Vaia and Gratianus, in ibid., pp. 262, 282–3.
279 'Summarium': 4th Censured Proposition, including his reply, in ibid., pp. 300–301.
280 'Summarium': 9th Proposition, in ibid., p. 304; Bruno, *De la causa, principio et uno*, Fifth Dialogue, p. 90.
281 Schoppe to Rittershausen, 17 February 1600, in *Macchiavellizatio*, p. 32; Silvestris, Mocenigo, Celestino, Iulius and Gratianus, in Firpo, *Il processo di Giordano Bruno*, pp. 268, 272–3.
282 Ibid., pp. 299–300, 269–70.
283 Schoppe to Rittershausen, 17 February 1600, in *Macchiavellizatio*, p. 32; Mocenigo, Celestino, Iulius, Vaia, Gratianus and Silvestris, in Firpo, *Il processo di Giordano Bruno*, pp. 267–9. 'Summarium': 10th Proposition, in ibid., p. 304.
284 Celestino and Gratianus, in ibid., pp. 284, 270. 'Summarium': 10th Proposition, in ibid., p. 304.
285 'Summarium': 8th Proposition, including Bruno's reply, in ibid., pp. 285, 304.
286 Schoppe to Rittershausen, 17 February 1600, in *Macchiavellizatio*, p. 32. Celestino (recalling a conversation with Bruno, Iulium, Vaia and Matthew de Orio), Gratianus and Silvestris, in Firpo, *Il processo di Giordano Bruno*, pp. 284–5.
287 Mocenigo and Silvestris, in ibid., pp. 283–5.

288 Gratianus and Silvestris, in ibid., pp. 284–5.
289 'Summarium': 1st and 7th Propositions, including Bruno's replies, in ibid., pp. 299, 303.
290 'Summarium': 5th Proposition, including Bruno's reply, in ibid., pp. 302–3.
291 Ibid., p. 270. Also, 'Summarium': 6th Proposition, including Bruno's reply, in ibid., p. 303.
292 Bruno (3rd and 14th Depositions), in ibid., pp. 254–5; Schoppe to Rittershausen, 17 February 1600, in *Macchiavellizatio*, p. 32.
293 'Summarium': 3rd Proposition, including Bruno's reply, in Firpo, *Il processo di Giordano Bruno*, p. 300.
294 'Giornale' dell'Arciconfraternita di San Giovanni Decollato (Rome), busta 8, num. 16 (5/1/1598–9/1/1602), Archivio di Stato di Roma. Summaries of records for the missing months and years are available in the inventory by Luigi De Santis and Giovanni Ricci Parracciani, 'Nomi dei Giustiziati assistiti negli ultimi momenti a dall'Archiconfraternita di S. Giovanni Decollato in Roma', 285/11 (1878), also titled 'Repertorio dei Giustiziati', Archivio di Stato di Roma.
295 Mercati, *Sommario*, pp. 10–13.
296 Yates, *Giordano Bruno and the Hermetic Tradition*, p. 355. Frances Yates, 'Bruno, Giordano', in *Encyclopedia of Philosophy*, ed. Paul Edwards, I (New York, 1967), pp. 405–8.
297 Finocchiaro, 'Philosophy versus Religion and Science versus Religion', pp. 78, 81–2.
298 Mayer, *The Roman Inquisition on the Stage of Italy*, p. 124. Francesco Beretta, 'Giordano Bruno e l'Inquisizione Romana: Considerazioni sul proceso', *Bruniana & Campanelliana*, VII (2001), pp. 15–49.
299 Bellarmino, *Disputationes: De controversiis Christianae fidei adversus huius temporis haereticos*, 2nd edn (Ingolstadt, 1588), pp. 145, 359; Bellarmino, *Disputationum: De controversiis Christianae fidei adversus huius temporis haereticos*, IV (Naples, 1858), pp. 33, 122, 136, 161, 164, 278, 384, 398, 639, 654, 678, 683.
300 Tedeschi, *Prosecution of Heresy*, pp. 94–5.
301 Schoppe to Rittershausen, 17 February 1600, in *Macchiavellizatio*, p. 32.
302 Herbert Jaumann, *Kaspar Schoppe (1576–1649), Philologe im Dienste der Gegenreformation* (Frankfurt am Main, 1998), p. 202.
303 J.W.S., 'Scioppus, Shoppius, or Schoppe, Caspar', *The Imperial Dictionary of Universal Biography*, vol. III (Glasgow, 1857), p. 746.
304 Frances Huemer, *Rubens and the Roman Circle* (New York, 1996), p. 11.
305 For example, Gregory noted that *fovebat aquas* is another translation of *ferebatur super aquas* (without referring to Bruno or Schoppe); Gregory, *Anima Mundi*, p. 126.
306 Marsilii Ficini, *In Plotini epitomae, seu argumenta, commentaris & annotationes*, in Ficini, *Operum* [*Opera*]: *in quo compraehenduntur ea, quae ex Graeco in Latinum Sermonem doctrissime transtulit, exceptis Platone atque Plotino Philosophis*, II (Basel, 1561): 'In librum De coelo [Plotinus], Comment. Summa Totius Libri', p. 1597.
307 In an unpublished manuscript, Bruno wrote Virgil's words: 'Principio coelum et terras, camposque liquentes, Lucentemque globum lunae,

Titaniaque astra, Spiritus intus alit, totamque infusa per arctus, Mens agitat molem', followed by the biblical line: 'Spiritus Domini replevit orbem terrarium et hoc quod contient omnia.' See Lutoslawski, 'Jordani Bruni Nolani opera inedita', p. 539.

308 Irenaei, *Adversus haereses*, I, chap. 25, sec. 4, pp. 264–5.

309 Paula Fredriksen, *Sin: The Early History of an Idea* (Princeton, NJ, 2012), pp. 1, 99–100.

310 Jerome, *Epistle to Pammachius*.

311 'The Anathematisms of the Emperor Justinian Against Origen', in The Fifth Ecumenical Council, Second Council of Constantinople (553 CE), in Callisti, *Ecclesiasticæ historiæ*, p. 893.

312 Schoppe to Rittershausen, 17 February 1600, in *Macchiavellizatio*, p. 34.

313 Martínez, 'Giordano Bruno and the Heresy of Many Worlds', p. 362.

314 Philosophers often used other expressions, such as: 'multos mundos', 'mundos alios', 'pluribus mundis', 'plures esse mundos', 'mundos asserit innumerabiles'; and in Italian, 'innumerabili mondi', 'molti mondi', 'infiniti mondi', and so on.

315 Schoppe to Rittershausen, 17 February 1600, in *Macchiavellizatio*, p. 31.

316 For example, Crowe, *Extraterrestrial Life Debate* (1986), p. 8. Crowe's important book focused on the period from 1750 onwards, so he did not take into account the evidence presently discussed; he mainly echoed Dick, *Plurality of Worlds*, p. 69. However, Dick later moved away from that opinion: Steven Dick, *Extraterrestrial Life and Our World View at the Turn of the Millennium* (Washington, DC, 2000), p. 9; and Dick, *Life on Other Worlds: The 20th-century Extraterrestrial Life Debate* (Cambridge, 1998), p. 10.

317 Pope Zacharias (748 CE), paraphrased in Caesare Baronio, *Annales ecclesiastici*, IX (Rome, 1600), p. 191.

318 Joannis Heckij [Joannes Heckius], 'Super Plinij. ii. Historias Na[tura] les' (Archivio Linceo, Rome, MS. 21), p. 11 verso. The title page seems to superimpose two dates: 'Incepi die i9a, Septembris anno i600' or 'i60i'.

319 Stobaeus wrote: 'Metrodorus [of Chios], the teacher of Epicurus', spoke of the production of the infinite, and Stobaeus noted that 'Anaximander, Anaximenes, Archelaus, Xenophanes, Diogenes [of Apollonia], Leucippus, Democritus, Epicurus, [posited] infinitely many in infinity according to the entire circuit.' See Ioannis Stobaei, *Eclogarum physicarum et ethicarum* (Göttingen, 1792), I, pp. 497–8; see also pp. 57, 293, 491. Also, Metrodorus had reportedly argued: 'It would be strange if a single ear of corn grew in a large plain or if there were only one world in the infinite,' according to Simplicius, quoted in Francis Cornford, 'Innumerable Worlds in Presocratic Philosophy', *Classical Quarterly*, XXVIII/I (1934), p. 13.

320 See also Iain Gardner and Samuel Lieu, *Manichaean Texts from the Roman Empire* (Cambridge, 2004), pp. 12, 16, 57, 85, 88, 199–208, 218, 221, 224.

321 Heckij, 'Super Plinij', p. 18 verso, emphasis added.

322 Ibid., p. 38 verso.

323 Lucio Vero Clarimontano [pseudonym], *Nova Apocalypsis inqua Innocentia Cæsaris Augusti et Fidelium Imperio Principum et Noxa Rebellium* (Luxembourg, 1626), p. 160.

324 *The Works of Aurelius Augustine*, V: *Writings in Connection with the Manichaean Heresy*, trans. Richard Stothert (Edinburgh, 1872), pp. 110–11, 116, 191, 277, 281.
325 Epiphanius, *Panarion*, secs II 4:3, 5:4, V 2:6, XXV 2:1, and XXVI 3:6, 9:9–10:0. Filastrii [Philaster], *Diversarum hereseon liber*, in Fridericus Marx, ed., *Corpus Scriptorum Ecclesiasticorum Latinorum*, XXXVIII (Prague, 1898), p. 18.
326 Ibid., Scholion 24(b), and p. 236.

2 Aliens on the Moon?

1 Several outstanding books provide such a comprehensive account, for example: Annibale Fantoli, *Galileo: For Copernicanism and for the Church*, trans. George Coyne (Notre Dame, IN, 1994); William Shea and Mariano Artigas, *Galileo in Rome* (Oxford, 2003); J. L. Heilbron, *Galileo* (Oxford, 2010); David Wootton, *Galileo, Watcher of the Skies* (New Haven, CT, 2010); Maurice Finocchiaro, *The Galileo Affair: A Documentary History* (Berkeley, CA, 1989). Much of the material that will now be discussed is not found in any of those works, and vice versa.
2 Francisci Piccolominei, *Librorum ad scientiam de natura attinentium, partes quinque* (Frankfurt, 1597), part 2: 'De Coelo', chaps 5–10, pp. 420–30.
3 Ibid., chap. 13, p. 521.
4 Ibid., chap. 5, p. 506, emphasis added.
5 Ibid., chap. 29, p. 463.
6 [Anon.], *c.* May–July 1600, quoted in Tiziana Provvidera, 'A New English Document on Giordano Bruno', *Bulletin of the Society for Renaissance Studies*, XIX (2002), p. 25. The spelling has here been modernized.
7 Guilielmi Gilberti, *De magnete, magneticisque corporibus, et de magno magnete tellure* (London, 1600), V, chap. 12, pp. 208–9.
8 Ibid., p. 209.
9 Ibid., p. 228.
10 Ibid., p. 210.
11 Ioannis de Pineda, *Commentariorum in Iob libri tredecim* (Seville, 1598), chap. 9, p. 461 [462]. See also Pineda, *Commentaria in Job* (Antwerp, 1612), p. 416.
12 Augustino Petreto, *D. O. M. turris Babel, et totius civitatis Babilonis haeresos vasta destructio* (Reggio nell'Emilia, 1601), n.p., item 52.
13 Ibid., item 250.
14 Ibid., item 165.
15 Ibid., item 25.
16 Ioh. Keppleri, *Somnium* [*c.* 1609], *seu Opus posthumum de astronomia lunari, divulgatum à Ludovico Kepplero filio* (Żagań and Frankfurt, 1634).
17 Max Caspar, *Kepler*, trans. C. Hellman (London, 1959), pp. 58–60.
18 Ibid., p. 62.
19 Ibid., p. 63.
20 Max Caspar, *Johannes Kepler*, 4th edn (Stuttgart, 1995), p. 109.
21 Caspar, *Kepler* (1959), p. 93.

22 Ioanne Keplero, *Prodromus dissertationum cosmographicum, contiens mysterium cosmographicum, de admirabili proportione orbium coelestium* (Tübingen, 1596), pp. 1, 4, 6, 23. According to the *Placita* (see above, chap. 1, n.15), Pythagoras taught that the world was formed of the five regular solid figures. *De placita philosophorum*, 1, chaps 9 and 24; 11, chap. 6.
23 Kepler, letter to Maestlin, quoted in Caspar, *Kepler* (1959), p. 69.
24 Keplero, *Mysterium cosmographicum*, p. [ii].
25 Kepler to Duke Frederick of Württemberg, 17 February 1596, in Johannes Kepler, *Gessamelte Werke*, ed. G. Dyck and Max Caspar, XIII (Munich, 1938), p. 218.
26 For a discussion of Bruce's friendship with Kepler and Galileo, see Massimo Bucciantini, *Galileo e Keplero, filosofia, cosmologia e teologia nell'età della Contrariforma* (Turin, 2005), pp. 93–116.
27 Edmund Bruce to Kepler, 5 November 1603, in Kepler, *Opera Omnia*, ed. Christian Frisch, vol. II (1923), p. 568. Kepler, *Gesammelte Werke*, ed. Max Caspar, IV (Munich, 1938), p. 450; trans. Robert Westman.
28 Robert Westman, *The Copernican Question* (Berkeley, CA, 2011), p. 375.
29 Joannis Keppleri, *De stella nova in pede serpentarii* (Prague, 1606), chap. 24, p. 121, emphasis in the original. Also in Kepler, *Gesammelte Werke*, I, p. 267.
30 Kepler to David Fabricius, 11 October 1605, in *Gesammelte Werke*, XV, p. 258.
31 Corn. Gemma, *De naturæ divinis characterismis* (Antwerp, 1575), p. 34. For discussion, see Hiro Hirai, *Cornelius Gemma: Cosmology, Medicine, and Natural Philosophy in Renaissance Louvain* (Rome, 2008), p. 24.
32 Ibid., p. 77.
33 Cornelii Gemmae, *De arte cyclognomica*, III (Antwerp, 1569), pp. 91–2.
34 Kepler to Johannes Brengger, 30 November 1607, in *Opera Omnia*, II (Frankfurt, 1859), pp. 591–2.
35 Kepler to Brengger, 5 April 1608, in ibid., p. 592.
36 Kepler, *Astronomia nova de motibus stellæ Martis ex observationibus Tychonis Brahe* (Prague, 1609), trans. William Donahue as *New Astronomy* (Cambridge, 1992), p. 66.
37 Nicolai Serarii, *Iosue, ab utero ad ipsum usque tumulum* (Paris, 1610), chap. 10, p. 1006.
38 It has become common to translate *Sidereus Nuncius* as *Starry Messenger*. The word 'messenger' is a possible translation for *nuncius*, but other translations are 'news' or 'message'. Albert Van Helden and others choose to read it as 'messenger' partly because Kepler titled his reply: *Dissertatio cum Nuncio Sidereo* (1610). Kepler wrote about his 'Discussion with the Starry Messenger', namely Galileo. Evidently Kepler meant messenger by *nuncio*, yet the word he used was different from *nuncius* in Galileo's book title. Yet following the principle that I do not use later sources to explain previous sources, I therefore don't agree that Galileo titled his own book *Starry Messenger*; instead, news (novelties) or message from the stars seems more appropriate to its contents since it is not about any messenger. Accordingly some translators, such as William Shea, have recently rendered the title as *A Sidereal Message*. For

discussion, see Edward Rosen, 'The Title of Galileo's *Sidereus Nuncius*', *Isis*, XLI/3–4 (December 1950), pp. 287–9. Rosen notes that Galileo's early drafts and later writings show that his book's purpose was 'simply to report the news about recent developments in astronomy, not to pass himself off solemnly as an ambassador from heaven'.

39 In 1609 Cosimo II became Grand Duke of Tuscany. By portraying the four satellites of Jupiter as emblems of the rising Medici dynasty, corresponding to Cosimo II and his three brothers, Galileo gained Cosimo's favour and advanced his career: from being a mere mathematician at the University of Padua he became appointed as philosopher of the noble Medici court. For analysis, see Mario Biagioli, *Galileo, Courtier* (Chicago, IL, 1993), pp. 103–57.

40 Bruno, *De immenso et innumerabilibus, seu de universo et mundis*, III, in Jordani Bruni, *Opera latine conscripta*, ed. F. Fiorentino, vol. I/2 (Naples, 1879), p. 324.

41 Galileo Galileo [Galilei], *Sidereus Nuncius* (Venice, 1610), title page.

42 Bruno, *De immenso*, I and II, in *Opera*, vol. I, pp. 210, 269.

43 Galileo, *Sidereus Nuncius*, p. 28 reverso.

44 Giordano Bruno, *De l'infinito universo et mondi* ([Venice], 1584), p. 66.

45 Galileo, *Sidereus Nuncius*, p. 28 reverso.

46 Bruno, *De l'infinito universo et mondi*, pp. 71, 87, 111.

47 Giordano Bruno, *De gli heroici furori* (Paris, 1585), Fifth Dialogue (n.p.).

48 Galileo, *Sidereus Nuncius*, p. 9. See also Galilei, *Sidereus Nuncius, or The Sidereal Messenger*, trans. Albert Van Helden (Chicago, IL, 1989), p. 43.

49 Giordano Bruno, *La cena delle Ceneri* (London, 1584), pp. 10–11, 55, 65, 69–70.

50 Galilei, letter to Giuliano de' Medici, 1 January 1611, in Antonio Favaro, ed., *Le opere di Galileo Galilei. Edizione Nazionale*, 20 vols (Florence, 1890–1909, repr. 1929–38), XI, p. 12.

51 Ioannis Kepleri, *Dissertatio cum Nuncio Sidereo nuper ad mortales misso a Galilaeo Galillaeo* (Prague, 1610), p. 1. For an account of Wackher and Kepler's friendship, see 'Kepler and Bruno on the Infinity of the Universe and of Solar Systems', *Journal for the History of Astronomy*, XXXIX (2008), pp. 469–95.

52 Martin Hasdale to Galileo Galilei, 15 April 1610, in *Le opere di Galileo Galilei, Prima edizione completa*, 15 vols (Florence, 1842–56), VIII (1851), p. 59.

53 Kepler, *Dissertatio cum Nuncio Sidereo*, pp. 25, 33.

54 Ibid., pp. 2–3.

55 Ibid., p. 25.

56 A couple of historians who have not neglected this point are Saverio Ricci, *La fortuna del pensiero di Giordano Bruno, 1600–1750* (Florence, 1990), and Laura Varanini, 'La *Dissertatio cum Nuncio Sidereo* fra Galileo e Bruno', *Bruniana & Campanelliana*, IX/1 (2003), pp. 207–15.

57 Ibid., p. 22.

58 Ibid., p. 24.

59 Kepler, *Dissertatio cum Nuncio Sidereo*, trans. E. Rosen, *Kepler's Conversation with Galileo's Sidereal Messenger* (New York, 1965), pp. 27–8.

60 Michaele Maestlino, *Disputatio in multivariis motuum planetarum in*

coelo apparentibus irregularitatibus (Tübingen, 1606), quoted in Kepler, *Dissertatio cum Nuncio Sidereo*, p. 18.

61 Martini Horky, *Brevissima peregrinatio contra Nuncium Sidereum* (Modena, 1610), reprinted in Favaro, ed., *Opere di Galileo Galilei: Edizione Nazionale*, III/1, p. 141.

62 General Inquisitor at Modena, Michael Angelus Lerrius, 18 June 1610, in ibid., p. 134.

63 John Wedderburn, *Quatuor problematum quae Martinus Horky contra Nuntium Siderum de quatuor planetis novis disputanda proposuit. Confutatio* (Padua, 1610), reprinted in Favaro, ed., *Opere di Galileo Galilei: Edizione Nazionale*, III/1, p. 162.

64 T., 'Campanella (Thomas)', in John Bernard et al., *A General Dictionary, Historical and Critical*, IV (London, 1736), p. 85.

65 Bernardino Telesio, *De rerum natura iuxta propriis principiis* (Naples, 1586).

66 Tommaso Campanella, *Philosophia sensibus demonstrata* (Naples, 1591).

67 Ibid., pp. 322–3. Ficino worked on his Latin translation of the *Enneads* of Plotinus, with commentaries, from 1484 until 1490, which was finally published in 1492, in Florence. Previously the works of Plotinus had been unavailable since late antiquity.

68 Marsilii Ficini, *Librum tertium de vita coelitus comparanda* [1489], *compositus ab eodem inter Commentaria eiusdem in Plotinum* [1490], in Marsilii Ficini, *Opera & quae hactenus extitêre*, I (Basel, 1561), III, p. 535, [on Plotinus] chap. 2. For the words by Iarchas, see Philostratus, *The Life of Apollonius of Tyana* (Cambridge, 1948), III, chap. 42, p. 322.

69 Luigi Firpo, *Il processo di Giordano Bruno* (Rome, 1993), p. 253.

70 Campanella, *Philosophia sensibus demonstrata*, ed. Luigi de Franco (Naples, 1992), pp. 523–5. For discussion see Germana Ernst, *Tommaso Campanella: The Book and the Body of Nature*, trans. David Marshall (New York, 2010), pp. 12–13.

71 Bernardini Telesii, *Quod Animal vniversvm ab vnica Animae substantia gubernatur, adversvs Galenvm liber vnicus*, ed. Antonio Persio (Venice, 1590).

72 Tommaso Campanella, 'Proemio. A Gasparo Schioppo' (1608), excerpt in *Oeuvres choisies de Thomas Campanella*, ed. Louise Colet (Paris, 1844), p. 24. See also Germana Ernst, ed., *L'ateismo trionfato: overo riconoscimento filosofico della religione universale contra l'antichristianesmo Macchiavellesco*, I (Pisa, 2004).

73 Léon Blanchet, *Campanella* (Paris, 1920), p. 25. Campanella had published this argument in his book *De sensitiva rerum facultate*, now lost.

74 Thomas Mayer, *The Roman Inquisition on the Stage of Italy, c. 1590–1640* (Philadelphia, PA, 2014), p. 47. Mayer provides a thorough account of the political and bureaucratic events by which Campanella was processed.

75 Ernst, *Tommaso Campanella*, p. 75.

76 Campanella, 'Proemio. A Gasparo Schioppo' (1608), in *Oeuvres choisies*, pp. 23–4.

77 Ernst, *Tommaso Campanella*, p. 77. Campanella did not specify his source for the story that Pythagoras faked his resurrection by hiding

underground but that story had been criticized by ancient writers: by the Athenian satirist Hermippus, by Celsus, who argued that Jesus was not unique and was perhaps a magician, by Diogenes, the biographer of philosophers, and by Tertullian, the Christian critic of false prophets. For example, see Tertullian, 'XVII. De Anima' (*c.* 215 CE), ed. J. H. Waszink, in *Tertulliani Opera*, II, ed. E. Dekkers, Corpus Christianorum, Series Latina (Turnhout, 1954), chap. 28, pp. 824–5. This story is a distortion of an ancient story by the historian Herodotus (*c.* 430 BCE), who mentioned that Salmoxis, a deceitful man who claimed to have returned from death, had originally been a slave of Pythagoras of Samos.

78 Ernst, *Tommaso Campanella*, pp. 76–7.

79 Ibid.

80 Campanella, 'Proemio. A Gasparo Schioppo' (1608), *Oeuvres choisies*, p. 24.

81 Mayer, *The Roman Inquisition on the Stage of Italy*, p. 52.

82 Ibid., p. 55.

83 [Gian Vittorio Rossi], 'Frater Thomas Campanella', in Jani Nicii Erithraei [pseudonym], *Pinacotheca imaginum, illustrium, doctrinæ vel ingenii laude, virorum* (Cologne, 1643), p. 43.

84 Tommaso Campanella, 'Madrigal VI', (*c.* 1608), in Alessandro D'Ancona, ed., *Opere di Tommaso Campanella*, I (Turin, 1854), p. 159.

85 Christopher Black, *The Italian Inquisition* (New Haven, CT, 2010), p. 86.

86 Peter Godman, *The Saint as Censor: Robert Bellarmine between Inquisition and Index* (Leiden, 2000), p. 175. Luigi Firpo, *I processi di Tommaso Campanella*, ed. E. Canone (Rome, 1998), pp. 256–8.

87 For discussion, see Eugenio Canone, 'L'editto di proibizione delle opere di Bruno e Campanella', *Bruniana & Campanelliana*, I/1–2 (1995), pp. 43–61.

88 Io. Brisichellen, 'Magistri Sacri Palatij', 7 August 1603, in Alexandri VII, ed., *Index librorum prohibitorum* (Rome, 1664), pp. 321–2.

89 Jill Kraye, 'Teaching Stoic Moral Philosophy: Kaspar Schoppe's *Elementa philosophiae moralis* (1606)', in *Scholarly Knowledge: Textbooks in Early Modern Europe*, ed. Emidio Campi (Geneva, 2008), p. 250.

90 Campanella, 'Proemio. A Gasparo Schioppo' (1608), in *Oeuvres choisies*, p. 23.

91 Ludovico delle Colombe, *Discorso contro il moto della terra* (1611), repr. in *Opere di Galileo Galilei, Prima edizione completa*, II, p. 370.

92 Ibid., p. 373.

93 Ibid.

94 Ibid., p. 378.

95 Tommaso Campanella to Galilei, Ides of [13] January 1611, in Campanella, *Lettere*, ed. Vincenzo Spampanato (Bari, 1927), p. 165.

96 Ibid., p. 167.

97 Origen, *De Principiis*, in *The Writings of Origen*, X of *Ante-Nicene Christian Library* (Edinburgh, 1869), II, chap. 1, sec. 3.

98 Ibid., chap. 7, secs 2–4.

99 Origen, *Contra Celsum* (248 CE), ed. Henry Chadwick (Cambridge, 1965), VI, sec. 9, p. 322; I, sec. 20, p. 21; VI, sec. 21, pp. 333–4.

100 'The Anathemas against Origen', in the 5th Ecumenical Council,

the 2nd Council of Constantinople (553 CE), in Nicephori Callisti, *Ecclesiasticæ historiæ libri decem & octo* (Basel, 1553), p. 893.

101 Marco Antonio Genevese, *Praxis archiepiscopalis curiae Neapolitanae* (Naples, 1602), p. 208; Roberto Bellarmino and Valenti Gozanga, *Riposta del Car. Bellarmino all difesa delle otto propositioni di Giovanni Marsilio Napolitano* (Naples, 1606), p. 16.

102 Andrea Vega, *Tridentini decreti, de iustificatione expositio, et defensio, libris XV* (Alcalá de Henares, 1564), Question 6, p. 709.

103 Pope Clement VIII, quoted in James Brodrick, *Robert Bellarmine, Saint and Scholar* (Westminster, MD, 1961), p. 156.

104 Stefania Tutino, *Empire of Souls: Robert Bellarmine and the Christian Commonwealth* (Oxford, 2010), e.g., p. 280.

105 Richard Blackwell, *Galileo, Bellarmine, and the Bible* (Notre Dame, IN, 1991), p. 46.

106 Godman, *The Saint as Censor*, pp. 163–4, 209–11, 301–3.

107 Justus Lipsius, *Physiologia Stoicorum libri tres* (Antwerp, 1604), pp. 121–2.

108 Ibid., pp. 120 and 70, respectively.

109 Ibid., pp. 126–8.

110 Ibid., p. 109.

111 Scioppii, 'Epistola', in Firpo, *Il processo di Giordano Bruno*, p. 355. This part of his letter was omitted in the version published in 1621.

112 Thomas F. Mayer, *The Roman Inquisition: A Papal Bureaucracy and its Laws in the Age of Galileo* (Philadelphia, PA, 2013), pp. 74, 417.

113 Ubbo Emmius to Sibrando Lubberto, 19 September 1608, in *Briefwechsel des Ubbo Emmius*. ed. H. Brugmans and F. Wachter, II (The Hague, 1923), pp. 51–2. Emmius was rector of the Latin school in Groningen. Lubbertus was a theologian. Emmius attributed the ideas in question to Simon Stevin. Actually, Stevin had merely asked what Earth would look like if someone were to observe it from the Moon.

114 Anton. Possevini, *Apparatus sacer ad scriptores Veteris & Novi Testamenti*, I (Cologne, 1608), p. 362.

115 Regarding their relationship, see Godman, *The Saint as Censor*, pp. 3–4, 120–21, 156–8, 230–31.

116 Possevini, *Apparatus sacer ad scriptores*, p. 839.

117 Petro Gregorio, *Syntaxeon artis mirabilis, in libros XL digestarum*, II (Cologne, 1610), pp. 79–80. Pierre Grégoire's encyclopaedic work incorporated sciences and mathematics with astrology and demonology, from a devoutly Catholic perspective, yet it was placed on the Index of Forbidden Books.

118 Placido Padiglia, *David penitente lezzioni sopra il cinquantesimo salmo di David* (Rome, 1610), p. 143.

119 Bruno's 3rd Deposition, in Firpo, *Il processo di Giordano Bruno*, p. 254. See also Wisdom 1:7.

120 Bellarmine's copy is at the Biblioteca Nazionale Centrale di Roma, item 6.36.E.11, which came from the Library of the Collegio Romano.

121 Bruno (14th Deposition), in Firpo, *Il processo di Giordano Bruno*, p. 270; also see 'Summarium': Sixth Censured Proposition, including Bruno's reply, in ibid., p. 303; Psalms 19:1, or 18:2 in the Latin Vulgate.

122 Padiglia, *David penitente lezzioni*, p. 143.

123 Ibid., p. 144.
124 Thomas Digges, 'A Perfit Description of the Caelestiall Orbes according to the most aunciente Doctrine of the Pythagoreans, latelye reuiued by Copernicus', in Leonard Digges, *A Prognostication Everlastinge of Righte Good Effecte*, ed. Thomas Digges (London, 1576), p. 43, English words modernized.
125 Malchus [Porphyry], *De Vita Pythagoræ*, ed. Cunrado Rittershusio (Altdorf, 1610), n.p.
126 Gasp. Scioppii, *Ecclesiasticus, auctoritati serenissimi D. Iacobi Magnæ Britanniæ Regis oppositus* (Augsburg, 1611), p. 264.
127 [John Donne], *Conclave Ignati: sive eius in nuperis inferni comitiis inthronisatio* (n.p., 1611); this first Latin edition was recorded in the Stationer's Register on 24 January 1611 (it was printed in London and Europe); pp. 8, 13, italics in the original.
128 Ibid., pp. 7–9. There has been some debate as to whether Donne's references to Copernicus and Galileo somehow suggest that Donne appreciated their works and used them to belittle the Jesuits. But actually, a close reading of Donne's text shows that he disdained the new astronomers. I agree with the systematic analysis presented by Chris Hassel, 'Donne's *Ignatius His Conclave* and the New Astronomy', *Modern Philology*, LVIII/4 (May 1971), pp. 329–37, which shows that Donne disliked the Copernican astronomers' arrogance and vainglory. As Hassel wrote, 'To Donne the astronomers and the Jesuits are equally foolish and equally dangerous.' Another scholar remarks that Donne had a 'characteristically negative attitude to innovation' and he thus viewed the Jesuits as 'guilty of innovations in religion'; see Achsah Guibbory, *Returning to John Donne* (Farnham, 2015), p. 27.
129 Donne, *Conclave Ignati*, p. 28.
130 Ibid., p. 5.
131 Ibid., p. 8.
132 Ibid., pp. 11–13.
133 Ibid., pp. 28–9.
134 Ibid., pp. 30–32.
135 Ibid., p. 34.
136 Ibid., pp. 16–17, 22–3.
137 John Donne, 'An Anatomy of the World' (1611), in *The Works of John Donne*, ed. Henry Alford, VI (London, 1839), p. 491.
138 Did Donne know Bruno's works? Edmund Gosse conjectured that Donne might have had some exposure to Bruno's works during a visit to Italy sometime between 1592 and 1596: Edmund Gosse, *The Life and Letters of John Donne*, I (London, 1899), pp. 56, 269. Lindsay conjectured that Donne's use of the circle metaphor as well as lines 7–10 in his 'Love's Alchymie' were drawn from Bruno: Jack Lindsay, 'Donne and Giordano Bruno', *Times Literary Supplement* (20 June 1936), p. 523. Ince conjectured that perhaps Donne learned of Bruno's works through Henry Percy, 9th Earl of Northumberland: Richard Ince, 'Donne and Giordano Bruno', *Times Literary Supplement* (27 June 1936), p. 544. Consequently, Yates agreed with Lindsay and Ince that Bruno influenced Donne: Frances Yates, 'Donne and Giordano Bruno', *Times*

Literary Supplement (4 July 1936), p. 564. Furthermore, Lindsay provided additional similarities to even argue that Donne's 'originality' stemmed from the influence of Bruno's works: Lindsay, 'Donne and Giordano Bruno', *Times Literary Supplement* (11 July 1936), p. 580.
139 Bellarmino to the Mathematicians of the Collegio Romano, 19 April 1611, in Favaro, ed., *Opere di Galileo Galilei: Edizione Nazionale*, XI, p. 87.
140 Cristoforo Clavio, Christoforo Grienberger, Odo Malcotio and Gio. Paolo Lembo to Roberto Bellarmino, 24 April 1611, in ibid., XI, p. 93.
141 Annibale Fantoli, *The Case of Galileo: A Closed Question?* (Notre Dame, IN, 2012), pp. 57–60.
142 Christophori Clavii, *Operum mathematicorum*, vol. III: *Commentarium in sphaeram Ioannis de Sacro Bosco & astrolabium* (Mainz, 1611), p. 15.
143 Ioannis Kepleri, *Narratio de observatis a se quatuor Iovis satellitibus erronibus, quos Galilaeus Galilaeus mathematicus Florentinus iure inventionis Medicaea Sidera Nuncupauit* (Florence, 1611), p. 1.
144 Iulio Caesare La Galla, *De phenomenis in orbe lunæ novi Telescopii Usu a D. Gallileo Gallileo* (Venice, 1612), p. 6. Also Galileo, *Sidereus Nuncius*, p. 9.
145 Ibid., pp. 5, 9, 14.
146 Ibid., p. 41.
147 Ibid., pp. 19, 21.
148 Ibid., p. 21.
149 Ibid., p. 22.
150 Ibid., p. 25.
151 Ibid., p. 27.
152 Ibid., p. 49.
153 Roberto Bellarmino, *Dichiaratione del Simbolo*, 2nd edn (Naples, 1605), p. 12. Previously, Bellarmine had also complained that some of the ancients worshipped the Sun as a god; see Bellarmino, *De gratia & libero arbitrio*, IV/2, in Bellarmino, [*Disputationes, de controversiis Christianae Fidei*] *tertia Controversia generalis, de reparatione gratiae per Iesum Christum Dominum Nostrum* (Ingolstadt, 1593), p. 278.
154 Roberto Bellarmino, *Explanatio in Psalmos* (Rome, 1611), p. 292; 2nd edn (Lyon, 1612), p. 292.
155 Roberti Bellarmini, 'Haeresiarchae, id est, qui haereses invenerunt vel suscitarunt, vel dices haereticorum fuerunt' (1592); in Godman, *The Saint as Censor*, p. 286; see also p. 167.
156 Godman, *The Saint as Censor*, pp. 182, 213, 219.
157 Roberto Bellarmino, *De amissione gratiae et statu peccati libri sex* (Heidelberg, 1613); reissued in Bellarmino, *Disputationum Roberti Bellarmini Politiani* S. J., S. R. E. *Cardinalis, De controversiis Christianae Fidei adversus huius temporis haereticos*, IV (Naples, 1858), IV, chap. 11, p. 161.
158 Ibid., p. 162.
159 Bellarmino, *Explanatio in Psalmos*, p. 775.
160 Ibid., p. 903.
161 Nicolai Mullerii, *Tabulae Frisicae lunae-solares quadruplices* (Alkmaar, 1611), chap. 1, p. 318.

162 Nicolai Nancelii, *Analogia microcosm ad macrocosmon; id est, relatio & proportio universi ad hominem* (Paris, 1611), p. 212.
163 Ioannis Kepleri, *Dioptrice sev demonstratio* (Augsburg, 1611), p. 18.
164 Joan. Fabro [Joannes Faber], *Praescriptiones Lynceae Accademiae* (Terni, 1624), p. [1].
165 Niccolò Lorini to Galileo, 5 November 1612, in Favaro, ed., *Opere di Galileo Galilei: Edizione Nazionale*, XI, p. 427.
166 Galileo to Prince Federico Cesi, 5 January 1613, in ibid., p. 461.
167 Prior to Galileo, it seems that Thomas Harriot had discovered sunspots (his manuscript was not published until 1833). Years later Fulgenzio Micanzio noted that Galileo first mentioned sunspots in August 1610. On March 1611 the Jesuit astronomer Christopher Scheiner, with an assistant, also discovered the spots, as Scheiner later reported. Also Johann Fabricius published his own independent discovery of sunspots before anyone else, in 1611. Galileo himself did not claim to be the first to discover sunspots, but the author of his book's preface gave him the credit, which led Scheiner to carry out a priority dispute and debate about sunspots. See, for example, Fantoli, *The Case of Galileo*, pp. 62–7.
168 Galilei, *Istoria e dimostrazioni intorno alle macchie solari e loro accidenti* (Rome, 1613), p. 14. See also Galileo, 'Letters on Sunspots', in *Discoveries and Opinions of Galileo*, ed. and trans. Stillman Drake (New York, 1957), p. 94.
169 Their works were not supported by powerful patrons; see Biagioli, *Galileo, Courtier*, pp. 71–2.
170 Galilei, *Istoria e dimostrazioni*, p. 133. Also in Drake, ed., *Discoveries and Opinions of Galileo*, p. 137.
171 [Christopher Scheiner], *De maculis solarib. Et stellis circa Iovem errantibus* (Augsburg, 1612), p. 26.
172 Simone Mario, *Mundus Iovialis anno M.DC.IX. detectus ope perspicilli Belgici* (Nuremberg, 1614).
173 Ioannes Georgius Locher, *Disquisitiones mathematicae, de controversiis et novitatibus astronomicis* (Ingolstadt, 1614), pp. 16–17. The last phrase reads: 'pro cuius rei stabilimento'.
174 Westman, *The Copernican Question*, pp. 324–8, 376–84, 487.
175 Galileo, Letter to the Grand Duchess Christina (1615), in Finocchiaro, *The Galileo Affair*, p. 97.
176 Benedetto Castelli to Galileo, 12 March 1615, in Favaro, ed., *Opere di Galileo Galilei: Edizione Nazionale*, XII, p. 154.
177 Paolo Foscarini to Sebastiano Fantone, 6 January 1615, in *Lettera del r. p. m. Paolo Antonio Foscarini Carmelitano. Sopra l'opinione de' Pittagorici, e del Copernico. Della mobilità della terra, e stabilità del sole, e del nuovo Pittagorico sistema del mondo* (Naples, 1615); trans. in Blackwell, *Galileo, Bellarmine, and the Bible*, pp. 217–51.
178 Foscarini, *Lettera*, pp. 4–5, 8.
179 Ibid., p. 14.
180 Ibid., pp. 9–10, emphasis added.
181 Ibid., p. 27. Blackwell's translation of this passage omits the plural of suns and moons.
182 Ibid., p. 29.

183 Bellarmino to Foscarini, 12 April 1615, Finocchiaro, *The Galileo Affair*, p. 67.
184 It is a common mistake to say that Sfondrato was head of the Index.
185 Lorini's Complaint to the Inquisition, 7 February 1615, in Finocchiaro, *The Galileo Affair*, p. 135.
186 Ximenes, in ibid., p. 143.
187 Attavanti, paraphrasing or quoting Caccini, in ibid., p. 145.
188 Giovanni Ciampoli to Galileo, 27 February 1615, in Favaro, ed., *Opere di Galileo Galilei: Edizione Nazionale,* XII, p. 146; Shea and Artigas, *Galileo in Rome*, pp. 64–5.
189 Piero Dini to Galileo, 14 March 1615, in ibid., p. 155.
190 Fantoli, *Galileo: For Copernicanism*, p. 449.
191 Roberti Bellarmini, *De ascensione mentis in Deum per scalas rerum creaturum opusculum* (Cologne, 1615), p. 181. Bellarmine's dedication specifies September.
192 Petreto, *Turris Babel*, n.p., item 25, p. 165.
193 Bellarmine's copy is at the Biblioteca Nazionale Centrale di Roma, item MISC.B.705.1, which came from the Library of the Collegio Romano.
194 Bellarmini, *De ascensione*, p. 277.
195 Bellarmino, *Secunda controversia generalis; De Christo capite totius ecclesiae* (Ingolstadt, 1585), I, chap. 3, p. 38.
196 Galileo, 1st Deposition, 12 April 1633, in Finocchiaro, *The Galileo Affair*, p. 256. See also ibid., p. 359, and Thomas F. Mayer, *The Trial of Galileo, 1612–1633* (Toronto, 2012), p. 157.
197 In some records Taverna also appears as Consultor of the Inquisition on 4 June 1596, and on January 1602. See Mayer, *The Roman Inquisition*, p. 270.
198 Lorenzo Cardella, *Memorie storiche de' Cardinali della Santa Romana Chiesa*, VI (Rome, 1793), pp. 110–11; Niccolò del Re, *Monsignor Governatore di Roma* (Rome, 1972), p. 97.
199 Some writers give vague or erroneous dates for when Sfondrato became an Inquisitor, as explained by Mayer, *The Roman Inquisition*, pp. 61, 270. Mayer said Sfondrato 'was an Inquisitor by 17 February 1600', which seems wrong since he appeared alongside other cardinals earlier, apparently all as Inquisitors, during the Roman proceedings against Bruno.
200 Danielle Bartoli, *Della vita di Roberto Cardinal Bellarmino* (Rome, 1678), p. 329. Note also that Sfondrato had unsuccessfully supported Bellarmine to become Pope in 1605, instead of Borghese.
201 There exists some confusion. Godman remarks that Sfondrato is 'often (and wrongly) described in the secondary literature as prefect of the Congregation for the Index'. Godman, *The Saint as Censor*, p. 216.
202 Records of these four meetings are reproduced in Sergio Pagano, ed., *I documenti del processo di Galileo Galilei* (Vatican City, 1984), pp. 219–22.
203 These complaints were by Lorini, Caccini, Ximenes, Attavanti and the Consultor's Report on Galileo's Letter to Castelli; Finocchiaro, *The Galileo Affair*, pp. 134–44.
204 Thomæ Campanellæ, *Apologia pro Galileo, mathematico Florentino* (Frankfurt, 1622), p. 5. Richard J. Blackwell, ed., *Defense of Galileo* (Notre Dame, IN, 1994), pp. 19–24, 39, 133–4.

205 Thomæ Campanellæ, *Disputationum in quatuor partes suæ Philosophiæ Realis,* Part I: *Quæstiones Physiologicæ* (Paris, 1637), Question 11, Article 4, p. 106. See also Blackwell, ed., *Defense of Galileo*, p. 24.
206 Piero Guicciardini to the Grand Duke Cosimo II, 4 March 1616, in Favaro, ed., *Opere di Galileo Galilei: Edizione Nazionale*, XII, p. 242.
207 Finocchiaro, *The Galileo Affair*, p. 149.
208 Consultors' Report, 24 February 1616. There are variations in published translations of this passage, but Finocchiaro rightly pointed out that the original manuscript includes a semicolon after the phrase 'foolish and absurd in philosophy.' See Finocchiaro, *The Galileo Affair*, p. 344. For discussion, see Christopher Graney, 'The Inquisition's Semicolon: Punctuation, Translation, and Science in the 1616 Condemnation of the Copernican System', available at https://arxiv.org, accessed 14 August 2017.
209 For discussion, see Finocchiaro, *The Galileo Affair*, pp. 30–31.
210 Cardinal Millini, Inquisition Minutes, 25 February 1616, in ibid., p. 147.
211 Special Injunction, 26 February 1616; see also Inquisition Minutes, 3 March 1616, in ibid., p. 148.
212 For example, see Biagioli, *Galileo, Courtier*, pp. 31–57.
213 Godman, *The Saint as Censor*, p. 219.
214 Piero Guicciardini to Curzio Picchena II, 5 December 1615, Finocchiaro, *The Galileo Affair*, XII, p. 207.
215 Galilei to Giacomo Muti, 28 February 1616, in ibid., pp. 240–41. For discussion, see Shea and Artigas, *Galileo in Rome*, pp. 91–2.
216 Inquisition meeting with the Pope, 3 March 1616, in Mayer, *The Trial of Galileo*, p. 95.
217 'Decretum Sacrae Congregationis', 5 March 1616, Rome.
218 Galileo, 'A Madama Cristina' (1615), in Favaro, ed., *Opere di Galileo Galilei: Edizione Nazionale*, V, p. 321. Finocchiaro, *The Galileo Affair*, p. 97, translates 'setta' as 'school'. Galileo referred to 'the sect' of Pythagoras again in his 'Considerazioni circa l'opinione Copernicana' (n.d.), in Favaro, ed., *Opere di Galileo Galilei: Edizione Nazionale*, V, p. 352; and Galileo, *Dialogo* (1632), p. 3, in Drake, ed., *Discoveries and Opinions of Galileo*, p. 12.
219 For example Bellarmini, *De controversiis Christianæ fidei, aduersùs huius temporis hæreticos*, II (Lyon, 1610), Præfatio, pp. 117, 609, 1022–3; Bellarmini, *Dispvtationvm Roberti Bellarmini de controversiis Christianæ fidei, aduersus huius temporis hæreticos*, I (Lyon, 1596); on Chrysostom, see pp. 6, 44, 110, 181, 224, 235, 343, etc.; on Hippolytus, see pp. 644, 647, 651, 657, 659, 663, 667. Bellarmine did not know the works of Hippolytus directly, knowing them through the texts of Tertullian and pseudo-Tertullian, from the *De praescriptione haereticorum*; see Edward Ryan, *The Historical Scholarship of Saint Bellarmine* (Louvain, 1936), p. 113. On Tertullian, see, for example, Bellarmini, *De controversiis Christianæ fidei*, III (Paris, 1613), pp. 17, 110, 200, 227, 323.
220 Bellarmino, *De aeterna felicitate sanctorum, libri quinque* (Cologne, 1616), III, chap. 6, p. 177. This book was approved for publication on 29 December 1615 and printed in March 1616.

221 Bellarmino, *Sexta controversia generalis, de ecclesia, quae est in Purgatorio* (Ingolstadt, 1587), II, chap. 4, p. 114. See also Luke 16:23.

222 Bellarmini, *De controversiis Christianæ fidei*, I (1596), chap. 8, p. 170; chap. 12, p. 186. See also Bellarmini, *Illvstrissimi et reverendissimi D. Roberti Bellarmini, S.R.E. Cardinalis, conciones habitæ* (Cologne, 1615), p. 135. Bellarmini, *De scriptoribvs ecclesiasticis, liber unus* (Paris, 1617), p. 491.

223 Bellarmini, *De reliquiis et imaginibus sanctorum*, chap. 16, in Bellarmino, *Septima controversia generalis, de ecclesia triumphante* (Ingolstadt, 1587), p. 220.

224 Bellarmini, *Iudicium, de libro, quem Lutherani vocant concordiae* (Ingolstadt, 1585), p. 141. For the full passage from *Against Vigilantius*, see Stefan Rebenich, *Jerome* (New York, 2013), p. 106.

225 Bellarmini, *Conciones* (Cologne, 1615), Concio VII, pp. 805–6.

226 Epistola Sancti Hieronymi ad Paulinvm (*c.* 394 CE), in *Biblia, ad vetustisima exemplaria nunc recens castigata* (Antwerp, 1570), p. x. Also in *Nicene and Post-Nicene Fathers*, 2nd ser., VI, trans. W. Fremantle et al. (Buffalo, NY, 1893), Letter 53.

227 Philostratus, *Life of Apollonius*, II, chaps 26, 28, 43: pp. 185, 189, 228.

228 Bruno, 'Oratio valedictoria', Academia Witenbergensi, 15 March 1588, in Bruni, *Opera latine conscripta*, ed. F. Fiorentino, vol. I/1, p. 22.

229 Lactantius, *Divine Institutes*, III, chap. 18, in *The Works of Lactantius*, XXI/I of *Ante-Nicene Christian Library* (Edinburgh, 1871), pp. 182–3.

230 Bellarmini, *Conciones*, *Concio* VII, p. 640.

231 Roberto Bellarmino, *De amissione gratiae et statu peccati libri sex* (1613); repr. in *Disputationum Roberti Bellarmini, de controversiis Christianae fidei adversus huius temporis haereticos*, IV (Naples, 1858), chap. 11, p. 161.

232 For example, see Jerome, letters to Avitus and to Marcellinus and Anapsychia, in *St Jerome: Letters*, VI of *Select Library of Nicene and Post-Nicene Fathers*, 2nd ser., pp. 240–41, 252.

233 On Porphyry, see Bellarmini, *De controversiis Christianæ fidei, Aduersus huius Temporis Hæreticos*, vol. I (1596), pp. 14, 25, 210, 668; Bellarmini, *De controversiis Christianæ fidei*, vol. IV (Ingolstadt, 1601), p. 357; Bellarmini, *Conciones habitae* (Cologne, 1615), Concio I, p. 7; Bellarminus, *De scriptoribvs ecclesiasticis philologica et historica dissertatio* (Paris, 1660), p. 58.

234 Galileo to the Tuscan Secretary of State, 6 March 1616, in Finocchiaro, *The Galileo Affair*, pp. 150, 151.

235 Ibid., p. 150.

236 Bruno to the Venetian Inquisition, Third Deposition, 2 June 1592, in Firpo, *Il processo di Giordano Bruno*, pp. 167, 269.

237 Bellarmino, 26 May 1616, in Finocchiaro, *The Galileo Affair*, p. 152.

238 Galileo to Piero Dini, 23 March 1615, in Favaro, ed., *Opere di Galileo Galilei: Edizione Nazionale*, V, p. 301. Also in Finocchiaro, *The Galileo Affair*, pp. 63–4.

239 Favaro, ed., *Opere di Galileo Galilei: Edizione Nazionale*, p. 302: Galileo wrote: '[Spiritus Dei] *foventem aquas seu incubantem super aquas*.'

240 Ibid., p. 305.

241 Galilei to Madama Cristina, 1615, in Favaro, ed., *Opere di Galileo Galilei:*

Edizione Nazionale, v, p. 345. See also Dionysius the Areopagite, *De divinis nominibus* (*c.* 500), in *Pseudo-Dionysius: The Complete Works*, trans. Colm Luibheid (Mahwah, NJ, 1987), pp. 75, 101–2. Ficino had written a *Commentary on De divinis nominibus*.

242 Bruno, *De gli heroici furori*, n.p., end of the Fourth Dialogue.

243 Cardinal Decio Carafa to Cardinal Giovanni Millini, 2 June 1616; Millini to Carafa, 9 June 1616, both in Pagano, ed., *I documenti del processo di Galileo Galilei*, p. 104.

3 The Enemies of Galileo

1 Nicolaus Mulerius, *Institutionum astronomicarum libri duo: Quibus etiam continentur Geographie principia, nec non pleraque ad artem navigandi facientia* (Groningen, 1616, repr. 1649), pp. x, xi–xii. Mulerius's preface is dated 6 August 1616 in the Julian calendar, which corresponds to 26 August in the Gregorian calendar.

2 Matthieu Riccius, *Histoire de l'expedition Chrestienne au royaume de la Chine entreprinse par les PP. de la compagnie de Iesus, comprinse en cinq livres*, ed. Nicolas Trigault, French trans. S.D.F. de Riquebourg-Trigault (Lyon, 1616), p. 176. See also Matthæi Riccii, *De Christiana expeditione apud Sinas suscepta ab Societate Ieesu, libri v*, ed. Rincolao Trigautio (Lyon, 1616), p. 624.

3 Antonii Ruvio, *Commentarii in libros Aristotelis Stagiritæ de coelo & mundo ortu* (Cologne, 1617), p. 54.

4 Kepler, 'Notes on the Dream', in *Kepler's Somnium: The Dream, or Posthumous Work on Lunar Astronomy*, ed. Edward Rosen (New York, 1967), p. 38.

5 Johannes Kepler, *Harmonices mundi libri v* (Linz, 1619); Kepler, *The Harmony of the World*, trans. and ed. E. Aiton et al. (Philadelphia, PA, 1997), IV, pp. 362–76.

6 Ibid., III, pp. 134–5.

7 Ibid., I, p. 12. By 460 CE the pagan theologian Proclus claimed that Pythagoras had discovered 'the structure of the cosmic figures', the five regular solids. Proclus, *A Commentary on the First Book of Euclid's Elements*, trans. Glenn Morrow (Princeton, NJ, 1970), p. 53.

8 Kepler, *Harmony of the World*, IV, p. 284.

9 Ibid., p. 304.

10 Kepplero, *Epitome astronomiae Copernicanae* (Linz, 1618), p. 125.

11 Ibid., pp. 125–6.

12 Ibid., pp. 109, 124.

13 Bellarmino and Franciscus Capiferreus, 'Decretum, Sacræ Congregationis Indicis', 18 May 1619, in *Index librorum prohibitorum* (1664), pp. 310–11.

14 *Edictum librorum qui post Indicem felicis recordationis Clementis VIII prohibiti sunt* (Rome, 1619), quoted in Maurice Finocchiaro, *Retrying Galileo, 1633–1992* (Berkeley, CA, 2005), p. 20.

15 Ioannis Tarde, *Borbonia sidera, id est planetæ qui Solis limina circvmvolitant motv proprio* (Paris, 1620), p. 16. See also Iean Tarde, *Les astres de Borbon, et apologie pour le Soleil* (Paris, 1627), p. 19.

16 Franciscus Capiferreus, 'Monitum Sacrae Congregationis ad Nicolai Copernici lectorem' (1620); this document is reprinted in Latin in William Roberts, *The Pontifical Decrees against the Doctrine of the Earth's Movement and the Ultramontane Defence of Them* (Oxford, 1885), pp. 122–4. One book that carefully analyses the emendations is Finocchiaro, *Retrying Galileo*, pp. 20–25.
17 The original page numbers for the censored sentences are: Copernicus, *De Revolutionibus* (1543), pp. iiij reverso, 3 verso, 6 verso, 7 verso, 9 verso, 10 verso, 122 verso.
18 This point is missing in Finocchiaro, *Retrying Galileo*. Finocchiaro's translations use the expression 'heavenly bodies' instead of 'stars', for example rendering *horum trium siderum* as 'these Three Heavenly Bodies', instead of 'these three stars'.
19 Kepler, *Epitome astronomiae Copernicanae* (1620); or *Epitome of Copernican Astronomy and Harmonies of the World*, trans. Charles Wallis (Amherst. NY, 1995), IV, p. 119.
20 Ibid., pp. 56, 119.
21 Ibid., pp. 56–7. Kepler argued that the Sun was probably the seat of the soul of the world, just as the heart is the seat of the human soul. See Ioannis Kepleri, *Pro suo opere Harmonices Mundi apologia* (Frankfurt, 1622), p. 24. For discussion, see Patrick Boner, *Kepler's Cosmological Synthesis: Astrology, Mechanism and the Soul* (Leiden, 2013), pp. 160–61.
22 Tommaso Campanella, *De sensu rerum et magia* (Frankfurt, 1620), pp. 214–15. This book was drafted in Italian as *Del senso delle cose e della magia*, in 1604 at the Sant'Elmo prison, and published in Italian in 1625; see Germana Ernst, *Tommaso Campanella: The Book and the Body of Nature*, trans. David Marshall (New York, 2010), pp. 114–15, 123.
23 Campanella, *De sensu rerum*, p. 217.
24 Ibid., p. 167.
25 Ibid., p. 194.
26 Ibid., p. 43.
27 Ibid., pp. 145–6.
28 Ibid., pp. 168–9.
29 Ibid.
30 Ibid., p. 231.
31 Io. Lavrentii, *De natvra dæmonvm*, in *Mallei maleficarvm, tractatvs aliqvot novi ac veteres*, II/2 (Lyon, 1620), p. 96.
32 Antonio Rusca, *De Inferno, et statv dæmonvm ante mvndi exitivm, libri qvinqve* (Milan, 1621), III, chap. 8, pp. 295–9.
33 Gaspar Schoppe to Konrad Rittershausen, 17 February 1600, printed in Gaspari Scioppii, 'Epistola, in qua haereticos jure infelicibus lignis cremari concludit' [also titled: 'Epistola, in qua sententiam de Lutheranis tanquam haereticis atram Romae fieri asserit & probat'], in *Macchiavellizatio* (Saragossa, 1621), p. 32.
34 Marta Fattori, '"*Qua epistola cum nimium utilis, et fructuosa sit, potius laude quam censura est digna*": un nuovo documento sulla lettera di Gaspare Scioppio', *Nouvelles de la République des Lettres*, 1–2 (2003), pp. 191–200.
35 Hilary Gatti, *Essays on Giordano Bruno* (Princeton, NJ, 2011), p. 313.

36 Christiano Longomontano, *Astronomiæ Danicæ appendix de asscititiis coeli phænomenis* (Amsterdam, 1622), p. 3.
37 Iulii Caesaris Lagalla, *De immortalitate animorum, ex Arist. sententia. Libri III* (Rome, 1622), pp. 353–4, 485–6.
38 Ibid., pp. 96, 292, 309–10, 313–14, 355, 373–5.
39 Bellarmini, *Disputationum, de controversiis Christianae Fidei adversus huius temporis Haereticos. Epitome*, ed. I. B. Desbois, I (Paris, 1602), p. 149 reverso.
40 Francisci Collii, *De animabus paganorum* (Milan, 1622), p. 2.
41 Ibid., p. 371.
42 Ibid., pp. 377–90.
43 Nicolao Romaeo, *Iohan Calvini Noviodunensis nova effigies centum coloribus ad vivum expressa* (Antwerp, 1622), pp. 588 verso, 682 verso; see also pp. 127 and 247.
44 Ibid., pp. 23, 558 reverso.
45 Barptolomaeo [*sic*] Agricola, *Symbolum Pythagoricum; sive de justitia in forum reducenda*, 2 vols (Naples, 1619). Another book on Pythagorean matters that had previously been banned was Viti Amerbachii, *Commentaria in Pythagorae et Phocylidis poëmata* (Strasbourg, 1539), censored in Rome in 1564 (and in Spain in 1559); see Jesús Martínez de Bujanda, ed., *Index de l'Inquisition Espagnole: 1583, 1584* (Québec, 1993), p. 557.
46 Agricola, *Symbolum Pythagoricum*, I, p. 121.
47 Ibid., I, p. 35.
48 Maffeo Barberini and Franciscus Capiferreus, 'Decretum, Sacræ Congregationis Indicis', 22 December 1622, *Index librorum prohibitorum* (1664), pp. 321–2.
49 Tommaso Campanella, 'Sopra i colori delle vesti: Sonetto' (undated), in Gio. Orelli, ed., *Poesie filosofiche di Tommaso Campanella* (Lugano, 1834), p. 111; I have written 'metal' for 'lead'.
50 Thomæ Campanellæ, *Apologia pro Galileo, mathematico Florentino* (Frankfurt, 1622); Richard J. Blackwell, ed., *Defense of Galileo* (Notre Dame, IN, 1994), p. 41.
51 Ibid., p. 43.
52 Ibid, p. 96; and Campanellæ, *Apologia pro Galileo*, pp. 40–41.
53 Blackwell, ed., *Defense of Galileo*, p. 48.
54 Ibid., p. 88.
55 Ibid., p. 89.
56 Ibid., p. 105.
57 Gregorius Magnus [Pope Gregory I], *Dialogorum libri IV, de vita et miraculis patrum Italicorum, et de aeternitate animarum* (Basel, 1496), IV, chap. 42. I doubt that Campanella referred to Gregory of Nyssa, a bishop who had written a literal commentary on Genesis (*c.* 375 CE).
58 Blackwell, ed., *Defense of Galileo*, pp. 89–90. Campanellæ, *Apologia pro Galileo*, p. 36. Campanella referred also to Ovid's *Metamorphoses*.
59 Blackwell, ed., *Defense of Galileo*, p. 96.
60 Ibid., p. 93.
61 Ibid., p. 101.
62 Ibid., p. 118.

63 Ibid., p. 119.
64 Ibid., p. 45.
65 Ibid., p. 53.
66 Gabriele Pennotto, *Generalis totius sacri ordinis clericorum canonicorum historia tripartita* (Rome, 1623), pp. 144, 16.
67 Campanellæ, *Apologia pro Galileo*, pp. 109–10.
68 Ibid., p. 111.
69 Ibid.
70 Ibid., pp. 112–13.
71 Ibid., pp. 62–3, 71.
72 Ibid., pp. 83, 85.
73 Ibid., pp. 63–4.
74 Ibid., p. 71.
75 Ibid., p. 80.
76 Ibid., pp. 78–9.
77 Ibid., pp. 59, 86.
78 Ibid., p. 49.
79 Ibid.
80 Ibid., p. 120.
81 Ibid., p. 49.
82 Ibid.
83 Galilei, *The Assayer* (1623); excerpts in *Discoveries and Opinions of Galileo*, ed. and trans. Stillman Drake (New York, 1957), pp. 237–8.
84 *Raccolta de libri prohibiti* (Milan, 1624). This prohibition was included in 'the Raccolta of 1624, the Elenchus of some years later, and the succeeding Indexes up to the time of Benedict XIV', the Index of which finally omitted it in the 1740s. See George Putnam, *The Censorship of the Church of Rome: and its Influence upon the Production and Distribution of Literature*, I (New York, 1906), p. 129.
85 Cardinal Zollern, quoted in Galileo to Federico Cesi, 8 June 1624, in *Le opere di Galileo Galilei. Edizione Nazionale*, ed. Antonio Favaro, 20 vols (Florence, 1890–1909, repr. 1929–38), vol. XIII, p. 182.
86 Urban VIII, quoted in ibid.
87 Augustino Oregio, *De Deo uno tractatus primus* (Rome, 1629), p. 195.
88 Marin Mersenne, *L'impieté des déistes, athées, et libertins de ce temps*, I (Paris, 1624), pp. 220–21.
89 Ibid., II, p. 299.
90 Ibid., I, pp. 229–31.
91 Miguel Granada, 'Mersenne's Critique of Giordano Bruno's Conception of the Relation between God and the Universe: A Reappraisal', *Perspectives on Science*, XVIII/1 (2010), pp. 33–5.
92 Mersenne, *L'impieté des déistes*, II, pp. 363–4.
93 Ibid., II, pp. 210, 365–8.
94 Marini Mersenni, *Quaestiones celeberrimae in Genesim* (Paris, 1623), col. 1079. Scholars had not realized that Mersenne was wrong; for example, see the otherwise excellent article by Granada, 'Mersenne's Critique', p. 30.
95 Mersenni, *Quaestiones celeberrimae*, col. 1078.
96 Marin Mersenne, *La verité des sciences* (Paris, 1625), p. 121.

97 In 1615 King James I pressured Donne to enter the Anglican Ministry by declaring that Donne could no longer be employed outside of the Anglican Church. In 1616 Donne was appointed Royal Chaplain. Despite his dislike of the Jesuits, Donne sought unity within Christendom. In his earlier book *Pseudo-Martyr* (1610) he insisted that Catholics in England should subscribe to the Oath of Allegiance to the king.
98 John Donne, Sermon CXIII, preached to the Earl of Exeter and his Company in his Chapel at St John's, 13 June 1624, in *The Works of John Donne*, IV (London, 1839), p. 576.
99 Ibid., pp. 108, 402, 432, 491, 494, 535.
100 Justo Justino Justinopolitano [Ludwig Camerarius], *Mysterium iniquitatis, sive Secreta secretorum Turco-Papistica secreta, contra libellum famosum, sub titulo Secreta Calvino-Turcica* (Justinopoli [Amsterdam?], 1625), p. 180.
101 Lucio Vero Clarimontano [pseudonym], *Nova Apocalypsis in qua Innocentia Cæsaris Augusti et fidelium imperio principum et noxa rebellium* (Luxembourg, 1626), p. 160.
102 Campanella, quoted in Peter Rietbergen, *Power and Religion in Baroque Rome: Barberini Cultural Policies* (Leiden, 2006), p. 121.
103 Caesar Longinus, *Trinvm magicvm, sive secretorum magicorvm opvs* (Frankfurt, 1630), pp. 45, 373, 385–91; Henning Grosse, ed., *Magica de spectris et apparitionibus spiritu* (n.p., 1656), pp. 186–7.
104 Panfilo Persico, *Del segretario libri quattro* (Venice, 1629), II, p. 171; Eng. trans. from Mario Biagioli, *Galileo, Courtier* (Chicago, IL, 1993), p. 315.
105 *Kepler's Somnium*, ed. Rosen, p. 5.
106 Lucæ Holstenii, 'De vita & scriptis Porphyrii philosophi dissertatio' (1630), I; in Lucas Holstenius, *Liber de vita Pythagorae. Eiusdem sententiae ad intelligibilia ducentes. De antro nympharum* (Rome, 1630).
107 Rietbergen, *Power and Religion in Baroque Rome*, p. 266.
108 Holstenius, *Liber de vita Pythagorae*, pp. 9, 65–6.
109 Ibid., p. 9.
110 Ibid., pp. 10, 8.
111 Ibid., p. 10, also on pp. 8, 16, 22.
112 Ibid., chap. 3, pp. 19–24; chap. 10, pp. 81–91.
113 Ibid., pp. 99, 120.
114 The original name of the Barberini family was Tafano: Italian for horsefly. The modest family heraldry included three horseflies. However, the family became wealthy. By 1607 Maffeo Barberini had changed the family emblem to three bees. Thomas Mayer, *The Roman Inquisition on the Stage of Italy, c. 1590–1640* (Philadelphia, PA, 2014), p. 101.
115 Francesco Barberini, quoted in Castelli to Galileo, 9 February 1630, in Favaro, ed., *Opere di Galileo Galilei: Edizione Nazionale*, XIV, p. 78. See Jules Speller, *Galileo's Inquisition Trial Revisited* (Frankfurt am Main, 2008), pp. 222–4.
116 Bruno, 3rd Deposition, 2 June 1592, in Luigi Firpo, *Il processo di Giordano Bruno* (Rome, 1993), pp. 167, 269. Bruno (3rd, 12th and 14th depositions) affirmed having such Pythagorean beliefs: Earth is a star, stars are worlds, infinitely many worlds exist; see ibid., pp. 268–9. 'Summarium': Tenth Censured Proposition, in ibid., p. 304.

117 John Heilbron, *Galileo* (Oxford, 2010), pp. 300–301.
118 Niccolò Riccardi to the Inquisitor in Florence, 24 May 1631, in Giorgio de Santillana, *The Crime of Galileo* (Chicago, IL, 1955), p. 317.
119 Liberti Fromondi, *Ant-Aristarchus, sive Orbis-Terrae Immobilis. Liber unicus. In quo decretum S. Congregationis S.R.E. Cardinal. an. CIↃ.IↃC.XVI. aduersus Pythagorico-Copernicanos editum defenditur* (Antwerp, 1631), p. 1.
120 Ibid., pp. 19–24.
121 Joannes Aventinus, *Annalium Boiorum* (Ingolstadt, 1554), III, p. 297. Aventinus died in 1534. His *Annals* were published posthumously and abridged; a more complete edition appeared in Basel in 1580. Although Aventinus remained a Catholic throughout his life, he sympathized with aspects of the Reformation, and described the struggle of Ludwig IV, Holy Roman Emperor and Duke of Bavaria, against the Papacy in the fourteenth century. Therefore, some Catholics viewed him as a Lutheran.
122 Fromondi, *Ant-Aristarchus*, p. 21.
123 Ibid., p. 29.
124 Ibid., pp. 2, 26.
125 Ioanne Baptista Morino, *Famosi et antiqui problematis de telluris motu, vel quiete* (Paris, 1631), p. 1.
126 Ibid., pp. 4, 5, 11.
127 Ibid., p. 138; see also pp. 12, 57, 136, where they are again called 'obstinate' and 'temerarius'.
128 Ibid., p. 57.
129 Ibid., pp. 12–13.
130 Ibid., p. 57.
131 Ibid., pp. 43–50, 105–6.
132 Ibid., p. 136.
133 Ibid., p. 139.
134 Galileo to Elia Diodati, 15 January 1633, in Favaro, ed., *Opere di Galileo Galilei: Edizione Nazionale*, vol. XV, pp. 23–4; see also VII, pp. 16–17. Also in Finocchiaro, *The Galileo Affair*, p. 224.
135 Ibid., p. 214. Stillman Drake, ed. and trans., *Discoveries and Opinions of Galileo* (New York, 1957), p. 5.
136 This similarity was pointed out by Isabelle Pantin; see David Wootton, *Galileo, Watcher of the Skies* (New Haven, CT, 2010), pp. 55, 276.
137 Pietro Redondi, *Galileo Heretic*, trans. Raymond Rosenthal (Princeton, NJ, 1989), p. 229.
138 Ferdinando Gregorovius, *Urbano VIII e la sua opposizione alla Spagna e all' Imperatore* (Rome, 1897), pp. 46–59.
139 Redondi, *Galileo Heretic*, pp. 230–31.
140 Rietbergen, *Power and Religion in Baroque Rome*, p. 266.
141 Galilei, 'Dialogo Secondo', in *Dialogo, Doue ne i congressi di quattro giornate si discorre sopra i due massimi sistemi del mondo Tolemaico, e Copernicano* (Florence, 1632), p. 183; as Galileo, *Dialogue Concerning the Two Chief World Systems*, ed. Stillman Drake (Berkeley, CA, 1962), p. 219.
142 'Dialogo Primo', in *Dialogo*, p. 3; Drake, ed., *Dialogue*, p. 11.
143 Ibid., pp. 53–4; Drake, ed., *Dialogue*, p. 71.
144 Wootton, *Galileo, Watcher of the Skies*, p. 212.

145 Galileo, *Dialogo*, Day 3, p. 324; also in Favaro, ed., *Opere di Galileo Galilei: Edizione Nazionale*, VII, p. 354; Drake, ed., *Dialogue*, p. 380.
146 Galileo, *Dialogo*, Day 1, p. 52; Day 2, pp. 110, 113; Day 3, pp. 323, 361–2; Drake, ed., *Dialogue*, pp. 139, 428.
147 Biagioli, *Galileo, Courtier*, pp. 322, 330–38.
148 Heilbron, *Galileo*, pp. 303–4.
149 Riccardi to the Inquisitor of Florence, 25 July 1632, in Redondi, *Galileo Heretic*, p. 239.
150 Campanella to Galilei, 5 August 1632, in Campanella, *Lettere*, ed. Vincenzo Spampanato (Bari, 1927), p. 241; his original phrase in Italian reads: 'ch'erano degli antichi pitagorici e democratici e di Vostra Signoria'.
151 Redondi, *Galileo Heretic,* p. 238.
152 Hippolytus, *Refutatio*, I, sec. 13, pp. 72–3.
153 Thomae Aqvinatis, 'Distinction of Things in General', *Svmmae Theologiae*, I (Venice, 1593), Question 47, Art. 3, fol. 165.
154 Marsilio Ficino, *Theologiae Platonicæ, de immortalitate animorum*, IX, in Marsili Ficini, *Opera & quæ hactenus extitêre, & quæ in lucem nunc primum prodiêre omnia* (Basel, 1561), p. 311.
155 Riccardi to Niccolini, 1632, quoted in Heilbron, *Galileo*, p. 307.
156 Pope Urban VIII, quoted by Francesco Niccolini to Andrea Cioli, 5 September 1632, in Favaro, ed., *Opere di Galileo Galilei: Edizione Nazionale*, XIV, pp. 383, 384 (Finocchiaro, *The Galileo Affair*, p. 229).
157 Urban VIII, quoted in Niccolini to Cioli, 13 November 1632, in ibid., XIV, p. 429 (Finocchiaro, p. 239).
158 Urban VIII, quoted in Niccolini to G. Gondi, 25 January 1642, in ibid., XVIII, p. 379.
159 Urban VIII, quoted in Niccolini to Cioli, 27 February 1633, in ibid., XV, p. 56 (Finocchiaro, p. 245).
160 Niccolini to Cioli, 11 September 1632, in ibid., XIV, p. 388 (Finocchiaro, p. 232).
161 Niccolini to Cioli, 13 March 1633, in ibid., XV, p. 68 (Finocchiaro, p. 247).
162 Niccolini to Cioli, 9 April 1633, in ibid., XV, p. 85 (Finocchiaro, p. 249).
163 Heilbron, *Galileo*, p. 308.
164 Bellarmino to Foscarini, 12 April 1615, in Favaro, ed., *Opere di Galileo Galilei: Edizione Nazionale*, XII, p. 171.
165 Clergymen rejected Galileo's arguments about what is 'probably' true; see Finocchiaro, *The Galileo Affair*, pp. 24–5.
166 Francesco Barberini, quoted in Niccolini to Cioli, 27 February 1633, in Favaro, ed., *Opere di Galileo Galilei: Edizione Nazionale*, XV, p. 56. Finocchiaro, p. 246, translates *dogma fantastico* as 'imaginary dogma'.
167 Galilei, *Dialogo*, p. 53; Drake, ed., *Dialogue*, p. 71.
168 [Anon.], *c.* May–July 1600, quoted in Tiziana Provvidera, 'A New English Document on Giordano Bruno', *Bulletin of the Society for Renaissance Studies*, XIX (2002), p. 25.
169 One commentator construes the expression 'fantasticall fellow' as a compliment. However, it was an insult. I have found more than fifty other instances of the word *fantasticall* and *fantastical* in dozens of books from 1590 to 1630, and all are negative. For example: 'he was

burned in Smithfield for obstinacy in new fantasticall opinions,' in N. D. [Robert Parsons], *An Examen of the Calendar, or the Catalogue of Protestant Saincts, Martyrs, and Confessors* (n.p., 1604), p. xl. Also, 'Margaret Ellis Confessor. This was a willfull woman of the towne of Bellirica [Billericay], who being condemned by the B. of London for divers fantasticall opinions obstinately held & defended by her, she died in Newgate' [May 1556], in ibid., p. civ. In Italian the words *fantastico* and *fantasticando* had thoroughly negative meanings.

170 Holste to Niccolò Peiresc, 1633, quoted in Rietbergen, *Power and Religion*, p. 267.

171 Holste to Peiresc, 7 March 1633, in Favaro, ed., *Opere di Galileo Galilei: Edizione Nazionale*, xv, p. 62.

172 Galileo, 1st Deposition, 12 April 1633, in ibid., xix, pp. 339–40 (Finocchiaro, *The Galileo Affair*, p. 259).

173 Niccolini to Andrea Cioli, 11 September 1632, in ibid., xiv, p. 389 (Finocchiaro, p. 233).

174 Ibid.

175 Zaccaria Pasqualigo, 17 April 1633, in Finocchiaro, *The Galileo Affair*, p. 271.

176 For biographies of Inchofer, see Dezsö Dümmerth, 'Les combats et la tragédie du Père Melchior Inchofer S.J. à Rome (1641–1648)', *Annales Universitatis Scientiarum Budapestinensis de Rolando Eötvös Nominatae. Sectio Historica xvii* (Budapest, 1976), pp. 81–112; and Thomas Cerbu, 'Melchior Inchofer, "un homme fin & rusé"', in *Largo campo di filosofare: Eurosymposium Galileo 2001*, ed. José Montesinos and Carlos Solís (La Orotava, 2001), pp. 587–611.

177 Melchior Inchofer, 17 April 1633, in Finocchiaro, *The Galileo Affair*, p. 264. See also Sergio Pagano, ed., *I documenti del processo di Galileo Galilei* (Vatican City, 1984), pp. 139–48.

178 Inchofer, in Finocchiaro, *The Galileo Affair*, p. 267.

179 Ibid., p. 268.

180 D'Ascoli was absent from Rome until late April, so he participated in Inquisition meetings from 4 May onward. See Pierre-Noël Mayaud, 'Les "Fuit congregatio sancti officii in . . . coram . . ." de 1611 à 1642: 32 ans de vie de la Congrégation du Saint Office', *Archivum Historiae Pontificiae*, xxx (Rome, 1992), pp. 231–89. Two other cardinal Inquisitors participated in 1616, Gallamino and Zapata, but they were not in Rome in 1633. See Annibale Fantoli, *Galileo: For Copernicanism and for the Church*, trans. George Coyne (Notre Dame, in, 1994), p. 448.

181 Ibid., p. 453.

182 For evidence of Cardinal Barberini's kindness toward Galileo, see Fantoli, *The Case of Galileo*, pp. 184–9.

183 Riccardi, quoted in Filippo Magalotti to Mario Guiducci, 7 August 1632, in Favaro, ed., *Opere di Galileo Galilei: Edizione Nazionale*, xiv, p. 370.

184 Final Report to the Pope, May or June 1633, in Finocchiaro, *The Galileo Affair*, p. 282.

185 Galileo's 4th Deposition, 21 June 1633, in ibid., p. 287.

186 Giacomo Bouchard to Fulgenzio Micanzio, 29 June 1633, in Favaro, ed., *Opere di Galileo Galilei: Edizione Nazionale*, xv, p. 166.

187 For a discussion of different kinds of heresy in relation to Galileo's trial, see Jules Speller, *Galileo's Inquisition Trial Revisited* (Frankfurt am Main, 2008), e.g., pp. 21–50. See also Finocchiaro, *Retrying Galileo*, pp. 12–16, 47, 272–4.
188 Fantoli, *The Case of Galileo*, p. 206.
189 The day after his condemnation to 'imprisonment', Galileo was allowed to stay instead at the Tuscan Embassy; the next week the Inquisition let him move to Siena to stay at the house of his friend, Archbishop Piccolomini. After six months Galileo was allowed to return under house arrest to his own home at Arcetri, then outside Florence. See Fantoli, *Galileo: For Copernicanism*, p. 464.
190 'Vesta' was an ancient name for the fire at the universal centre: 'the figure of the whole universe, in the centre of which the Pythagoreans placed the fire, namely Vesta, and they called it unity. They held that the Earth is not immobile and is not located in the centre of the circumference, but is carried in a circle around the fire.' Plutarch, 'Numa Pompilius' (*c.* 75 CE), in Plutarchi, *Summi et philosophi et historici parallela*, p. 45.
191 Liberti Fromondi, *Vesta, sive Ant-Aristarchi Vindex . . . In quo Decretum S. Congregationis S.R.E. Cardinalium anno M.DC.XVI. & alterum anno M.DC.XXXIII. aduersus Copernicanos terraæ motores editum, iterum defenditur* (Antwerp, 1634), [p. 19].
192 Cerbu, 'Melchior Inchofer', p. 591.
193 Ibid., p. 590. Melchiore Inchofer, *De epistola B. Virginis Mariae ad Messanenses conjectatio* (Rome, 1631).
194 Cerbu, 'Melchior Inchofer', p. 592.
195 Leone Allacci to Fabio Chigi, in ibid., p. 591.
196 Ibid., p. 598.
197 For discussion, see William Shea, 'Melchior Inchofer's "Tractatus Syllepticus": A Consultor of the Holy Office Answers Galileo', in *Novità celesti e crisi del sapere*, ed. Paolo Galluzzi (Florence, 1984), pp. 283–92.
198 Richard Blackwell, *Behind the Scenes at Galileo's Trial* (Notre Dame, IN, 2008), pp. 45–6.
199 Melchioris Inchofer, *Tractatvs syllepticvs, in quo, qvid de terrae, solisq. motv, vel statione, secundum S. Scripturam, & Sanctos Patres sentiendum* (Rome, 1633), pp. 77–8, 86; trans. in Blackwell, *Behind the Scenes*, quotations on pp. 108, 123, 167.
200 Campanella, *Apologia pro Galileo* (1622); *Defense of Galileo*, trans. Blackwell, pp. 50, 48.
201 Nicolai Mulerii, 'Isagoge in tabulas Frisicae', *Tabulae Frisicae Lunae-Solares quadruplices* (Alkmaar, 1611), chap. 1, p. 318.
202 Nicolaus Mulerius, *Institutionum astronomicarum libri duo* (Groningen, 1616), p. [x].
203 Inchofer, *Tractatus syllepticvs*, p. 92.
204 Diogenes, *Lives of Eminent Philosophers*, ed. B. Hicks (Cambridge, 1972), bk 1, sec. 13, p. 74.
205 Inchofer, *Tractatus syllepticvs*, p. 92.
206 Ibid., p. 80.

207 For discussion, see Michael Gorman, 'A Matter of Faith? Christoph Scheiner, Jesuit Censorship, and the Trial of Galileo', *Perspectives on Science*, IV/3 (1996), p. 298.
208 Inchofer, *Tractatus syllepticvs*, Imprimatur by Lucas Waddingus, [p. iv].
209 Galileo to Elia Diodati, 25 July 1634, in *Le opere di Galileo Galilei, Prima edizione completa*, 15 vols (Florence, 1842–56), VII, pp. 49–50. This letter is missing in Favaro, ed., *Opere di Galileo Galilei: Edizione Nazionale*, XVIII.
210 See Melchiorre Inchofer, *Vindiciarum S. Sedis Apostolicae, sacrorum tribunalium et authoritatum adversus neo-Pythagoraeos terrae motores, et solis statores* [Libri Duo], 1635, Rome, Biblioteca Casanatense, MS 182.
211 Domenico Berti, *Il processo originale di Galileo Galilei* (Rome, 1876), pp. xci–xcii.
212 Mariano Artigas, 'Un nuovo documento sul caso Galileo: EE 291', *Acta Philosophica*, X/2 (2001), pp. 199–214.
213 For discussion on the issue of transubstantiation, see Lucas Mateo-Seco, 'Galileo e l'Eucaristia. La questione teologica dell'ACDF, *Index, Protocolli*, EE, f. 291r–v', *Acta Philosophica*, X/2 (2001), pp. 243–56; and William Shea, 'Galileo e l'atomismo', *Acta Philosophica*, X/2 (2001), pp. 257–72.
214 Rafael Martínez, 'Il Manoscritto ACDF, *Index, Protocolli*, vol. EE, f. 291r–v', *Acta Philosophica*, X/2 (2001), pp. 215–42.
215 Francesco Beretta notes that G3 might instead be from as late as 1642. See Beretta, 'Melchior Inchofer et l'hérésie de Galilée: Censure doctrinale et hiérarchie intellectuelle', *Journal of Modern European History*, III/1 (2005), p. 44.
216 Ibid., pp. 41–2.
217 Philippi Lansbergii, *Commentationes in Motum Terræ Diurnum, & Annuum*, trans. Martin van der Hove (Middelburg, 1630).
218 Inchofer, *Vindiciarum S. Sedis Apostolicae*, p. 5; Augustine, *Confessions*, V, chap. 5.
219 Inchofer, *Vindiciarum S. Sedis Apostolicae*, pp. 6, 17.
220 Galileo, *Dialogo*, Day 3, p. 324; also in Favaro, ed., *Opere di Galileo Galilei: Edizione Nazionale*, VII, p. 354. Drake, ed., *Dialogue*, p. 380.
221 Inchofer, *Vindiciarum S. Sedis Apostolicae*, p. 15.
222 Ibid., p. 16.
223 Ibid., p. 17.
224 Ibid., p. 34. Augustine, *City of God*, XVI, chap. 9.
225 Inchofer, *Vindiciarum S. Sedis Apostolicae*, p. 35.
226 Ibid., pp. 11, 88, 92, 94.
227 Dionysius of Alexandria, 'Against the Epicureans', in *The Works of Gregory Thaumaturgus, Dionysius of Alexandria and Archelaus*, XX of *Ante-Nicene Christian Library*, ed. A. Roberts and J. Donaldson (Edinburgh, 1871), p. 171.
228 Inchofer, *Vindiciarum S. Sedis Apostolicae*, p. 36.
229 Ibid. Plutarch had written that Cleombrotus asserted 183 worlds. Plutarch, 'Quare Expirauerint Oracula', in Plutarchi Chaeronei, *Ethica, sive Moralia, Opera quae extant, omnia* (Basel, 1573), p. 533.
230 Thomas Gaisford, ed., *Theodoreti Episcopi Cyrensis Graecarum Affectionum Curatio* (Oxford, 1839), IV, p. 162.

231 Inchofer, *Vindiciarum S. Sedis Apostolicae*, p. 36.
232 Siméon bar Yohay, *K'abbalae Denudatae*, II: *Liber Sohar Restitutus* (Frankfurt, 1684), pt 2, Third Treatise: 'Idra Suta, seu Synodus Minor', sec. 19: 'De Labiis Microprosopi & ore ejus', p. 588.
233 Inchofer, *Vindiciarum S. Sedis Apostolicae*, p. 37.
234 Ibid., p. 36. Tertullian, 'Annotationes Iacobi Pamelii in librum de pallio', in Tertullian, *Opera quae hactenus reperiri potuerunt omnia* (Paris, 1616), chap. 2, p. 11.
235 Hieronymus, 'Epistola ad Avitum', quoted in Ludovicus Lucius, *Historiae ecclesiasticae*, I (Basel, 1624), p. 185. Jerome there referred to Origen's *Peri Archon*, in *Origenis Adamantii Opera* (Paris, 1619), II, pp. 436–7, III, p. 463.
236 Inchofer, *Vindiciarum S. Sedis Apostolicae*, p. 37.
237 Ibid.
238 Ibid., p. 39.
239 Ibid., p. 42.
240 Ibid., p. 43.
241 Ibid., p. 45.
242 Ibid., p. 46.
243 Ibid., p. 51.
244 St Augustine, *The Manichean Debate*, I (New York, 2006), pp. 20, 48, 108, 118.
245 St Augustine, 'Answer to Felix, a Manichean', in ibid., p. 286.
246 Inchofer, *Vindiciarum S. Sedis Apostolicae*, p. 56.
247 Ibid., p. 81.
248 Ibid., p. 84.
249 Ibid., p. 98. Friedrich Marx, ed., *Sancti Filastrii episcopi Brixiensis Diversarum hereseon liber*, Corpus Scriptorum Ecclesiasticorum Latinorum 38 (Vienna, 1898), p. 76.
250 Inchofer, *Vindiciarum S. Sedis Apostolicae*, pp. 110–22.
251 Ibid., p. 125.
252 Ibid., pp. 128–9.
253 Ibid., p. 137.
254 Ibid., p. 149.
255 Ibid.
256 Diogenes, 'Anaxagoras', in *Lives of Eminent Philosophers*, ed. Hicks, II: 3, secs 6–8, pp. 153–4.
257 Guilielmi Gilberti, *De Mundo nostro Sublunari Philosophia Nova. Opus posthumum* (Amsterdam, 1651), pp. 165, 199–201. See Hilary Gatti, *Giordano Bruno and Renaissance Science* (Ithaca, NY, 1999), pp. 33–7.
258 Gilberti, *De Mundo nostro Sublunari Philosophia Nova*, pp. 84, 124–6, 158, 165, 217–18, 246.
259 Galileo, *Dialogo*, pp. 393–406; Drake, ed., *Dialogue*, pp. 467–71, 477–9.
260 Galileo, *Dialogo*, pp. 396–7; Drake, ed., *Dialogue*, pp. 467–8.
261 Gilbert, *De Magnete* (1600), p. 209.
262 Inchofer, *Vindiciarum S. Sedis Apostolicae*, p. 156.
263 Ibid.
264 John of Damascus, *De Fide Orthodoxa* (*c.* 730), in Philip Schaff and Henry Wace, eds, *St Hilary of Poitiers / John of Damascus*, Select Library

of Nicene and Post-Nicene Fathers of the Christian Church, 2nd ser., vol. IX (Oxford, 1899), II, chap. 6, p. 22.
265 Inchofer, *Vindiciarum S. Sedis Apostolicae*, p. 158.
266 Nicephori Callisti, *Ecclesiasticæ historiæ libri decem & octo* (Basel, 1553), p. 893.
267 Inchofer, *Vindiciarum S. Sedis Apostolicae*, p. 159.
268 Ibid.
269 Ibid.
270 Ibid., p. 162.
271 Ibid., p. 168.
272 Ibid., p. 188. Martiani Capellae, *De nuptiis philologiae et Mercurii et de septem artibus liberalibus*, bk 8, ed. Ulrich Friedrich Kopp (Frankfurt am Main, 1836), p. 670.
273 Inchofer, *Vindiciarum S. Sedis Apostolicae*, p. 184.
274 Ibid., p. 186.
275 Ibid., p. 187.
276 Ibid., p. 190.
277 Ibid., p. 194.
278 Ibid., p. 195.
279 Franciscus Magdalenus, 'Monitum S. Congregationis ad Nicolai Copernici Lectorem, eiusque emendatio, permissio & correctio' (1620), reissued in Antonio Schyrleo de Rheita, *Oculus Enoch er Eliae sive Radius sieromysticus* (Antwerp, 1645), Preface: unnumbered pages [p. xxi].
280 Inchofer, *Vindiciarum S. Sedis Apostolicae*, p. 197.
281 Ibid., p. 198.
282 Ibid., p. 200; Fromondi, *Ant-Aristarchus, sive Orbis-terrae Immobilis*, pp. 28, 29.
283 Inchofer, *Vindiciarum S. Sedis Apostolicae*, p. 209.
284 For example, see the work by another Jesuit theologian, Adami Tanneri, *Theologiae scholasticae*, III (Ingolstadt, 1627), col. 465 [p. 294].
285 Inchofer, *Vindiciarum S. Sedis Apostolicae*, p. i.
286 Beretta, 'Melchior Inchofer et l'hérésie de Galilée', p. 42.
287 Inchofer, *Vindiciarum S. Sedis Apostolicae*, p. 210.
288 Melchiore Inchofer, *Epistolae B. Virginis Mariae ad Messanenses Veritas vindicata* (Messina, 1629), p. 205. This book was republished in 1631 and 1632.
289 Ibid., p. 205.
290 Ibid., e.g. pp. 22–4, 46, 48, 145.
291 Inchofer, *Vindiciarum S. Sedis Apostolicae*, p. 43.
292 Ibid., p. 165.
293 Tommaso Campanella, *Sensu rerum et magia*, p. 168.
294 [Giordano Bruno] Nolano, *Cabala del cavallo Pegaseo. Con l'aggiunta dell'Asino Cillenico* (Paris, 1585), Second Dialogue [unnumbered pages].
295 Inchofer, *Vindiciarum S. Sedis Apostolicae*, p. 168.
296 See Zachariae Pasqualigi, *Disputationes metaphysicae*, 2 vols (Rome, 1634–6), I, p. 11; II, col. 352.
297 Enrico Carusi, 'Nuovi documenti del processo di Giordano Bruno', *Giornale Critico della Filosofia Italiana*, VI (1925), p. 134. Carusi notes that the statement on the slip of paper was apparently written by the same hand that wrote this copy of that 'Decreto della Congregazione'.

298 Firpo, *Il processo di Giordano Bruno*, p. 337.
299 Rietbergen, *Power and Religion in Baroque Rome*, p. 125.
300 Royal Letter to Ambassador Cardinal Borgia, 2 April 1635, in Luigi Amabile, *Fra Tommaso Campanella ne'Castelli di Napoli, in Roma ed in Parigi*, II (Naples, 1887), p. 217.
301 See the correspondence between Francesco Barberini and Giorgio Bolognetti, and between Cardinals Antonio and Francesco Barberini with Giulio Mazarini; in Amabile, *Fra Tommaso*, II, docs 255–98, pp. 218–32.
302 Cardinal Francesco Barberini to Giorgio Bolognetti, 23 October 1634, in ibid., p. 218.
303 Barberini to Bolognetti, 6 November 1634, in ibid., p. 218.
304 Amabile, *Fra Tommaso*, p. 96.
305 Ibid., p. 124.
306 Thomas Campanella, *De sensu rerum, et magia*, 2nd edn (Paris, 1637), p. iv.
307 Thomæ Campanellæ, *Defensio libri sui De sensu rerum*, in ibid., p. 33.
308 Ibid., p. 21.
309 Ibid., p. 51.
310 Ibid., p. 34.
311 Ibid., pp. 21–2.
312 Ibid., pp. 94–5.
313 Ibid., p. 22.
314 Most English translations of the Latin Vulgate Bible do not convey the literal words of Genesis, which Campanella repeatedly quoted as 'Producant aquæ reptilia animæ viventis, et volatilia', and 'Producat Terra animam viventem'. See Campanella, *Defensio*, pp. 13, 37, 46, 56, 67; and *De sensu rerum*, 2nd edn, p. 2.
315 Campanella, *Defensio*, p. 21.
316 Campanella, *De sensu rerum* (1637), ii, chap. 32, pp. 114–15.
317 Athanasii Rhetoris, *Anticampanella in Compendium redactus adversus Librum de sensu rerum et magia* (1638), pp. 7–8, in Athanasii Rhetoris, *Antipatellarus, Epistola de unione Ecclesiarum ad Alexandrinum et Hierosolymorum patriarchas, Anticampanella in Compendium Redactus* (Paris, 1655).
318 Ibid., pp. 15–16, 20.
319 Ibid., pp. 18–19.
320 Ibid., p. 28.
321 N. Chorerius, quoted in Camille Latreille, *De Petro Boessatio (1603–1662): ac de conditione litteratorum virorum in Delphinatu eadem aetate* (Vienna, 1894), pp. 46–7.

4 Worlds on the Moon and the Stars

1 John Wilkins, *The Discovery of a World in the Moone; or, A Discourse tending to Prove that 'tis Probable there may be Another Habitable World in that Planet* (London, 1638), p. 80.
2 Ibid., p. 34; see also pp. 12–13.
3 Bellarmino to Foscarini, 12 April 1615, in Maurice Finocchiaro, *The Galileo Affair: A Documentary History* (Berkeley, CA, 1989), p. 67.

4 Federico Cesi to Galileo Galilei, 12 January 1615, in *Le opere di Galileo Galilei, Prima edizione complete*, VIII (Florence, 1851), p. 340.
5 Attavanti, paraphrasing or quoting Caccini, in Finocchiaro, *The Galileo Affair*, p. 145.
6 Ximenes, in ibid., p. 143.
7 Piero Guicciardini to Grand Duke Cosimo II, 4 March 1616, in *Le opere di Galileo Galilei. Edizione Nazionale*, ed. Antonio Favaro, 20 vols (Florence, 1890–1909, repr. 1929–38), XII, p. 242.
8 Alessandro Tassoni to Albertino Barisoni, 5 March 1616, in Alessandro Tassoni, *Lettere*, ed. Pietro Puliatti, I (Bari, 1978), p. 259.
9 Antonio Querengo to Cardinal Alessandro d'Este, 5 March 1616, in Favaro, ed., *Opere di Galileo Galilei: Edizione Nazionale*, XII, p. 243.
10 Liberti Fromondi, *Ant-Aristarchus, sive Orbis-Terrae Immobilis. Liber unicus. In quo decretum S. Congregationis S.R.E. Cardinal. an.* CIƆ.IƆC.XVI. *aduersus Pythagorico-Copernicanos editum defenditur* (Antwerp, 1631), p. 29.
11 Ibid., p. 21.
12 Nicolai Caussini, *De eloquentia sacra et humana, libri XVI*, 3rd edn (Paris, 1630), p. 1005.
13 Dominico Gravina, *Catholicarum praescriptionum adversus nostri temporis hæreticos*, III/I (Naples, 1630), p. 38.
14 Ibid., pp. 38–9.
15 For example, he praised Pope Urban VIII and the following cardinals (who participated in Galileo's trial): Francesco Barberini, Guido Bentivoglio, Gaspar Borgia, 'Felici Centino' (d'Ascoli), Desiderio Scaglia and Laudivivio Zachia. See Dominico Gravina, *Pro sacro deposito fidei catholicae et apostolicae, fideliter à Romanis pōtificibus custodito, Apologeticus. Adversus novatorum calumnias* (Naples, 1629), p. [i]; and Gravina, *Vox Turturis seu de florenti usq. ad nostra tempora, S S. Benedicti, Dominici, Francisci, et aliarum sacrarum Religionum statu* (Cologne, 1627), pp. 297, 393.
16 Ioannes de Solorzano Pereira, *Disputationem de Indiarum Iure sive De iusta Occidentalium Indiarum inquisitione, acquisitione, et retentione* (Madrid, 1629), pp. 158–9.
17 For example, Jerome, 'Ad Pammachium et Marcellinum Apologia Hieronymi adversum Ruffinum', II (402 CE), in Sancti Hieronymi Stridoniensis, *Opera Omnia*, ed. Mariani Victorii (Paris, 1624), p. 511.
18 Gabriele Vazquez, *Commentariorum ac Disputationum in Primam Partem Sancti Thomæ*, I (Lyon, 1631), p. 107. Vazquez was a Jesuit theologian.
19 Philaster, 'Catalogus Haereseon ex Epiphanii Eusebijque commentarijs collectus', in *Divi Aurelii Augustini liber de haeresibus ad Quodvultdeum. Item S. Philastrii ejus coætanei Catalogus hæresium pœnè omnium* (Helmstedt, 1621), heresy 77 [unnumbered pages].
20 *Decretum Gratiani emendatum et notationibus illustratum, Unà cum glossis, Gregorii XIII. Pont. Max. iussu editum* (Rome, 1582) pt 2, Causa 24, Question 3, col. 1895 [unnumbered p. 1011]; *Corpus Iuris Canonici emendatum et notis illustratum: Gregorii XIII. Pont. Max. iussu editum*, with new indices and appendices by Pauli Lanceloti (Lyon, 1591 ['Cum Licentia']), pt 2, Cause 24, Question 3, col. 877 [unnumbered pages];

Corpus Iuris Canonici (Lyon, 1605), col. 877; *Corpus Iuris Canonici* (Lyon, 1622), col. 877.

21 Antonii Ruvio, *Commentarii in libros Aristotelis Stagiritæ de coelo & mundo* (Cologne, 1617); repr. in Lyon (1620 and 1625), and Brescia and Cologne (1626).

22 Johann Heinrich Alsted, 'Chronologia, hæresium, sectarum, schismatum', in Johannis-Henrici Alstedii, *Thesaurus chronologiæ, in quo Universa temporum & historiarum series in omni vitae genere ita ponitur ob oculos*, 2nd edn, rev. and augmented (Herborn, 1628), p. 380.

23 Johannis Henrici Alstedii, *Theatrum Scholasticum, in quo consiliarius philosophicus proponit et exponi* (Herborn, 1610), p. A iv. See Pauli Bolduani, *Bibliotheca Philosophica, sive Elenchus scriptorum philosophicorum* (Jena, 1616), p. 147: 'Jordani Bruni artificium perorandi, communicatum à Joan. Henrico Alstedio. Francof. 1612. A. in 8.'

24 Vincentii Burgundi, *Bibliotheca Mundi, seu Speculi maioris*, II (Douai, 1624), pp. 10–11, 1565.

25 Recall that in 1599 Bruno told his Inquisitors that he would revoke his propositions only 'if the Apostolic See and the Holiness of Our Lord held these eight propositions as definitely heretical'. The Pope replied that Bruno should be informed that, indeed, 'these propositions are heretical, and not only heretical as now declared, but by the most ancient Fathers of the Church and the Apostolic See.' Luigi Firpo, *Il processo di Giordano Bruno* (Rome, 1993), pp. 315, 340–41.

26 'The Anathemas against Origen', in The Fifth Ecumenical Council, The Second Council of Constantinople (553 CE), Third Anathema.

27 Francesco Ingoli to Cardinal Bonifacio Caetani, 9 August 1613, in Archivio Caetani, Rome, Fondo Generale, no. 140664. Quoted in M. Bucciantini, 'Teologia e nuova filosofia. Galileo, Federico Cesi, Giovambattista Agucchi e la discussione sulla fluidità e corruttibilità del cielo', in *Sciences et religions de Copernic à Galilée (1540–1610), Actes du colloque international organisé par l'École française de Rome* (Rome, 1999), p. 412.

28 Livio Galante, *Christianae theologiae cum Platonica comparatio* (Bonn, 1627), p. 269. Galante credited Augustine for reproving this heresy, and he attributed it to the Manichaeans and to the men 'who committed wicked errors and were excommunicated and condemned in Paris'.

29 Cardinal Carolus Pius and Franciscus Magdalenus Capiferreus, 'Sacrae Congregationis Indicis. Decreto', 4 February 1627, reproduced in [Pope] Alexandri VII, ed., *Index Librorum Prohibitorum* (Rome, 1664), p. 333.

30 Roberto Fludd, *Utriusque cosmi maioris scilicet et minoris metaphysica, physica atque technica historia*, I (Oppenheim, 1617), pp. 51, 121–2, 146, 167; see also pp. 25, 39, 98, 107. In another book, Fludd argued that Jesus Christ is the 'Platonic soul of the world', a soul contained in a divine spirit that emanates from the Sun and which is infused in all bodies, and vivifies them; see Roberto Fludd, *Anatomiae amphitheatrum effigie triplici* (Frankfurt, 1623), pp. 36, 254–6, 304.

31 For example, the Jesuit professor Cornelius Lapide commented: 'Pythagoras decided that souls transmigrate from one body to another; this error also spread among the Hebrews.' See Cornelio Lapide,

Commentaria in Acta Apostolorum, Epistolas canonicas et Apocalypsin (Antwerp, 1627), p. 88. Consider another example: 'Christ refuted two errors, one by the Gentiles, the other by the Jews. The Gentiles believed the error of the transmigration of souls, which were placed by God as in prisons, in ugly bodies for previous sins, and therefore he said that he did not sin. Because given that God raised the soul in that very subject, and previously it had no being, He could not imprison them there for previous faults.' See Geronimo de la Cruz, *Defensa de los estatutos y noblezas españolas: destierro de los abusos, y rigores de los informantes* (Zaragoza, 1637), II, p. 166.

32 One example: 'some heretics believed in the transmigration of souls, that those of the ones who died returned and were placed in other bodies, and were born, that the soul of Adam had entered the body of St Paul, thus he could say that he was the first sinner. But this is a manifest heresy.' According to the Bishop of Barbastro, Geronymo Batista de Lanuza, *Las Homilias sobre los Evangelios de la Quaresma*, III (Zaragoza, 1636), Homily 38, p. 638. Another example: 'it was opportune to remove a pernicious heresy: it was the error of Pythagoras, followed by malicious people, saying that souls exited their bodies, and entered into others.' Bartolomé Cayrasco de Figueroa, *Tercera parte del Templo militante, flos santorum, y triumpho de sus virtudes* (Lisbon, 1618), p. 145; published previously in 1600 and 1609.

33 Likewise, in 1627 Galante noted that 'what Pythagoras said' about transmigration 'is utterly deceptive and condemned by all as legally deserved'. Galante, *Christianae theologiae*, p. 25.

34 *Corpus Iuris Canonici* (1622), col. 876 [unnumbered pages]. Not everyone thought that Tertullian himself should be considered a heretic. For example, Augustine argued that by declaring that the soul is immortal and that it is corporeal, and even that God is corporeal, Tertullian did not deserve to be considered a heretic because he might have merely been using the word corporeal as a way to describe that which is not nothing.

35 Gabriel Naudé to Ismaël Boulliau, 15 August 1640, in René Pintard, *Le libertinage érudit dans la première moitié du XVIIe siècle* (Geneva, 2000), pp. 473–4.

36 Gabriel Naudé, *Apologie pour tous les grands personnages qui ont esté faussement soupçonnez de magie* (Paris, 1625); 2nd edn (The Hague, 1653), p. 331.

37 Bellarmino, *De aeterna felicitate sanctorum, libri quinque* (Cologne, 1616), IV, p. 240.

38 Giordano Bruno, *De l'infinito universo et mondi* ([Venice], 1584), p. [xxix].

39 Bellarmino, *De aeterna felicitate sanctorum*, IV, pp. 227–8; the margin cites 'Psalm 57', presently known as Psalm 58:10.

40 John Donnelly and Roland Teske, eds, *Robert Bellarmine, Spiritual Writings* (New York, 1989), p. 26.

41 C. E. of the Society of Jesus [Edward Coffin], *A True relation of the last sickness and death of Cardinal Bellarmine, who dyed in Rome the seaventeenth day of September, 1621, and of such things as have happened in, or since his buriall* ([Saint-Omer], 1622). The following quotations are

from excerpts of Coffin's book, copied in: [Thomas] Chalmers, 'Death of Bellarmine', *The Youth's Instructer and Guardian*, new ser., XVII (March 1853), pp. 112–18.

42 Chalmers, 'Death of Bellarmine', pp. 113–14.

43 Vincenzo Renieri to Galileo, 6 March 1641, in Favaro, ed., *Opere di Galileo Galilei: Edizione Nazionale*, XVIII, p. 304; Paganini Gaudentii, *De Pythagorea animarum transmigratione* (Pisa, 1641), p. 8: 'Non desunt hodie qui asserant, telescopium sive tubum opticum a Galilaeo, viro notissimi nominis, primum manasse.'

44 Gaudenzi, 'In Morte del Famosissimo Galileo, Tre Sonetti' (Pisa, 1642).

45 Paganini Gaudentii, *De errore sectariorum huius temporis labyrintheo* (Pisa, 1644), pp. 252–3.

46 John Heilbron, *Galileo* (Oxford, 2010), p. 356.

47 Pope Urban VIII, quoted in Francesco Niccolini to G. Gondi, 25 January 1642, in Favaro, ed., *Opere di Galileo Galilei: Edizione Nazionale*, XVIII, p. 379.

48 Francesco Barberini to Giovanni Muzzarelli, 25 January 1642, in ibid., p. 380.

49 Cornelio Lapide, 'Commentaria in Iosue' (1630s), in Corn. Cornelii a Lapide, *In Iosue, Iudices, et Ruth Commentarii* (Paris, 1642), chap. 10, p. 56.

50 Diary entry, 5 November 1646, in Balthasar de Monconys, *Iournal des Voyages de Monsieur de Monconys* (Lyon, 1665), p. 130.

51 David Wootton, *Galileo, Watcher of the Skies* (New Haven, CT, 2010), pp. 240–50; Castelli to Galileo, 7 June 1639, in Favaro, ed., *Opere di Galileo Galilei: Edizione Nazionale*, XVIII, pp. 57–8.

52 Anon., review of Samuel Brown's *The Tragedy of Galileo Galilei* (Edinburgh, 1850), in *The Eclectic Review*, XXVII (March 1850), pp. 272–3.

53 In the Gregorian calendar Galileo died on 8 January 1642 and Newton was born on 4 January 1643. In the Julian calendar Galileo died on 29 December 1641, whereas Newton was born on 25 December 1642.

54 Petrus Cazraeus to Pierre Gassendi, 3 November 1642, in Petri Gassendi, *Epistolae* [*Opera Omnia*], vol. VI (Lyon, 1658), p. 451.

55 Petri Gassendi, *Syntagmatis Philosophici*, pt II: *Physica*, in *Opera Omnia*, I (Florence, 1658), pp. 126–7.

56 Ibid., p. 142.

57 Ibid., p. 252.

58 Ibid., p. 142.

59 Ch. Jeannel, *Gassendi spiritualiste* (Montpellier, 1859), pp. 15–16, 20. Jeannel denounced the 'monstrous doctrine' in Virgil's *Aeneid*.

60 For discussion, see Gorman, 'A Matter of Faith', pp. 302–3.

61 This circumstance, and others, show that the Jesuits as a whole were not conspiring against Galileo. For discussion, see Michael Gorman, 'A Matter of Faith? Christoph Scheiner, Jesuit Censorship, and the Trial of Galileo', *Perspectives on Science*, IV/3 (1996), pp. 294–301; Richard Blackwell, *Behind the Scenes at Galileo's Trial* (Notre Dame, IN, 2008), p. 43; Annibale Fantoli, *The Case of Galileo: A Closed Question?* (Notre Dame, IN, 2012), pp. 211–14.

62 Jakob Bidermann, Giovanni Battista Rossi and Ioannes Alvarado, 'Judicium Revisorum Collegij Rom. de Vindicijs P. Melchioris

Inchoveri', 29 January 1636, in the Archivum Romanum Societatis Iesu (ARSI), Fondo Gesù 655, item 066, fol. 198*v*.

63 Inchofer, 'Responsio ad ea quae Patres Censores opponunt in meo libro', 6 February 1636, in ARSI, F. G., item 067, fol. 199*v*.

64 Ibid., item 067, fol. 200*v*. The Jesuit censors replied to Inchofer's complaints; see Jakob Bidermann, Franciscus Armatus, Giovanni Rossi and Ioannes Alvarado, 'Responsio ad Apologiam R. P. Melchioris Inchoveri', 4 March 1636, in ARSI, F. G., item 67, fol. 202*v*–203*v*.

65 Gorman, 'A Matter of Faith?', p. 305.

66 Blackwell, *Behind the Scenes*, p. 43.

67 Thomas Cerbu, 'Melchior Inchofer, "un homme fin & rusé"', in *Largo campo di filosofare: Eurosymposium Galileo 2001*, ed. José Montesinos and Carlos Solís (La Orotava, 2001), p. 592.

68 One of its authors was the former Jesuit Giulio Clemente Scotti; see Gorman, 'A Matter of Faith?', pp. 303–4.

69 Blackwell, *Behind the Scenes*, p. 44.

70 Natanael Jomtob [pseudonym of Antonio Puig y Blanch], *La Inquisicion sin máscara, ó disertación, en que se prueban hasta la evidencia los vicios de este tribunal, y la necesidad de que se suprima* (Cadiz, 1811), p. 348. The author, Puig y Blanch, based his account on quotations from the official report by a member of the Holy Office: Rodrigo Ruiz de Zepeda Martínez, *Auto General de la Fee . . . celebrado . . . en Mexico, à los 19 de Noviembre de 1659* (Mexico City, 1659).

71 Jomtob, *La Inquisicion sin máscara*, p. 349.

72 Jaime Contreras and Gustav Henningsen, 'Forty-four Thousand Cases of the Spanish Inquisition (1540–1700): Analysis of a Historical Data Bank', in *The Inquisition in Early Modern Europe: Studies on Sources and Methods*, ed. Gustav Henningsen and John Tedeschi (DeKalb, IL, 1986), p. 114. Comparable statistical figures do not yet exist for the Roman Inquisition, mainly because too many documents were lost, dispersed or destroyed. See John Tedeschi, *The Prosecution of Heresy: Collected Studies on the Inquisition in Early Modern Italy* (Binghamton, NY, 1991), pp. 23–45.

73 Dan. Georgi Morhofi, *Polyhistoris continuatio*, *tomum philosophicum & practicum*, ed. Johanne Möllero (Lübeck, 1708), p. 239. Morhof lived from 1639 to 1691; this volume was published posthumously. His *Polyhistor* volumes grew from lectures he gave at the University of Kiel around 1666.

74 Ibid., p. 26; see also pp. 303, 343.

75 Ibid., p. 260; see also p. 303.

76 Ibid.

77 Jo. Alberti Fabricii, *Bibliotheca Græca* (Hamburg, 1705), I, chap. 20, pp. 132–3.

78 Ibid.

79 Algernon Herbert, *Nimrod: A Discourse on Certain Passages of History and Fable*, IV/1 (London, 1829), pp. 140, 160; and IV/2 (London, 1830), pp. 178, 486.

80 Ibid., IV/2, p. 509.

81 Ibid., pp. 509–15.

82 Ibid., p. 514.
83 Ibid., pp. 479–97.
84 The gradual process by which the Catholic Church revised its views on Copernicanism and Galileo is discussed thoroughly in Maurice Finocchiaro, *Retrying Galileo, 1633–1992* (Berkeley, CA, 2005).
85 George Waddington, *A History of the Church from the Earliest Ages to the Reformation*, 2nd edn (London, 1835), p. 103.
86 Howell rightly makes this point, though he does not discuss Pythagorean religious beliefs. Kenneth J. Howell, *God's Two Books: Copernican Cosmology and Biblical Interpretation in Early Modern Science* (Notre Dame, IN, 2013), pp. 163, 185, 205.
87 Étienne Tempier, Condemnation of 1277, see 'Sequuntur errores annotati in rotulo, 1277'; reissued in Faculty of the University of Paris, and Henricus Deinfle, *Chartularium Universitatis Parisiensis*, I (Paris, 1889), p. 543.
88 Ibid.
89 Gaspar Schoppe to Konrad Rittershausen, 17 February 1600, printed in Gaspari Scioppii, 'Epistola, in qua haereticos jure infelicibus lignis cremari concludit' [also titled: 'Epistola, in qua sententiam de Lutheranis tanquam haereticis atram Romae fieri asserit & probat'], in *Macchiavellizatio* (Zaragoza, 1621), pp. 30–35, p. 34: 'imò Monstra', in Firpo, *Il processo di Giordano Bruno*, p. 352.
90 'Decretum Sacrae Congregationis', 5 March 1616 (Rome, 1616).
91 Roberti Bellarmini, *Disputationes de controversis Christianae fidei, adversus hujus temporis hæreticos*, I (Ingolstadt, 1581), 'Præfatio', repr. Venice, 1721, p. x.
92 Thomæ Campanellæ, *Apologia pro Galileo, mathematico Florentino* (Frankfurt, 1622); Richard J. Blackwell, ed., *Defense of Galileo* (Notre Dame, IN, 1994), p. 45.
93 Ibid., p. 151.
94 Galileo to Nicolò de Peiresc, 21 February 1635, in Favaro, ed., *Opere di Galileo Galilei: Edizione Nazionale*, XVI, p. 215.
95 John Chrysostom, *Homilies of S. John Chrysostom, Archbishop of Constantinople, on the First Epistle of S. Paul the Apostle to the Corinthians* (Oxford, 1845), Homily V (on 1 Corinthians 1:26–7), sec. 2, p. 53.
96 Galilei, 'Dialogo Primo', in *Dialogo, Doue ne i congressi di quattro giornate si discorre sopra i due massimi sistemi del mondo Tolemaico, e Copernicano* (Florence, 1632), p. 43; as Galileo, *Dialogue Concerning the Two Chief World Systems*, ed. Stillman Drake (Berkeley, CA, 1962), p. 58.

ACKNOWLEDGEMENTS

This book began from research for a chapter in one of my previous books, *Science Secrets*. It also benefited from feedback I received when writing another book, *The Cult of Pythagoras*. I therefore thank several colleagues, friends and anonymous reviewers who gave me suggestions in those earlier projects. In early 2009 I presented a keynote talk on Galileo and the Roman Inquisition at the University of Puerto Rico, for the Fourth Centenary of Galileo's discoveries. I did not then anticipate that I would subsequently spend eight years working on a book about the topics I explored in that talk. I thank the Honors Program of the University of Puerto Rico, and in particular, Professor Idsa Alegría, for kindly inviting me.

In the autumn of 2012 the College of Liberal Arts kindly interrupted my teaching duties so that I could take care of my father, who had become disabled, and work also on the present project. For months I struggled to help my father, Ronald Martínez Cuevas, trying to enable him to talk and walk again, and to regain his health, but unsuccessfully. During the nights I read about Galileo.

I thank two anonymous reviewers for the *Journal for the History of Astronomy* for helpful comments on a draft article that I submitted back in 2012. At that point I could either revise it or expand it into a book. I warmly thank Frances Rivera Avilés for insisting that it could better become a book. Had it not been for her insistence, I would not have found the surprising material that I later discovered.

I thank the Department of History and the College of Liberal Arts of the University of Texas at Austin for funding my research trips to Italy. I thank the following libraries, and their archivists, for enabling me to access their rare manuscripts and books: the Biblioteca Casanatense in Rome, the Archivio di Stato di Roma, the Archivio Storico Capitolino in Rome, the Archivum Romanum Societatis Iesu, the Biblioteca Vallicelliana in Rome, the Archivio Storico of the Academia Nazionale dei Lincei in Rome, the Galileo Museum in Florence, and the Biblioteca Nazionale Centrale di Roma. In particular, I want to thank Maria Temide Bergamaschi, Laura Giallombardo, Michele Di Sivo, Alessandra Marrone, Brian Mac Cuarta SJ and Elisabetta Caldelli. Plus, Frances Rivera Avilés kindly helped me to gather information from certain manuscripts in Rome. I also thank the librarians of the Perry-Castañeda Library of the University of Texas, the Harry Ransom Humanities Research Center, the Library of the University of Notre Dame, the Biblioteca Cervantina of the Instituto Tecnológico y de Estudios Superiores de Monterrey, México, and finally, the Archivo General de la Nación, in México City.

Next, I must thank scholars who have made enormous, valuable contributions to the study of Galileo and Bruno, and who kindly responded to my inquiries. First, I thank Richard J. Blackwell for warmly and enthusiastically discussing Galileo and Melchior Inchofer with me, and I thank his son Thomas for putting us in contact. I thank John L. Heilbron for kindly reading an early version of my manuscript and giving me encouragement and advice on how to shape it and improve it. I also give thanks to William R. Shea for his attentive reading of my manuscript, and his helpful, expert and genuinely caring comments. Bill's humanity shines warmly even through the impersonal medium of email. I also thank him for putting me in touch with Hilary Gatti. Consequently, Gatti kindly agreed to read my manuscript, and I am enormously grateful for her diligent and generous labour. I greatly learned and benefited from her meticulous corrections and helpful suggestions, especially in her main area of scholarship, Giordano Bruno. Next, Jole Shackelford read my manuscript too, and I'm glad that he enthusiastically agreed with my findings about Bruno, since he had written about the Bruno myths. Plus, he shared his expertise on science in the Renaissance. Next, Maurice Finocchiaro kindly read parts of my manuscript, and he expertly pinpointed how to better characterize some of Galileo's claims. It was a pleasure discussing Galileo with him. I also thank scholars who, although they lacked the time to read my manuscript, did kindly give me some useful pointers, namely Ingrid Rowland and David Wootton.

I presented some of my research findings in four talks at UT Austin, in 2011, 2013, 2015 and 2017. I thank the audience in those talks, especially Bruce J. Hunt for multiple comments and suggestions. Miriam Bodian oriented me on some of the literature on the Roman Inquisition. Martha Newman clarified historical questions about sins, heresies and Catholic religious orders. Susan Boettcher explained to me the rise of the Protestant Reformation and especially Martin Luther's beliefs. Jorge Cañizares-Esguerra gave me suggestions about competing notions of sin and souls in the history of Christianity. H. Darrel Rutkin kindly shared his insights about Plotinus, Ficino and the notion of the soul of the world. Alexander P. D. Mourelatos gave me advice on the literature about the Presocratics and the Pythagoreans. Stephen Case kindly discussed with me his work on Platonic notions of the nature of stars.

In early 2015 I discussed myths about Bruno in a talk I presented at the March Meeting of the American Physical Society. I thank the session organizer, Catherine Westfall, as well as the participants for various interesting comments. In June 2015 I presented a talk about Bruno and Galileo at the 12th Biennial Conference on the History of Astronomy at the University of Notre Dame. I thank the conference organizers, along with the participants, for their useful feedback. I especially thank Michael J. Crowe and Steven J. Dick for kindly discussing the history of the belief in many worlds with me. I am indebted to their pioneering publications on the history of the extraterrestrial life debate, and am very flattered that they appreciated and agreed with my findings about Bruno. I also thank Owen Gingerich for additional helpful information, praise and encouragment. Sarah Reynolds discussed Kepler's work on the Moon with me. Eric C. Young offered insights about the early Church Fathers' opinions on celestial lights, to complement my research on their views on heresies.

I presented two talks on Bruno and Galileo at the annual conference of the History of Science Society in 2015 and 2017. I thank the participants for helpful discussions and questions, especially Paula Findlen, Mordechai Feingold, Jole Shackelford, Michael Kerze and Rienk Vermij. I also discussed Galileo and Bruno with Michael Shank and John Lisle, who gave me useful comments. I should also express my gratitude to editors and anonymous reviewers supporting academic journals for which I had submitted articles on Giordano Bruno, namely *Annals of Science*, *Bruniana & Campanelliana*, and *Early Science and Medicine*. Furthermore, Miguel A. Granada kindly emailed me useful pointers in response to a couple of articles I had published about Giordano Bruno. And I thank James H. Dee and Jole Shackelford for help with several Latin translations.

I thank the Institute for Historical Studies at the University of Texas at Austin for granting me a Fellowship to enable me to polish the final manuscript during 2016–17. I also thank the participants at my IHS talk for their helpful feedback, especially A. Azfar Moin. I gratefully acknowledge a University of Texas at Austin Subvention Grant awarded by the Office of the President. I also thank the Office of the Vice President for Research.

I thank my mother Lillian Montalvo Conde for happily being around. Brett M. Bennet kindly suggested that I submit my manuscript to his editor Ben Hayes at Reaktion Books. Hence, I warmly thank Ben Hayes for adopting the project. I also appreciate the suggestions of the anonymous reviewers, and the work of the copy editor David Rose. Amy Salter kindly and meticulously worked as text editor. So I thank them and the staff at Reaktion Books for helping me share this brand new story about Bruno and Galileo.

PHOTO ACKNOWLEDGEMENTS

The author and publishers wish to express their thanks to the below sources of illustrative material and/or permission to reproduce it. Some locations and titles of works are given here for the sake of brevity.

Archivio di Stato di Roma: p. 76; photos by the author: pp. 32, 76, 89, 95, 119, 144, 186, 202, 249, 250, 252, 261; Stefano della Bella, etching for Galileo Galilei, *Opere di Galileo Galilei linceo nobile Fiorentino . . .* (Bologna, 1656): p. 111 (photo Metropolitan Museum of Art [open access]); Metropolitan Museum of Art: p. 143 (Harris Brisbane Dick Fund, 1936 – open access); Musei Capitolini, Rome: p. 256; church of S. Ignazio, Rome: pp. 144, 249, 250; basilica of St Peter, Rome: p. 186; from Antonio Tempesta, *Recens prout hodie iacet almæ urbis Romæ cum omnibus viis aedificiis que prospectus accuratissime delineatus* [1593] (Rome, 1645): pp. 46, 72, 75, 201 (photos Metropolitan Museum of Art [open access]); Francesco Villamena, *Galileo Galilei linceo filosofo e matematico del Ser.mo Gran Duca di Tosc.a . . .* ([Rome], 1613): p. 138 (photo Metropolitan Museum of Art [open access]).

INDEX

Page numbers in *italics* refer to illustrations